DIGITAL SIGNAL PROCESSING IN TELECOMMUNICATIONS

BT Telecommunications Series

The BT Telecommunications Series covers the broad spectrum of telecommunications technology. Volumes are the result of research and development carried out, or funded by, BT, and represent the latest advances in the field.

The series includes volumes on underlying technologies as well as telecommunications. These books will be essential reading for those in research and development in telecommunications, in electronics and in computer science.

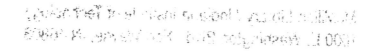

DIGITAL SIGNAL PROCESSING IN TELECOMMUNICATIONS

Edited by

F.A. Westall

Head of the Applied Signal Processing Section
Data Communications Division
BT Laboratories
Martlesham Heath, UK

and

S.F.A. Ip

Senior Member of Professional Staff
Applied Signal Processing Section
Data Communications Division
BT Laboratories
Martlesham Heath, UK

CHAPMAN & HALL
London · Glasgow · New York · Tokyo · Melbourne · Madras

Published by Chapman & Hall, 2–6 Boundary Row, London SE1 8HN

Chapman & Hall, 2–6 Boundary Row, London SE1 8HN, UK

Blackie Academic & Professional, Wester Cleddens Road, Bishopbriggs, Glasgow G64 2NZ, UK

Chapman & Hall Inc., 29 West 35th Street, New York NY10001, USA

Chapman & Hall Japan, Thomson Publishing Japan, Hirakawacho Nemoto Building, 6F, 1–7–11 Hirakawa-cho, Chiyoda-ku, Tokyo 102, Japan

Chapman & Hall Australia, Thomas Nelson Australia, 102 Dodds Street, South Melbourne, Victoria 3205, Australia

Chapman & Hall India, R. Seshadri, 32 Second Main Road, CIT East, Madras 600 035, India

First edition 1993

© 1993 British Telecommunications plc

Printed in Great Britain by Alden Press, Oxford.

ISBN 0 412 47760 2

Contents

Contributors

D K Anthony	Formerly Signal Processing, BT Laboratories
P Barrett	Signal Processing, BT Laboratories
P R Benyon	Decision Support Systems, BT Laboratories
N J Billington	Copper Access, BT Laboratories
I Boyd	Signal Processing, BT Laboratories
P Branch	Subjective Performance, BT Laboratories
A P Breen	Signal Processing, BT Laboratories
A P Clark	Late Professor of Telecommunications, Loughborough University of Technology
N G Cole	Copper Access, BT Laboratories
J M Connell	University of Manchester
J Cook	Copper Access, BT Laboratories
K T Foster	Copper Access, BT Laboratories
D K Freeman	Decision Support Systems, BT Laboratories
I Goetz	Mobile Systems, BT Laboratories
C D Gostling	Signal Processing, BT Laboratories
D R Guard	Voice Messaging, BT Laboratories
P J Hughes	Speech Platforms, BT Laboratories
S F A Ip	Signal Processing, BT Laboratories
R J Knowles	Customer Networks, BT Laboratories
A Lewis	Signal Processing, BT Laboratories
L F Lind	Professor, Electronic Systems Engineering, University of Essex
A Lowry	Signal Processing, BT Laboratories
N J Lynch-Aird	Core System Design, BT Laboratories

R M Mack — Signal Processing, BT Laboratories

D J Myers — Video Services, BT Laboratories

M Ogden — Customer Networks, BT Laboratories

J H Page — Signal Processing, BT Laboratories

D Pauley — Speech Platforms, BT Laboratories

D A Smee — Signal Processing, BT Laboratories

C B Southcott — Operator Automation, BT Laboratories

R G C Williams — Business and Mobile Systems, BT Laboratories

F A Westall — Signal Processing, BT Laboratories

M W Whybray — Video Systems, BT Laboratories

C S Xydeas — Professor of Electrical Engineering, University of Manchester

G Young — Copper Access, BT Laboratories

Preface

This book is dedicated to the subject of digital signal processing (DSP) in telecommunications. As the term 'DSP' means different things to different people, the chapters have been carefully selected to give the reader a broad perspective of the subject. The book is intended to provide an insight into the DSP technologies and techniques that are regularly deployed at BT Laboratories to solve practical telecommunications problems.

At the start of the information age, information was transmitted from place to place on telecommunications networks and received directly by another human being. As the age progressed it became possible to store information and effectively reduce the requirement for the sender and recipient to be simultaneously available. At the current time it is becoming increasingly possible to process the information before or after storage to detect, sense, extract, transform or convert information into a form more suitable for humans to comprehend or even to reduce the need for human interaction in some cases. It is at this human/machine interface where the real world phenomena of information as a stream of continuous analogue signals meets the discrete, discontinuous world of digital technology. It is here that the role of digital signal processing and the related specialised DSP devices provides the vital link between the two worlds.

DSP is a rich subject, and to do it justice would require much more space than is available here, though most chapters provide tutorial material to help those unfamiliar with the subject. In attempting to keep the mathematics to a manageable level, some of the elegance and rigour has been lost. For this the editors apologise in advance, though the ample references associated with each chapter should satisfy all but the most insatiable of academic appetites.

BT Laboratories have made notable contributions in the development and application of digital signal processing — from early work on PCM systems and C J Hughes' work on the conception of microprocessor structures in the 1960s through to work in the 1970s leading to a variety of practical fixed and adaptive filtering devices and to commercial systems for deployment in data transmission and switching applications. However, it was the advent of low-cost programmable DSP devices during the 1980s that allowed a wide range of practical and cost-effective systems to emerge from laboratories such as ours. After a decade of practical experience using DSP devices, the time is ripe for a retrospective review and a glimpse into the exciting future of this key technology.

This book brings together the work of many of the DSP practitioners based at BT Laboratories and concentrates on applications developed over the last three to four years. The introductory chapter provides a broad overview to set the scene and to put the subject into context. The second chapter addresses the important topic of fixed digital filtering and presents some new ideas as illustrations. Two tutorial chapters then cover adaptive filtering — a major component in the ubiquitous application of DSP technology within speech-band (3 kHz) systems. There then follow several chapters describing representative telecommunications applications in some detail — high-speed digital subscriber-loop systems, data transmission including PSTN modems and channel coding, speech coding, network emulation and noise suppression applications. In order to place the speech applications even more clearly in context, a further chapter provides a background to the speech production and perception processes that underpin the key speech applications. To round off the broad coverage of DSP applications, further developments of considerable potential are covered in two chapters on video coding and adaptive neural processing.

As DSP devices grow in complexity, so does the need for design methods and associated tools, to allow the dramatic potential of this technology to be economically unleashed. Recent advances in computer-aided development tools and environments to support DSP system design are described in one chapter, while several authors describe different aspects of the development process relevant to their particular applications, a good example being that of the DSP development method and tool employed for the Skyphone speech codec. Further advances in this area are anticipated, which will slow the growth of software costs and open up important new possibilities for DSP technology.

The last chapter illustrates the impact of DSP in the varied fields of network modelling and channel monitoring. These embryonic subjects will be among the main beneficiaries of progress in leading-edge DSP technology, illustrating the extent to which traditional boundaries between telecommunications and computing are blurring.

It can now truly be claimed that DSP has come of age. The authors hope that this snap-shot of the broad range of telecommunications applications for DSP has effectively communicated the excitement of working in this dynamic and rapidly expanding field.

Last, but not least, a considerable amount of effort goes into the production of a book like this, and the editors would like especially to thank all the authors and reviewers, too numerous to mention by name, for their dedication, skill and hard work. Additionally, one of us (FAW) has had the good fortune to have been associated with many of the DSP practitioners at the laboratories at some stage over the formative last two decades. To all of you, mentors and managers alike, thank you for your support.

F A Westall
S F A Ip

1

DIGITAL SIGNAL PROCESSING IN TELECOMMUNICATIONS

F A Westall and S F A Ip

1.1   INTRODUCTION

Over the last decade digital signal processing (DSP) has emerged as a key enabling technology in telecommunications. In this short period it has moved from being a research curiosity into use in a broad spectrum of practical applications within both public and private networks. It has heralded a resurgence of interest in signal processing, and could provide a key to the long-awaited convergence of telecommunications and computing [1]. It is therefore timely to review progress on DSP and it is to this end that this book is directed.

The term DSP means different things to different people. To many it is synonymous with a class of exceptionally fast microcomputers targeted at real time signal processing; whilst to others it relates to the algorithms or mathematical processes that accomplish the desired signal transformation.

There are other viewpoints too. A development engineer may be less concerned with the best attainable theoretical performance, than with achieving the specified goals within cost, timescale and marketing constraints. The concern will be with issues of documentation, testability, manu-facturability and product support, whilst at the same time trying to build in future-proof safeguards against advances in technology, loss of key staff, or customers' changing requirements.

Signal processing, as with other engineering disciplines, requires an accurate assessment and balancing of these often conflicting viewpoints to ultimately achieve success in the market-place. In order to appreciate the impact of DSP, it is necessary to be aware of the environment in which it operates (Fig. 1.1) and to consider the subject from the different viewpoints indicated above. It is this breadth of treatment that makes this book timely.

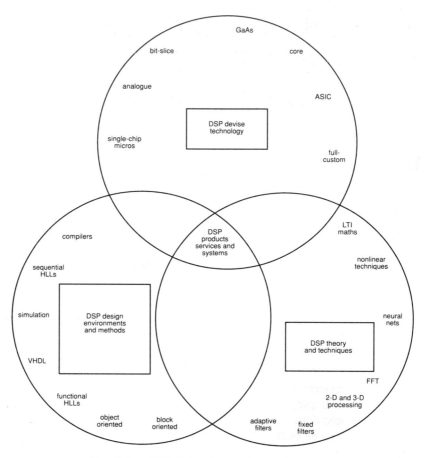

Fig. 1.1   Digital signal processing environment.

Although digital techniques have been pre-eminent in recent years, signal processing itself is not a new subject. Telegraphic transmission, now over 150 years old, was the forerunner of modern high-speed adaptive modems; early voice coders (such as Dudley's voice operation demonstrator — VODER) date back to the late 1930s and Von Kempelen made a voice synthesizer as early as 1791 [2]. The mathematics on which many current

DSP systems are based were established over 20 years ago, such as with the work of Widrow on adaptive filtering in the early 1960s [3].

However, it was the advent of the single-chip digital signal processing microcomputer in the late 1970s that helped to convert research into practical, cost-effective systems and has resulted in the upsurge of interest in new telecommunications applications.

Although impressive progress has been made over the last decade in silicon technology, engineers are still presented with some special problems when implementing DSP systems. Real time constraints and other hardware limitations all conspire to limit the performance that can be achieved in practice. The designer often has to work with fuzzy specifications, as with a speech codec, where the aim might be to achieve the best 'subjective' performance. Although some computer-based support tools are available to help, these generally still lack the necessary integration to ensure that designs are consistent and supportable, whilst at the same time achieving the real time requirements.

Despite these reservations DSP is now firmly established as a key technology, with many current applications giving a taste of the potential power of this technology for the future.

This chapter provides a broad overview of the subject. A review is given of the key technologies that underpin the field, followed by a snap-shot of current DSP-related activities at BT Laboratories. To conclude, some predictions are made on future trends and applications.

1.2   WHAT IS DSP?

1.2.1   A DSP system

A typical DSP system is shown in Fig. 1.2. The input signal can take a variety of forms. For example, it could represent a speech signal from a microphone, or be a modulated data signal from a telephone line. It could be a video signal from a camera to be encoded for transmission over a digital link, or to be stored on a computer.

The input signal is band-limited and sampled prior to conversion into an encoded bit stream using the analogue-to-digital converter. If information is not to be lost, then the sampling rate (f_s) should be at least twice the highest frequency of the input band-limited signal.

The digital signal processing device acts on the digitized input, sample-by-sample, and modifies the data in some way, such as via a sequence of multiplier-accumulator (MAC) operations. This processing of data is the key to DSP and is quite different from, for example, switching systems, where

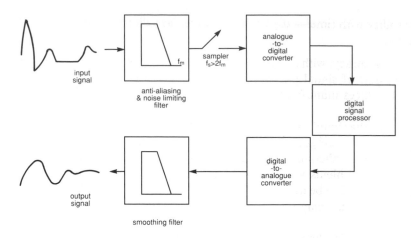

Fig. 1.2   Digital signal processing system.

the processor provides routeing only and does not substantially modify the data *en route*. The run-time constraints are therefore quite different, although both are often referred to as 'real time systems'.

Finally, after processing, the digitized samples are converted back to analogue values, and a subsequent filter provides interpolation, or smoothing, back to a continuous waveform.

It should be noted that this model, whilst being typical of many cases, is by no means universal. Speech recognizers, for example, do not produce continuous waveforms at the output, rather they produce decisions on which of a group of possible utterances the input speech most resembles. Sometimes the input signals are already in discrete form, as from a compact disc, thereby obviating the need for an analogue-to-digital converter.

1.2.2   DSP versus analogue — pros and cons

DSP has been one of the major beneficiaries of the dramatic reductions in cost, size and power consumption of silicon technologies in recent years. DSP systems are also compatible with other modern digital technologies such as transmission and switching networks, and computers. Interfacing to such systems is therefore much simpler than with analogue technology of equivalent functionality. Other advantages are:

● programmability — this gives the designer the flexibility to modify or upgrade the program late in the development, perhaps even in the cold light of field-trial experience;

- stability with time — digital signal processing does not drift or age with time;

- good accuracy, with excellent definability and repeatability — this extends the range of signal processing tasks that can be implemented reliably, and reduces manufacturing and testing costs.

On the other hand, for simple signal processing tasks, such as telephones interfacing to analogue exchange lines, the extra cost of DSP may be unacceptable. Furthermore, the high clock speeds associated with DSP devices can present problems with crosstalk and electromagnetic radiation. Such systems also tend to be more power-hungry than equivalent analogue systems. Other factors that may be grouped into the 'con' category include:

- rapid rate of technological change — there is often a steep learning curve associated with the introduction of each new family or generation of technology leading to 'family inertia' — a tendency to stick with one device manufacturer to minimize re-learning costs;

- more mathematical knowledge needed — DSP is a rich subject with solid mathematical foundations and, as there are often a variety of alternatives for progressing towards a performance target, some mathematical knowledge is required to reduce the bewildering choice of algorithms on which to concentrate for more detailed consideration;

- poor development and debug tools — although improving, current design tools lack the sophistication, integration and usability to allow the designer to exploit the full potential of the device technology.

Now that a basic DSP system has been defined, the means available for implementing the processing function in Fig. 1.2 are reviewed, starting with the most popular programmable technologies.

1.3   PROGRAMMABLE DSP DEVICES

1.3.1   Historical background

A major milestone for DSP came in 1979 with the announcement of the first commercial programmable device from Intel, the 2920. This device contained on-board analogue converters, but was limited in not containing a single-cycle multiplier. The credit for the first 'true' single-chip DSP microprocessor device[1] goes to AMI with the S2811, though production difficulties meant that the device did not become available in the UK until much later than the announcement date of 1978. In 1980 NEC launched the μPD7720 — the first commercially available DSP device with a dedicated multiplier. This product range is still being updated today, making it the most durable DSP device architecture. This was followed by the successful range of Texas Instruments (TI) products, beginning in 1982 with the TMS32010, and leading to the latest TMS320C40, which is capable of 50 million floating-point operations per second. Today TI is still the the market leader with more than 50% of the market for DSP devices, due in no small part to their emphasis on development support. The credits for producing the first complementary metal oxide semiconductor (CMOS) and floating-point device go to Hitachi for their HSP device of 1982, and a significant leap in processing throughput came in 1983 with the MB8764 from Fujitsu with a 120 ns cycle time and dual internal data buses. The first 'high-performance' floating-point device was the DSP32 from AT&T launched in 1984.

Motorola were relative late-comers with the MC56001 fixed-point device in 1986 and the MC96002 IEEE-compatible floating-point device in 1990.

Since 1980 the time taken for a multiply and accumulate operation (MAC) for DSP devices has reduced from 400 ns (TMS32010) to below 40 ns (MC56116 and TMS320C40) — a 10-fold increase in processing throughput. Pin counts have increased from a maximum of 64 in 1980 to a current maximum of over 200. This trend is significant in that the larger number of pins has provided considerable additional architectural flexibility in terms of external memory expansion and interprocessor communications. The key multiplier component now occupies under 5% of the die area compared with around 40% in 1980, and the die area itself has increased by a factor of 3 over this period [4]. On-chip RAM has increased over an order of magnitude (from 2.5 k to 32 kbytes), driven by advances in silicon technology which was based on 4 μm N-channel MOS (NMOS) in 1980, but now exploits sub-micron CMOS.

An interesting recent trend is indicated by the ADSP-21msp50 'mixed signal processor' from Analog Devices [5], which combines voice-band signal conditioning, 16-bit sigma-delta A-to-D conversion, D-to-A and a DSP device on a single CMOS chip. In spirit at least, the link is thereby completed with that first DSP device from Intel.

---

[1] The commonly known digital signal processor, DSP microcomputer and microprocessor will be generally referred to as 'DSP devices' throughout this book.

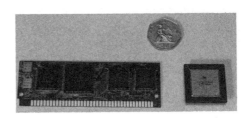

power dissipation, 250W   2.5W   0.25W
circa 1982   circa 1985   1989

Fig. 1.3   Evolution of LMS adaptive filter technology.

It is interesting to note that in the period up to 1986, DSP technology enjoyed a three order of magnitude reduction in cost, size, weight and power consumption [6]. To put this into perspective, Fig. 1.3 shows the evolution of an audio echo cancellation system developed at BT Laboratories, from the early transistor-transistor logic (TTL) prototype (1982), through a custom large-scale integration (LSI) multi-chip implementation (1985), to a modern single-chip DSP device of 'similar' functionality.

A summary of some single-chip DSP devices is provided in Table 1.1, and a more comprehensive review can be obtained from Lee [1].

Table 1.1   Single-chip programmable DSP devices.

Company	Part number	Date introduced	MAC time (nsecs)*	Number of bits fixed-point	Number of bits floating-point
AMI	S2811	1978	300	12/16	
NEC	μPD7720	1980	250	16/32	
	μPD77230	1985	150		32
Texas	TMS32010	1982	390	16/32	
Instruments	TMS32020	1987	200	16/32	
	TMS320C25	1989	100	16/32	
	TMS320C30	1989	60	24/32	32/40
	TMS320C40	1992	40	32	40
	TMS320C50	1990	35	16	
Motorola	MC56001	1986	75	24	
	MC96002	1990	50	32/64	32/44
	MC56002	1991	50	24/48	
AT&T	DSP32C	1988	80	16 or 24	32/40
	DSP16A	1988	25	16/36	
	DSP3210	1992	60	24	32/40
Analog	ADSP-2101	1990	60	16	
Devices	ADSP-21020	1991	40	32	32/40

* MAC = one multiply and accumulate operation.

1.3.2   Functionality of DSP devices

DSP devices are intended for real time applications involving exceptionally fast arithmetic intensive processing. They have most or all of the following characteristics:

- a hardware multiplier and adder producing a product/sum in a single cycle;

- simultaneous accessing of instructions and data;

- fast RAM on chip — often simultaneously accessible from two blocks via independent data buses;

- hardware to facilitate low (or zero) overhead looping and branching;

- fast interrupt handling and hardware I/O support;

- multiple hardware address generators that operate in single cycle;

- multiple operations performed in parallel;

- hardware for overlapping of instruction decoding and execution (pipelining) for raw speed;

- poor general-purpose functions — small stack, limited instruction set.

It should be noted that many of these features are shared with modern general-purpose reduced instruction set computers (RISC), indicative of a growing tendency to blur the distinctions between the different hardware technologies.

1.3.3   Selecting a DSP device

The DSP system of Fig. 1.2 can be implemented using one or more of the following architectures:

- fixed-point, single-chip devices (e.g. the MC56001, TMS320C20);

- floating-point, single-chip devices (e.g. the MC96002, TMS320C30);

- application-specific devices, such as specialist FFT, digital filter and speech recognition chips (e.g. Zoran 34325);

- high-speed, bit-slice devices, made up by interconnecting arithmetic and memory components in 'slices' of 2 or 4 bits (e.g. AMD2900 series).

Fixed-point DSP devices are the preferred choice for applications where component cost and low-power dissipation are important factors. Many DSP device manufacturers produce variants which feature much higher processing speeds (20-40 million operations per second) but at the expense of reduced precision (typically 16-bit) and increased power dissipation. Examples include the Motorola MC56116 and the Texas Instruments TMS320C50.

Floating-point devices provide an increase in dynamic range and precision relative to fixed-point technology, but are usually more expensive. They are more amenable to high-level language support, and may therefore be easier to program. Consequently, the reduced development timescales may compensate for the higher initial outlay for some low-to-medium volume applications.

For video or radar rate processing, high-speed bit-slice microprocessor technology may be the most appropriate choice, especially where cost and power dissipation issues are less critical.

Table 1.2   Device requirements for various DSP applications.

Application area	Sample rate (kHz)	Sample period	MACs available per sample		Number of DSP devices *	
			400 ns	40 ns	400 ns	40 ns
Voiceband	8	125 μS	312	3120	3	<1
Audio	44	22.7 μS	56	560	14	2
Video	5000	200 ns	<1	5	>1500	150+

*To implement a 250-tap LMS adaptive digital filter.

Table 1.2 illustrates the effect of bandwidth and hence sampling rate on the processing requirements for two DSP devices — 1st generation (circa 1980, 400 ns MAC time), and the latest with a 40 ns MAC time (such as the TMS320C40 mentioned later in section 1.9). The example given is of a 250-tap adaptive FIR filter based on the least-mean-squares (LMS) adaptive algorithm (see Fig. 1.10; $N = 250$). It is assumed that each filter tap contribution and update requires 3 MAC cycles. No provision has been made for I/O interprocessor communications in this simplified example.

The main advantages of DSP devices are that they provide a fast route to implement new applications, and that the programmability allows greater flexibility for product revisions and upgrades. On the other hand, the proprietary 'edge' of a product may be diminished since competitors have easy access to identical hardware, reverse engineering is simpler and patent or copyright protection of software can be both complex and costly.

Although this chapter focuses on applications targeted primarily at single-chip DSP devices, this technology is by no means the only option available to system designers [7]. Some of these alternatives are considered next.

1.4   ALTERNATIVE AND COMPLEMENTARY DSP DEVICE TECHNOLOGIES

1.4.1   Custom LSI

Early work in the late 1970s and early 1980s at BT Laboratories resulted in a variety of custom devices for implementing programmable or adaptive digital filters. An example was the filter and detect (FAD) device, which allowed digital filters of up to 16th-order to be implemented as a cascade of second-order biquadratic sections at a sampling rate of 8 kHz. Provision was also made on the chip to implement the level-detect function required for multifrequency tone detection. This successful development was incorporated into a variety of applications, including the System X exchange [8].

Early modems for the switched telephone network, such as the popular BT modem DM4962, made extensive use of in-house designed custom-LSI devices [9]. Figure 1.4(a) shows a photomicrograph of the adaptive filter chip, implemented in 5 μm NMOS.

A further development implemented a general-purpose digital adaptive filter chip-set for voice applications. The coefficient adaptation was provided using the power-normalized least mean squares (LMS) algorithm. This device was used in a number of research demonstrators of advanced audio-conferencing terminals, loudspeaking telephones [10] and network echo cancellers, though it was never exploited commercially outside BT. A photomicrograph of the coefficient updating component is shown in Fig. 1.4(b).

The advantage of implementing DSP using custom LSI devices is that it keeps the unit costs low, as well as allowing the silicon designer greater flexibility in trading off a number of competing performance issues. It may also be harder to 'reverse engineer' than a programmed DSP device. The main disadvantage is that the development timescales tend to be much longer than for implementing equivalent functionality on programmable devices, and the development costs are correspondingly greater. Thus it is not justifiable to use such technology for prototyping or demonstrators. A further problem is that designs must be frozen early in the development life cycle and will require costly device reworks if there is a need to modify the design due to product revisions or upgrades.

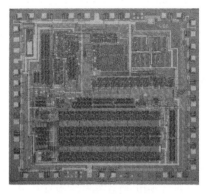

(a)   Modem adaptive equalizer.

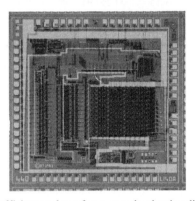

(b)   Coefficient updater for a speech-adaptive digital filter.

Fig. 1.4   Early examples of custom-ICs for DSP.

Although diminishing in significance in telecommunications, such technology still enjoys some popularity, especially for high-performance, high-reliability military applications, and for very cost-sensitive, high volume, consumer and domestic products.

1.4.2   Semi-custom and gate-array

The boundaries between semi- and full-custom silicon technologies are becoming less well-defined. DSP functions are available as standard cells from some application-specific integrated circuit (ASIC) vendors. Indeed several commercial high-speed DSP devices were themselves designed via this route. Software tools for ASICs are becoming more integrated with computer-aided engineering (CAE) tools for high-level design, offering the future prospect of converting DSP data-flow specifications directly into gate-array or standard cell ASIC implementations. Such technology provides the system designer with the option of reducing component costs, or reducing power supply requirements, but with development timescales more akin to those required for DSP devices. This trend is further fuelled by recent interest in hardware description languages, which are now versatile enough to be useful in designing special-purpose DSP chips [11]. VHDL (VHSIC hardware description language) — now an IEEE standard — is supported by a growing number of simulator, debugger, graphic interface and synthesizer tools [12].

ASIC devices are often incorporated into DSP system designs to mop up random logic, clock generators and other 'external' DSP functionality, in order to minimize chip count and hence size and cost.

Having reviewed the available hardware technologies, the chapter now considers some issues associated with the DSP system design process itself.

1.5   DEVELOPMENT TOOLS FOR PROGRAMMABLE DSP DEVICES

1.5.1   The DSP system design process

Currently there are no well-established formal design methods for DSP, though Fig. 1.5 shows a 'typical' system design process for a DSP application. During the conceptualization or specification stage, a signal processing requirement is expressed by an appropriate description technique, such as a data-flow diagram, a sequence of mathematical operations, formal notation or natural language. However it is achieved, a good specification technique should foster the development of a well-structured, consistent, clear and easy-to-understand statement of requirements.

For DSP applications, this is usually followed by a prototyping or experimentation phase where the specification is tested, normally via simulation using a high-level programming language (HLL). Simulation is employed to validate the specification, usually under a number of artificial mathematical assumptions. As a minimum, the prototyping phase should enable the designer to establish a target performance for the system using either simulated or real world signals for more accurate results. It also allows specific algorithmic and network interactions to be studied without loss of focus through considerations such as hardware limitations, real time constraints or even availability of the actual hardware. It facilitates the analysis, optimization and debugging of the specification with the aid of powerful computer-based graphical tools.

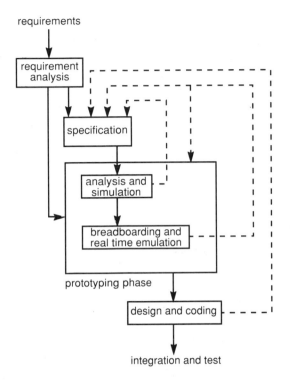

Fig. 1.5   Design process for real time DSP system.

Following high-level simulation, the restrictions of the hardware environment may be imposed by breadboarding and real time emulation using a DSP development system [13]. Key parts of the simulated system may be compiled and downloaded on to the target device for real time, or non-real time testing as appropriate, with all the analysis tools of the design environment available for system debugging, and for comparison with the simulated results [14-16]. Hence, early evidence can be provided to establish the feasibility and capability of the system.

Prototyping may also provide an opportunity to demonstrate various technical options in a format the customer can readily understand, and to build confidence in the technical solution on offer. It is vital to do this prior to embarking on the subsequent detailed design and development phases, where there will be progressively less flexibility, and more cost, in introducing changes to the system specification.

As shown in Fig. 1.5, the specification, design, simulation and emulation stages are likely to form an iterative cycle until the required performance goal is achieved.

The refined specification that emerges from the prototyping phase is then subjected to progressively detailed levels of specification, design and coding, following an appropriate development life-cycle model and method [17], until the system integration, conformance testing, field trial and engineering release of the product have been satisfactorily completed.

1.5.2   The need for support tools

Early work on DSP involved breadboarding systems using standard TTL or MOS parts prior to implementation on custom silicon (Fig. 1.3 (left)). Such prototypes were often cumbersome, inflexible, and very difficult to debug. The first few generations of programmable devices were not much better with development support limited to simple debuggers and sometimes a software simulator. Programmers could not usually afford the inefficiencies of compilers to obtain run-time code directly from the corresponding HLL. Consequently, DSP devices were usually hand-coded directly in assembler and some *ad hoc*, specialist techniques were developed to ensure that efficient (real time) execution rates were achieved. Furthermore, under pressure of timescales, programs were frequently not documented with sufficient detail, resulting in systems that were hard to maintain and even harder to modify. Code developed in this way tended to have a high design risk and also a low tolerance to changes in key staff.

Each new generation of DSP device required a considerable overhead in familiarization. Consequently, engineers tended to be reluctant to migrate to a different DSP device, even though in some cases it was clearly superior in (potential) performance terms. The availability of good support tools remains an important factor in selecting a DSP technology vendor.

As the emphasis in DSP has shifted over the last decade from hardware to software, so a number of important points have emerged which are well recognized by other software disciplines [17]:

- the cost of maintaining a complex DSP software system, i.e. bug fixing, porting to a new DSP device, and product enhancement, can be more than twice the initial development cost;

- the cost of undetected errors rises exponentially with the stage of the design life-cycle (a specification mistake, for example, could cost up to a hundred times more if it gets through to the final stages of development, than would be the case if it had been found earlier);

- too much time is spent on debugging DSP systems, the '90% complete' effect being as evident in DSP as elsewhere (the length, and cost, of the testing phase is usually underestimated — it can amount to as much as 30% of the initial development costs for a DSP product);

- there is a tendency in DSP system design to concentrate on ensuring that the functional requirements are met but, sometimes considered as incidental, the non-functional factors, such as reliability, maintainability, manufacturability and system costs, are just as important in ensuring the whole life costs are kept down to levels the market will tolerate;

- there is a shortage of real time DSP software engineers, a problem that is compounded by the fact that a certain mystique surrounds the subject, and is perpetuated by the myth that only a very small proportion of real time programmers have the 'right stuff' to be DSP system designers.

As the complexity of DSP techniques and devices has increased, so has the need for higher-level support tools. Modern devices use advanced hardware features such as parallelism and pipelining to achieve high throughputs, though these details are hidden from the programmer who has access to an HLL and associated compiler to generate run code. Ease of programming is however achieved at the expense of loss of control over these very hardware features that must be exploited to achieve efficient run time operation. There is no easy solution to this dilemma at present. Further problems are experienced when building systems which involve a mixture of heterogeneous devices, and which require coupling of concurrent synchronous (e.g. algorithmic) and asynchronous (e.g. state machine) processes.

It is clear from the preceding discussion that there is a growing need for design methods and associated environments to manage the increasing complexity of the DSP programming task, throughout the development life-cycle and beyond. To be effective, such methods and tools must embrace software and hardware integration, and need to take account of the requirement for design feedback and iteration. To the best knowledge of the authors, no such integrated tool-set for DSP system design yet exists, although prototyping environments, such as that described in Chapter 5, go some way to meeting these needs.

1.5.3   Current support tools for DSP devices

1.5.3.1   High-level language support

High-level languages, such as Fortran and C, have been popular for many years for simulation of signal processing systems, such programs being generally easier to write and maintain than assembler programs. Although commercial C compilers exist for many of the popular microcomputers, the reality is that few, if any, can currently generate run code that approaches the efficiency of hand-coding.

The sequential nature of conventional HLLs has long been recognized as a limitation in exploiting the inherent parallelism that exists in many DSP devices. To overcome compiler inefficiency, some global or local post-optimization can be applied, though to a very limited extent at present. Other recent attempts have been made to overcome the deficiencies of traditional HLLs; C has been extended to accommodate DSP requirements [18,19] and some functional languages such as Silage have been developed specifically to describe DSP applications [20]. Although employing some interesting techniques, they have yet to achieve any significant take up within the DSP community. In another approach, Spectron Microsystems (SPOX^{TM} and μSPOX^{TM}) has provided a set of optimized DSP assembler library functions for the TMS320C30 and MC56000/MC56001 which can be accessed directly via sub-routine calls from an application program written in C. This can subsequently be compiled for the target device.

1.5.3.2   CAE environments and tools for DSP

In recent years, a large number of computer-aided packages have been developed to assist the process of specifying, simulating, analysing, implementing and verifying algorithms and applications. Available tools range from digital filter design and signal analysis packages, such as FDAS^{TM} [21], Hypersignal^{TM} [22] and ILS^{TM} [23], to packages that encompass a large part of the DSP system development process [24].

Of particular interest are recent developments which allow algorithms to be specified at a high-level as a hierarchical block-oriented data-flow chart. Once specified in this way, the design can then be simulated directly without recourse to programming languages. Figure 1.6 gives a taste of this type of approach. These tools may incorporate a library of building blocks for frequently-used signal processing functions which can be used to hierarchically build more complex systems. If a required block is not available in the tool's library, the user can define a new one using an HLL (typically C or Fortran).

Commercial examples of such systems include the Signal Processing Worksystem (SPW^{TM}) from Comdisco and DSPlay^{TM} from Burr-Brown. In addition to the block diagram editor, both systems allow assembler programs for selected target DSP devices to be generated directly from the block-diagram specification. Non-commercial developments include the Gabriel system from the University of California, Berkeley, which also allows for multiple sampling rates and both automatic parallel code-generation and parallel hardware simulation [25].

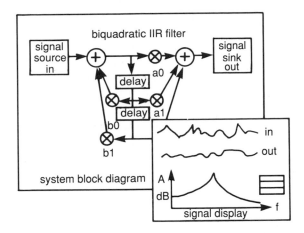

Fig. 1.6   Example of block-diagram based DSP design tool.

There are several advantages to this approach. The block diagram description of synchronous DSP systems is a natural one for most engineers, more familiar than high-level languages, and much more so than using an assembler language. Customized application modules may be put into a library to be reused by others, thereby preserving the original investment in the work. The approach encourages the use of structured programming and hierarchical decomposition, and is relatively independent of device technology. On the other hand, it must also be remembered that not all DSP systems conveniently fit the synchronous data-flow model. Difficulties can occur when modelling systems with multiple sample rates, with concurrent processing, and when handling asynchronous interrupts, although tools such as Gabriel [25] address some of these issues.

Having dealt with the technological and design issues, the following section gives a brief review of telecommunications applications for DSP.

1.6   TELECOMMUNICATIONS APPLICATIONS FOR DSP

In recent years there has been a blossoming of the use of DSP techniques, spurred on by the technological trends indicated above, and the reducing cost of the hardware. Although this chapter is mainly concerned with telecommunications, for completeness it is worth mentioning other applications which include:

- automotive — e.g. active suspension, noise reduction and 'intelligent' engine control;

- avionics and military — e.g. radar, virtual instruments, control of 'intentionally unstable' aircraft;

- hi-fi and domestic — e.g. tone controls, music synthesis, TI's Speak-&-Spell™ and voice recognition 'toys';

- medical — e.g. foetal heart monitoring, ECG processing;

- geophysical and underwater — e.g. sonar and seismic processing.

The range of telecommunications applications is very large and expanding. There follows a brief snap-shot of some active areas of DSP-related work at BT Laboratories.

1.6.1   Speech applications

Speech applications for DSP have matured considerably in recent years as a result of major advances in the following underlying techniques.

Speech recognition — this is a class of pattern recognition where features are compared with stored templates. The technology has progressed from speaker-dependent isolated-word recognition, to speaker-independent connected-word recognition systems [26, 2]. Current work focuses on sub-word recognition and large vocabulary speaker-independent recognition. Speaker verification is a related technology which provides validation of the identity of the speaker from the voice-print.

Speech synthesis — by text-to-speech using rules, or analysis-synthesis techniques (see Chapter 10).

Speech enhancement — cancellation of echoes on long international circuits to enable simultaneous two-way speech communications, or for use in loud-speaking telephones or conference terminals to prevent 'howl-round' (see Chapter 4). Interest has also grown in applications for adaptive noise suppression, especially for very noisy environments where the signal-to-noise ratio is low (see Chapter 14).

Speech coding — converts speech into digital bit streams for storage and transmission (see Chapter 11). It exploits redundancy in speech signals to reduce the transmitted bit rate, for radio-based systems such as aeronautical telephony, digital cellular and digital cordless systems. Speech codecs have also been applied in customers' private networks, e.g. for speech-and-data multiplexers. Figure 1.7 illustrates the trade-off between bit rate and quality for some 'popular' encoding techniques. More complex schemes require increasingly sophisticated generations of DSP devices for viable one-chip implementation (see Chapter 12). This is also illustrated in Fig. 1.7. Such techniques are also of current interest in encoding enhanced quality wideband speech (7 kHz bandwidth) at 64 kbit/s for ISDN applications, audio-conferencing or commentary channel transmission [27].

Interactive speech systems — using the above technologies to simplify the human interface to advanced network services for applications, such as replacement of human operators, interactive intelligent network services, voice messaging, language translation and voice-controlled information, and database services [26,28].

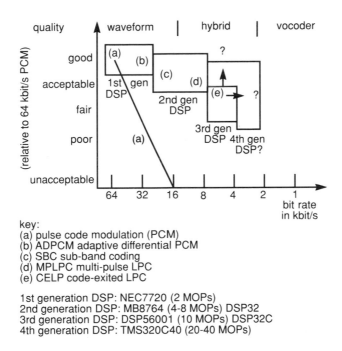

key:
(a) pulse code modulation (PCM)
(b) ADPCM adaptive differential PCM
(c) SBC sub-band coding
(d) MPLPC multi-pulse LPC
(e) CELP code-exited LPC

1st generation DSP: NEC7720 (2 MOPs)
2nd generation DSP: MB8764 (4-8 MOPs) DSP32
3rd generation DSP: DSP56001 (10 MOPs) DSP32C
4th generation DSP: TMS320C40 (20-40 MOPs)

Fig. 1.7   Speech coding quality versus bit rate.

1.6.2   Modems for the switched telephone network

Current high-speed modems are capable of operating at full-duplex rates up to 14.4 kbit/s on international two-wire PSTN circuits, and incorporate the latest adaptive filtering and error correction techniques (see Chapter 8).

Figure 1.8 shows an example of the latest PSTN modem from BT together with an expanded view of the data-pump module which incorporates all of the important CCITT Recommendations and Bell modes, and makes extensive use of advanced DSP techniques (see Chapter 7).

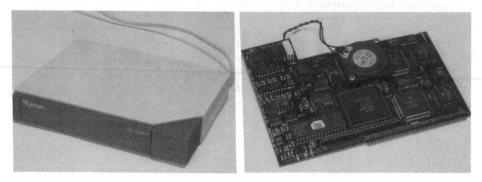

Fig. 1.8   BTL-designed DSP-based modem.

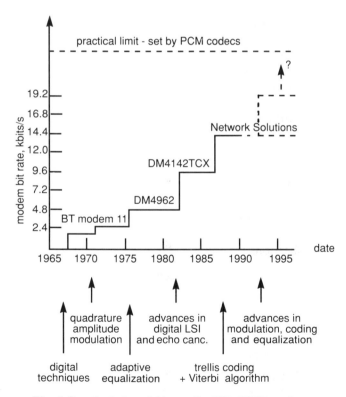

Fig. 1.9   Evolution of bit rate for BT's PSTN modems.

Figure 1.9 illustrates the evolution of practically achievable bit rates for BT PSTN modems, with an indication of some of the key technological milestones and the products in which they made their debut. The impact of DSP is evident in the fact that, since 1970, there has been an order of magnitude improvement in modem bit rates for the PSTN, whilst at the same time achieving similar reductions in cost, size and power dissipation.

1.6.3   Local network transceivers

Advances in DSP techniques are also leading to new developments for digital multichannel pair-gain systems at 2×64 kbit/s and above, over the copper loop between the customer's premises and the local exchange (see Chapter 6). Such high-rate digital subscriber line (HDSL) technology will enable broadband local access at rates up to 2 Mbit/s over unscreened copper pairs [29]. DSP technology can also be used for fault location and diagnosis on such circuits.

1.6.4   Radio access systems

DSP will play a highly significant role in new radio-access systems, as in channel coding for the emerging second-generation groupe speciale mobile (GSM) cellular radio system, second-generation digital cordless (CT2) telephones, and the emerging digital European cordless telephony standard (DECT) [30]. DSP techniques are also beginning to be deployed in digital HF, VHF and microwave radio-relay systems. In such areas new technologies such as bipolar CMOS, in combining high-bandwidth bipolar and high-density/low-power CMOS on one chip, may allow analogue and digital processing to co-exist in economic balance for large volume applications.

1.6.5   Image and video

Many visual applications are under development using DSP technology [31]:

- small-screen video-telephony for ISDN basic-rate access at rates up to 64 kbit/s;
- facsimile and still-picture processing for surveillance applications;
- video codecs for videoconference and broadcasting applications, at rates from 2 Mbit/s to 565 Mbit/s;
- 'wire-frame' videotelephones, allowing hearing-impaired people to communicate by sign language over normal telephone lines (see Chapter 15).

1.6.6 Other applications

There are also many other areas where DSP devices are being used:

- error correction and data-compression, for modems, mobile and aeronautical systems;

- digital encryption for secure data and telephony;

- port and bus interfacing for customer premises equipment;

- general-purpose mathematics and matrix processing for PCs and work-stations, including plug-in DSP development system boards [13];

- 2-D and 3-D graphics processing;

- decision-support systems, especially using artificial neural networks (see Chapter 16);

- line and interface transmission testing equipment (see Chapter 17).

1.7 DSP ALGORITHMS AND THEORY

Digital signal processing has the advantage of being built on sound mathematical foundations. The basic theory of digital filtering, both fixed and adaptive, was established long before practical implementations were available, and several decades before software techniques became popular with the single-chip DSP device.

1.7.1 What is an algorithm?

Named after Al-Kuwarizmi, a ninth century Persian mathematician, algorithms are the basic processes, procedures or sets of rules for processing signals. They often incorporate some aspect of generality, e.g. the linear-prediction algorithm, artificial neural networks and the least-mean-squares adaptive filtering algorithm. Algorithms can be 'fuzzy' and there are often many similar ways of potentially achieving the same task. A typical set of design trade-offs with reference to adaptive digital filtering algorithms is illustrated later in this section.

1.7.2   Overview

Following pioneering work on sampling by Nyquist (1924), theoretical work focused on digital equivalents of established analogue filter techniques, and on spectral analysis. Early notions of signal vectors, correlation and Toeplitz systems can be traced back to Yule (1907) and Albert Einstein (1914). The mathematics of linear-time invariant (LTI) systems (Weiner, Shannon, Gabor, Dirac and others) led to the discrete Fourier transform (DFT), followed by a plethora of 'fast' transforms, of which the fast Fourier transform (FFT) became the most widely adopted. Adaptive filtering became established from the late 1950s and early 1960s, with the Widrow-Hoff algorithm of 1959 becoming by far the most successful. This technique was subsequently extended to modems following work by Lucky [32] leading to the 'decision-directed' method of equalizer training. Further significant gains in usable bit rates for PSTN modems were made in recent years (Fig. 1.9) with the adoption of trellis-coded modulation (TCM) and the associated Viterbi algorithm (see Chapters 8 and 9).

Following the invention of pulse-code-modulation by Sir Alec Reeves in 1937, speech coding took off with the concepts of prediction-based waveform coding and linear predictive coding (LPC) (see Chapter 11). A breakthrough came in the late 1960s when the speech community rediscovered the Levinson-Durbin algorithm for solving systems of linear equations. The LPC algorithm has been extended in recent years with improved methods for excitation source modelling for bit rates between 4 and 9.6 kbit/s. Speech recognition became established over the last decade with dynamic time warping (DTW), hidden Markov modelling (HMM) and other pattern classification techniques. A variety of nonlinear signal processing techniques has been studied in recent years, of which the artificial neural network (see Chapter 16) [33] and homomorphic processing (see Chapter 10) [34] are currently favoured.

Since they are embodied in many practical DSP systems in tele-communications, it is worth briefly reviewing some key results from the extensive field of fixed and adaptive filtering.

1.7.3   Fixed digital filters

Fixed (non-adaptive) digital filters (see Chapter 2) can be broadly classified into two major groups:

- finite impulse response (FIR), where the sampled impulse response is time-limited;

● infinite impulse response (IIR), with an unbounded time response to an impulse.

Finite impulse response networks are generally based on the tapped-delay-line filter (Fig. 1.10 (a)), and as they do not employ feedback they are unconditionally stable. It is also simple to impose phase linearity on the frequency response of such networks, and it is relatively easy to predict the sensitivity of the design to finite-precision effects.

Infinite impulse response filters, on the other hand, often require less processing than FIR especially for designs which involve short transition bands or a tight bound on amplitude or phase response in the passband. Since they often employ feedback, i.e. their response contains both poles and zeros, care is needed in selecting the filter coefficients to ensure that the network remains stable. Other difficulties can occur in higher-order filters due to the build-up of quantization noise as a result of finite-precision arithmetic limitations.

A wide variety of techniques exists for designing digital filters, from the simplest windowing methods for FIR filters, to the more complex nonlinear optimization techniques for IIR designs. A number of specialist techniques, such as wave and lattice digital filtering, have also been proposed in recent years.

The interested reader is commended to a number of excellent teaching texts on the subject [34,35].

1.7.4   Adaptive digital filters

Adaptive digital filters have enjoyed considerable commercial success in telecommunications products. Most high-speed modems use adaptive equalization filters and many long-distance telephone circuits increasingly make use of adaptive echo cancellers to facilitate simultaneous two-way speech communications. Cowan and Grant provide a good overview of techniques and applications [36].

The basic principle of adaptive filtering is very simple. The aim is to modify the impulse response, and hence frequency response, of a network to minimize some measure of distortion. This is usually obtained by comparing, with a desired response, the output of the filter at regular intervals during adaption (see Chapter 4 for the definition of adaption). One popular technique is the LMS algorithm which selects the coefficients (a_N) to provide the best least-squares match to the desired response over time. This is shown diagrammatically in Fig. 1.10.

Such optimization can also be performed on IIR filters which employ feedback, though with the major difficulty of guaranteeing stability of the filter during adaption. For this reason most practical adaptive filters used in telecommunications are still based on the LMS FIR adaptive filter structure of Fig. 1.10.

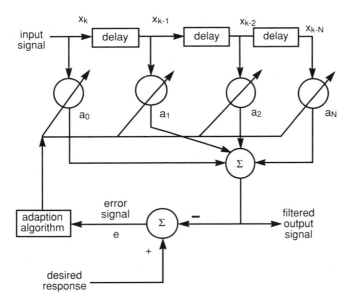

(a)   Tapped-delay-line digital adaptive filter.

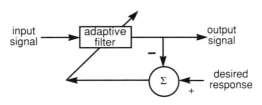

(b)   Symbolic representation.

Fig. 1.10   Adaptive digital filter.

This basic filter network can be applied to a wide variety of applications. Two examples are shown in Fig. 1.11 — the predictor, as used in waveform speech codecs (see Chapter 11) and the adaptive equalizer, as used in high-speed modems (see Chapter 7). Other applications include noise or echo cancellation (see Chapters 14 and 4), and channel modelling or estimation (see Chapter 17). Chapter 3 provides a more comprehensive review of adaptive digital signal processing techniques and Chapter 4 a review of telecommunications applications.

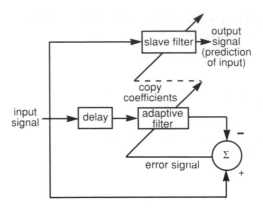

(a)   Adaptive predictor.

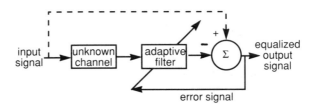

(b)   Adaptive channel equalizer.

Fig. 1.11   Examples of adaptive filter applications.

In recent years a considerable body of theoretical knowledge has been established on the subject of adaptive filtering [36]. Many of these techniques, such as fast-Kalman, recursive least squares (RLS) and lattice filters, have been studied principally with regard to their convergence and stability properties. For practical systems, however, the basic theoretical assumptions of no noise, linearity and stationarity do not usually apply and, consequently, it is usual to compare available adaptive algorithms by means of the following criteria:

- rate of convergence — the speed with which the filter settles to the desired response;

- misadjustment noise — the closeness of the match between the filter response and the desired response, after adaption;

- robustness — the sensitivity of the filter response to ill-conditioned input signals;

- tracking capability — the ability of the filter to track time variations in the channel response;

- computational complexity — the amount of processing that is required in unit time;

- numerical properties — the sensitivity of the design to wordlength limitations;

- structure and ease of implementation — the match between the processing requirements and the processor architecture.

In addition, each specific application will usually require further considerations relating to algorithm/network interactions. For an adaptive filter employed in a waveform speech codec, for example, the designer also needs to take account of factors such as:

- filter delay, which affects echo;

- transparency, which affects the ability to interwork with existing non-speech services;

- tandeming ability, which determines the degree to which distortion builds up after multiple encodings;

- robustness to link errors, which results in loss of synchronization between the encoder and decoder prediction filters;

- and last, but most importantly, the effect of the filter design on subjective quality as perceived by the customer.

This example illustrates how the design of an adaptive filter, or indeed most practical algorithms, is a complex trade-off between a number of conflicting attributes, and is very strongly dependent on the application. Such multi-dimensional optimization must rely heavily on the effective use of simulation and prototyping methods as described in section 1.5 to achieve a good balance of all the design trade-offs.

1.7.5   Other areas

Other active research areas in DSP theory include:

- higher-order spectral analysis;

- information theory and error control coding;

- nonlinear modelling;

- pattern processing and classification techniques;

- multidimensional filtering;

- motion estimation and transform coding;

- systolic processing for high-speed radar;

- linear algebra and seismic processing;

- fast-algorithms for adaptive filters.

Of particular current interest are highly-parallel DSP algorithms for targeting on the next generation 'parallel' DSP devices, and also efficient iterative algorithms for high-speed geometric and elementary function calculations, such as divide, square-root, trigonometric and vector rotations. In this context, the co-ordinate rotation digital computer (CORDIC) has been particularly prominent [37,38].

1.8   DSP AND NETWORKS

1.8.1   Network emulation, modelling and simulation

In Chapter 13 commercial DSP devices are described that were used to emulate in real time the operation of a network comprising telephone instruments and transmission plant with associated delays and losses. The advantage of this approach is that it allows new terminal or transmission plant to be subjectively evaluated under controlled and repeatable laboratory conditions. Since many transmission artefacts, such as echo, have an impact on the perceived quality of conversations, powerful DSP technology is needed to achieve duplex real time operation.

In another study (see Chapter 17), the tools associated with a DSP system design environment were used to study the interactions of a variable-rate speech codec with a packet-based network. Although the problem was complex and not amenable to real time operation, DSP devices were used to good effect as application accelerators, and to perform real time testing and debugging of elements of the system, such as the speech codec. The intention with this type of work is not to emulate the detailed operation of the network, but rather to model important specification aspects of the system, statistically or otherwise, using a combination of sample-driven and event-driven simulation.

1.8.2   International standardization

One further factor that has helped to foster interest in DSP is the emergence of stable international standards to support the use of DSP-based terminals.

A classic example of this is the V.series Recommendations for modems, promulgated by the International Telegraph and Telephone Consultative Committee (CCITT). The presence of these stable international standards has been a strong driver for the deployment of DSP in modems, and this in turn has created the market for more sophisticated DSP technology. The V.32 Recommendation for 9.6 kbit/s duplex modems, introduced in 1984, and the more recent V.32bis for 14.4 kbit/s, have been particularly influential in this respect (see Chapter 7).

Speech coding has also been subject to considerable international standardization activities in recent years (see Chapter 11). Standards now exist for 32 kbit/s adaptive differential PCM (G.721), 16 kbit/s low-delay code-excited LPC (G.728), wideband (7 kHz) speech coding (G.722), variable-rate speech coding for circuit multiplication equipment (G.724) and for wideband packet networks, amongst others. There has also been significant progress in standardization of both source and channel codecs for the second-generation cellular mobile systems (GSM), for aeronautical and maritime digital telephony, and for digital European cordless telephones (DECT).

International standards also exist for image coding, digital facsimile (Groups 3 and 4) and video and music codecs (ISO MPEG), all of which are important beneficiaries of DSP. Of particular current interest is the standardization debate on high-definition television.

1.9   FUTURE TECHNOLOGY TRENDS AND APPLICATIONS

1.9.1   Future device technology trends

The inexorable pace of silicon technology continues with an approximate doubling of transistor count every two years (Moore's Law). With today's most advanced microprocessor products touching two million transistors, one can predict single-chip programmable DSP devices with around 30 million transistors at feature sizes of less than 0.2 μm by the end of the decade. During this period, instruction cycle times may drop to below 10 ns, or 100 million instructions per second, perhaps resulting in well over 200 million parallel operations per second. Advances in interconnect technology will be required to allow full advantage to be taken of these higher speeds. Recent developments in asynchronous DSP device architectures using self-timed logic may provide one option for improving on the speed of conventional synchronous processors [39].

General-purpose microprocessor and DSP technologies continue to converge. RISC-type functionality, such as mixed integer/floating-point processing and high-speed data and instruction caches, are likely to be incorporated into future DSP devices (and vice versa) for embedded applications such as 3-D graphics and image processing.

An interesting recent development is the DSP core. These cores are programmable DSP devices which are surrounded by custom-specific logic. A good example is the MC56200 adaptive filter chip from Motorola which features a MC56001 fixed-point DSP core. The DSP56156 is another example of such a device with an on-board oversampled Sigma-Delta analogue converter combined with a core based on the 56116 fixed-point processor. A number of companies, in addition to Motorola, have announced plans to market such technology. A related trend, already evident, will be to high-performance DSP devices with scaled-down functionality, whereby very high throughputs are achieved at the expense of wordlength precision and other reductions in hardware capability.

The recently announced TMS320C40 from Texas Instruments is the first of a new wave of so called 'parallel DSP' devices combining multiple high-speed bi-directional communication processors (six at 20 Mbytes/s) and floating-point DSP functionality in a single device. Powerful arrays of such devices could be interconnected for applications requiring highly parallel processing. This significant development, in allowing networks such as those shown in Fig. 1.12 to operate in a highly co-operative fashion, may have the same kind of impact on the practical exploitation of the vast academic field of parallel algorithms that the stand-alone programmable DSP device had on sequential algorithms over the last decade [4,40]. While the TMS320C40 is intended to multiprocess with other TMS320C40s, the recently emerged low-cost AT&T DSP3210 is intended to multiprocess with Intel 80*86 or Motorola 680*0 devices. The DSP3210 together with the visible caching operating system (VCOS) and the multimedia module library are targeted at future multimedia applications running on a personal computer.

Commercial optical signal processing devices should emerge from research laboratories over the next few years. One early application for such technology could be the ubiquitous artificial neural networks [33] (see Chapter 16), where the nonlinearities and highly parallel interconnection requirements may be more readily achieved using optical detection and holographic techniques. Some combination of DSP and optical techniques may be appropriate, and possible, for such applications. Some companies are also believed to be developing Gallium Arsenide DSP chip sets, and an experimental 8-bit DSP device based on a Josephson junction has recently been reported which is capable of 1G operations per second [41].

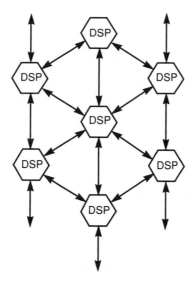

(a)   Hexagonal grid (e.g. image processing).

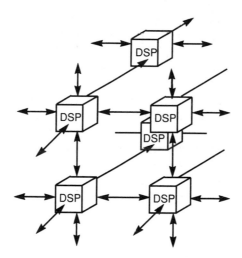

(b)   Hybercube (e.g. finite element processing).

Fig. 1.12   Parallel DSP arrays.

1.9.2   Future trends in DSP system design tools

Research and development of computer-aided tools and methods for DSP application design is an active and rapidly-growing area.

The need for assembler coding will diminish as the efficiency of automated code generation systems improve and the sophistication of DSP devices increases. Consequently, systems will increasingly be designed at a high level using powerful and integrated design environments. This trend will be spurred on by the rapid pace of technological change, which will make an investment in hand-crafted assembler code obsolete much quicker than an investment in a library of application-proven algorithms.

HLL compiler efficiency should continue to improve in stages, the biggest gains, perhaps, being realized when future microcomputer architectures and compilers are jointly designed and optimized.

A particular challenge will be the design of compilers and schedulers for the next generation of 'parallel' DSP devices featuring very high interconnection bandwidths. This technology will allow real time DSP to be extended into areas such as video and graphics processing. Such devices will be difficult to program in assembler language, so a new generation of development support tools will be necessary, allowing for both simulation and real time debugging of parallel and multi-DSP systems [42,43].

Recent studies [44] claim that some object-oriented programming (OOP) languages are suitable for solving signal processing problems. This area is likely to grow in significance as benefits of OOP, such as maintainability, portability and manageability, become more apparent to DSP system engineers [45]. It may be possible to combine new functional languages for DSP, such as Silage [20], with the block diagram approach to form an efficient design environment in the future.

Design tools and methods for DSP should become more integrated [16], perhaps allowing for a wider variety of hardware architectural options to be considered. By this means the designer may trade silicon for functionality, for example, by reducing the wordlength to what is actually needed. The Cathedral project at Katholieke Universiteit Leuven is an example of research effort directed at automatic synthesis of DSP device architectures [46].

The ability to combine analogue, DSP core and random-logic on one slice of silicon already exists. Making this facility widely available to DSP system designers will require much closer coupling of software and hardware design, simulation and test capability in emergent CASE tools. Particular issues that need addressing include design-for-test, test synthesis and combined software/hardware debugging techniques and tools.

Recently published work indicates that CAE support for multirate [47,48], multiprocessor [43], and heterogeneous systems can reasonably be expected in the foreseeable future [25]. Such tools may also make some provision for automatic task partitioning, run time scheduling and run-code optimization (see Chapter 5).

1.9.3   Future applications for DSP in telecommunications

As optical-fibre systems with virtually limitless bandwidth spread through terrestrial telecommunications networks, then the need for signal encoding and data compression may diminish on circuit-switched point-to-point connections. However, with the growth in corporate 'virtual' networking and wideband packet networks, the added flexibility afforded by variable-rate speech and image coding will be in demand for competitive bandwidth management schemes. Bandwidth will also remain under pressure in mobile and personal communications systems. In such areas, the demand for DSP technology will continue to grow strongly over the next decade, driven by international standardization, and by technical advances in the on-chip integration of digital and analogue functionality.

Demand for new interactive speech services within telecommunications networks will continue to grow, spurred on by increased customer expectations and competitive pressure to reduce core network operating costs. Advances in speech recognition, coding and synthesis and DSP technology will provide the catalysts, as in the past. Future developments could include advanced operator services and a variety of interactive voice-based database enquiry services.

As network digitalization progresses, then reduced circuit losses coupled with increasing delays at the periphery of the network will increase demand for more sophisticated echo control. This will be compounded by the anticipated growth in audio-conferencing systems, and in particular, hands-free telephony, with their increased potential for generating echo at the acoustic interface. DSP will be a key technology in controlling echo in such systems.

With the predicted rapid growth in visual-based telecommunications over the next decade, the need to process digital images will demand increasingly sophisticated DSP technology. Complex DSP products will also be required to facilitate bandwidth compression for future high definition television systems which are currently in the process of standardization. Depending on regulatory and commercial factors, new high quality audio, stereo and music-based services are also likely to emerge in telecommunications networks, as a consequence of the availability of enabling DSP technology.

As is already evident with the NEXT computer, future generations of workstation may include DSP devices as standard peripherals for a variety of purposes, including application accelerators, mathematics co-processors, audio, modem, facsimile, advanced graphics and video processing. Multimedia terminals for the ISDN should also benefit from advances in DSP for future applications such as teleconferencing and teleworking.

Other opportunities will be created in non-traditional DSP areas such as LAN, bus and disk controllers. Interest should also grow in the use of DSP devices as accelerators for numeric-intensive applications such as decision-support systems and network modelling/design, spurred on by emerging 'parallel' DSP technologies. Indeed, the on-chip convergence of DSP device, microcomputer, ASIC and analogue technology may mean that in the foreseeable future the distinctions between telecommunications and conventional computing may disappear altogether.

1.10   CONCLUSIONS

The emergence of the low-cost, programmable DSP device over the last decade has resulted in a resurgence of interest in signal processing. Many non-technical factors contribute to the current expansion in telecommunications applications for DSP — reducing cost of hardware, competitive and regulatory conditions, and international standardization activities are all important. However, DSP technologies do not just replace old practices — they transcend the limits of analogue methods making it possible to achieve results that were hitherto impossible or impracticable, the high-speed duplex modem operating at speeds up to 14.4 kbit/s on two-wire switched telephone lines being a good example.

The rapid pace of technological change presents problems for the engineer, as does the lack of an integrated design method to support the growing complexity of the programming task over the complete development life cycle. However, better tools are appearing which will slow the growth of software costs, and will improve both the quality and reusability of the design investment.

DSP is a dynamic, practically-relevant field, overlapping the traditional subjects of mathematics, electronics and computer science. Coming generations of DSP devices, in converging with high-performance microprocessor, ASIC and analogue technology, may hold the key to the much promised — but little realized — integration of telecommunications and computing.

The completion of the trunk network digitalization in the UK has led to increased customer expectations of quality and cost competitiveness. Digital signal processing has matured at the right time and has created many new opportunities in its wake. As it enters its second decade, DSP technology is well-positioned to exploit these opportunities.

REFERENCES

1. Lee E A: 'Programmable DSPs: a brief overview', IEEE Micro Magazine, pp 14-16 (October 1990).

2. Fallside F and Woods W A: 'Computer speech processing', Prentice Hall Inc, NJ (1985).

3. Widrow B and Stearns S D: 'Adaptive signal processing', Prentice Hall Inc, Signal Processing Series, NJ (1985).

4. Ahmed H M and Kline R B: 'Recent advances in DSP systems', IEEE Comms Magazine, $\underline{29}$, No 5, pp 32-45 (May 1991).

5. Davis H, Fine R and Regimbal D: 'Merging data converters and DSPs for mixed-signal processors', IEEE Micro Magazine, $\underline{10}$, No 5, pp 17-27 (October 1990).

6. Morris L R: 'Guest editor's introduction to digital signal processing microprocessor', IEEE Micro Magazine, $\underline{6}$, No 6, pp 6-8 (December 1986).

7. Higgins R J: 'Digital signal processing in VLSI', Prentice Hall Inc, Englewood Cliffs, NJ (1989).

8. Adams P F and Macmillan R: 'A MOS IC for digital filtering and level detection', IEEE Journal, $\underline{SC—16}$, No 3, pp 183-190 (June 1981).

9. Brownlie J D et al: 'Custom-designed integrated circuits for data modems', BT Technol J, $\underline{3}$, No 1, pp 14-19 (January 1985).

10. Lewis A V and South C R: 'Extension facilities and performance of an LSI adaptive filter', Proceedings of ICASSP, San Diego (March 1984).

11. Kumar K A and Petrasko B: 'Designing a custom DSP circuit using VHDL', IEEE Micro Magazine, $\underline{10}$, No 5, pp 46-53 (October 1990).

12. Harr R E: 'VHDL comes of age for designers', ASiC Technology & News, $\underline{2}$, pp 20-21 (June 1990).

13. Loughborough Sound Images: a range of PC-based and workstation DSP development systems.

14. Ip S F A, Knowles R J and Easton P W: 'Digital Signal Processing System Design Methodology', IEE Colloquium Digest No. 1990/120, IEE Computing and Control Division, pp 2/1-4 (1990).

15. Knowles R J, Ip S F A and Easton P W: 'Structured DSP design methodology', IASTED Int Conf on Signal Processing and Digital Filtering, Lugano (1990).

16. Ip S F A, Westall F A, Knowles R J, Benyon P R and Easton P W: 'A prototyping environment for digital signal processing systems', to be published in Computing & Control Engineering Journal.

17. Jackson L A: 'Software system design methods and the engineering approach', BT Technol J, $\underline{4}$, No 3, pp 7-23 (July 1986).

18. Leary K W and Waddington W: 'DSP-C: a standard high level language for DSP and numeric processing', Proceedings of ICASSP, 4, pp 1065-1068 (1990).

19. Leary K W and Cavigioli C: 'The ADSP-21020: an IEEE floating point and fixed point DSP for HLL programming', Proceedings of ICASSP, pp 1077-1080 (1991).

20. Genin D, Hilfinger P, Rabaey J, Sheers C, and De Man H: 'DSP specification using the SILAGE language', Proceedings of ICASSP, 4, pp 1057-1060 (1990).

21. Filter Design and Analysis System, Momentum Data Systems, Costa Mesa, USA (1988).

22. Hypersignal User Manual, Hyperception (1988).

23. Interactive Laboratory Systems Users' Manual, Signal Technology Inc (1986).

24. Signal Processing WorkSystem Users' Guide, COMDISCO Systems Inc (1989).

25. Lee E A, Ho W H, Goei E E, Bier J C and Bhattacharyya S: 'Gabriel: a design environment for DSP', IEEE Transactions on ASSP, 37, No 11, pp 1751-1762 (1989).

26. Wheddon C and Linggard R: 'Speech and language processing', Chapman and Hall (1990).

27. Jayant N S and Noll P: 'Digital coding of waveforms, principles and applications to speech and video', Prentice-Hall, Signal Processing Series, NJ (1984).

28. Strathmeyer C R: 'Voice in computing: an overview of available technologies', IEEE Computer Magazine, 23, No 8, pp 10-15 (August 1990).

29. Rowbotham T R: 'Local loop developments in the UK', IEEE Comms Mag, 29, No 3, pp 50-59 (March 1991)

30. Groves I: 'Personal communications — a vision of the future' BT Technol J, 8, No 1, pp 7-11 (January 1990).

31. Kenyon N: 'Audiovisual telecommunications: applications and networks', BT Technol J, 8, No 3, pp 7-14 (July 1990).

32. Lucky R W: 'Automatic equalisation for digital communications', Bell System Technical Journal, 44, pp 547-588 (April 1965).

33. Lippmann R P: 'An introduction to computing with neural nets', IEEE ASSP Magazine, pp 4-22 (April 1987).

34. Rabiner L R and Gold B: 'Theory and application of digital signal processing', Prentice-Hall Inc, NJ (1975).

35. Parks and Burrus: 'Digital filter design', Wiley (1987).

36. Cowan C F N and Grant P M: 'Adaptive Filters', Prentice-Hall, Signal Processing Series, NJ (1985).

37. Westall F A: 'Efficient DSP realisations for PSK modem receivers', Proc of IERE Int Conf on Digital Processing of Signals in Comms, Loughborough, pp 143-152 (April 1981).

38. Ahmed H M: 'Directions in DSP processors', IEEE J Sel Areas in Communications, 8, No 8, pp 1420-1427 (October 1990).

39. Jacobs G and Broderson R: 'A fully asynchronous DSP using self-timed logic', ISSCC Tech Digest, pp 150-151 (1990).

40. Simar R Jr: 'TMS320C40: a DSP for parallel processing', Proceedings of ICASSP, pp 1089-1092 (1991).

41. Kotain S et al: 'A 1 GOPs 8b Josephson DSP', ISSCC Technical Digest, pp 148-149 (1990).

42. Lauwereins R, Engels M, Peperstraete J, Steegmans E and Ginderdeuren J V: 'GRAPE: a CASE tool for digital signal parallel processing', IEEE ASSP Magazine, 7, No 2, pp 32-42 (1990).

43. Engels M, Lauwereins R and Peperstraete J A: 'Rapid prototyping for DSP systems with multiprocessors', IEEE Design & Test of Computers Magazine, 8, No 2, pp 52-62 (1991).

44. Karjalainen M: 'DSP software integration by object-oriented programming: a case study of QuickSig', IEEE ASSP Magazine, 7, No 2, pp 21-31 (1990).

45. Saunders J H: 'A survey of object oriented programming languages', Journal of object oriented programming, 1, No 6, pp 5-11 (1989).

46. De Man H, Rabaey J, Six P and Claesen L: 'Cathedral-II: a silicon compiler for digital signal processing', IEEE Design and Test, pp 13-25 (December 1986).

47. Buck J, Ha S, Lee E A and Messerschmitt D G: 'Multirate signal processing in Ptolemy', Proceedings of ICASSP, pp 1245-1248 (1991).

48. Barrera B and Lee E A: 'Multirate signal processing in Comdisco's SPW', Proceedings of ICASSP, pp 1113-1116 (1991).

2

FIXED FILTERING — A REVIEW AND SOME NEW IDEAS

L F Lind

2.1 INTRODUCTION

There are many books on fixed digital filters [1-4], but it is the intention of this chapter to cover three ideas that so far have probably not appeared in book form.

After developing some tutorial material on filter definitions, stability, and the bilinear design method, the chapter will deal with the first idea — all-pole digital filters. These filters have a constant numerator, which means that all of the calculation work is devoted to the denominator polynomial. In broad terms the denominator can then be of greater degree, which leads to some impressive gains in performance.

The second idea is a general method for designing FIR filters to meet both magnitude and phase specifications. The word filter is used loosely here, for the method also covers Hilbert transformers, phase equalizers, and so on. The advantage of the approach is that exact equations are used. This makes it fast, and easy to programme. A surprising result emerges — the best filter coefficients are equal to the eigenvector of the smallest eigenvalue of a special square matrix.

The final idea is for an important telecommunications task — the demodulation of a frequency shift keyed waveform. FIR filters are used in a very special way to achieve the result. The eye diagram that results is beautiful to behold, since it has zero telegraph distortion, and zero ripple horizontal paths from one eye crossing to the next. The demodulator can work at high speed, needing only one or two cycles of carrier per element period to achieve a high degree of perfection.

Throughout, the chapter attempts to give plausible explanations, rather than rigorous mathematical proofs which can be found by consulting the references at the end of the chapter.

2.2   DIGITAL FILTER REVIEW — BILINEAR AND ALL-POLE DESIGNS

A fixed digital filter satisfies the fixed coefficient linear difference equation:

$$y(k) = a_0 x(k) + a_1 x(k-1) + \ldots + a_p x(k-p) - b_1 y(k-1)$$
$$- b_2 y(k-2) - \ldots - b_q y(k-q), \ k = 0 \text{ to } \infty \qquad \ldots (2.1)$$

where p and q are fixed constants which determine the size of the filter. Minus signs are used in front of the b coefficients for a reason. These coefficients will become positive in the transfer function, as will be shown. The a and b coefficients are the lifeblood of the filter, determining its frequency and time domain performances. For fixed filters, the central problem is to find the best values for these coefficients.

Equation (2.1) can be calculated if there are a sequence of $(p+1)x$ values, the previous q y outputs, and the coefficients. The equation does not, however, explain how to get started. For $k=0$, the usual assumption is that all previous x and y values are zero, i.e. the memory has been reset. Otherwise, starting or boundary values must be stated to get equation (2.1) going.

If this equation is viewed as a filter structure, it has three main elements. There is the multiplier, the adder (or subtractor), and the delay storage element. Equation (2.1) can be drawn as the flow diagram of Fig. 2.1.

On Fig. 2.1 the multipliers are denoted by labels on the straight lines, the delay units by the boxes marked T, and summations by the arrowheads. The top portion, involving the a coefficients, is called the FIR (finite impulse response) part of the equation, whereas the b coefficients give rise to the IIR (infinite impulse response) part.

All of these elements are easy to implement with DSP devices. Traditionally the multiplication operation was the most troublesome, taking many clock cycles to execute. Now, however, special-purpose hardware circuitry exists to speed up the process.

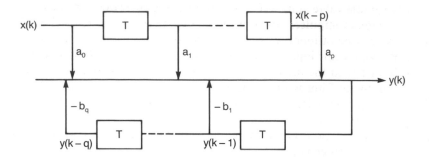

Fig. 2.1   Digital filter structure.

The memory requirement of equation (2.1) can be performed without much work. For example, it is not necessary to use shift registers. With a cyclic memory, as shown in Fig. 2.2, and the current $x(k)$ stored in 6, $x(k-1)$ in 1, and so on to $x(k-5)$ in 5, then the pointer shows where $x(k)$ is. When $x(k+1)$ arrives, it is put in 5, and the pointer moved to 5. Now the system has been adjusted so that the new $x(k)$ is in 5, $x(k-1)$ in 6, and finally $x(k-5)$ in 4. Shift register movement of the data is not necessary.

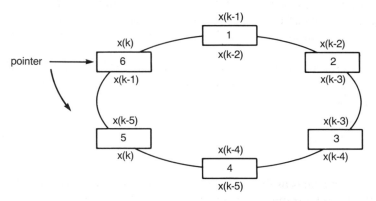

Fig. 2.2   A cyclic memory with a movable pointer.

2.2.1   Time response

The performance of equation (2.1) is usually tested by inputting a 'digital impulse', which is:

$$\delta(n) \text{ is defined as } \delta(0) = 1, \text{ and } \delta(k) = 0 \text{ for all } k \neq 0 \qquad \qquad \dots (2.2)$$

The sample value at $n = 0$ for this impulse looks strange when compared to an analogue impulse, which goes to infinite height at this point! This, however, can be rationalised on the basis that the output of a digital filter, when driven by this input, is the same as the filter's transfer function. The same is true with the analogue filter, driven by the analogue impulse.

If the digital filter is FIR, the b coefficients are zero. Figure 2.1 shows that, as a digital impulse sweeps through the delay stages, it successively causes each a coefficient to appear in turn at the output. This demonstrates the aptness of the name finite impulse response, since the output is zero when $k > p$. An FIR filter is always stable.

Next consider a pure IIR filter, where all the a coefficients except for a_0 are zero. Figure 2.1 shows that the b coefficients are in feedback loops. It is known that feedback can be dangerous, and can even cause instability, unless kept under control. This is the main worry with an IIR filter.

An unstable filter has an output that tends to infinity as k gets large. Here is an unstable filter:

$$y(k) = x(k) + 2y(k-1) \qquad \ldots (2.3)$$

For a digital impulse input, the growth in y is exponential, with the first few values $y = \{1\ 2\ 4\ 8\ 16\ \ldots\}$. In practice, infinite growth cannot take place. Sooner or later overflow will occur in the arithmetic. This is similar to an output voltage wanting to exceed the supply voltage in an active filter, causing limiting.

The coefficients of an IIR filter must be controlled so that this instability cannot occur. To see the link between coefficients and stability, the z-transform is used:

$$W(z) = \sum_{k=0}^{\infty} w(k)z^{-k} \qquad \ldots (2.4)$$

This transform is useful only if the $w(k)$ time sequence has some pattern. Then $W(z)$ can be found in closed form. If $w(k)$ consisted of, say, random numbers, then equation (2.4) cannot be computed. An infinite amount of work would need to be done to find the transform!

The z variable is in general a complex variable. It can be given any value such that equation (2.4) converges. Later, its link to the frequency domain will be seen.

Difference equations are difficult to solve. However, with the z-transform a difference equation is converted to a simple algebraic equation, which is easy to solve. For example, considering the pure IIR filter with zero initial conditions:

$$y(k) = x(k) - 0.9y(k-1) - 0.2y(k-2) \qquad \ldots (2.5)$$

and letting $x(k)$ be a digital impulse, then the transforms of $y(k)$, $y(k-1)$, and $y(k-2)$ are Y, $z^{-1}Y$, and $z^{-2}Y$, and $x(k)$ has the simple transform of 1. The transform of equation (2.5) is:

$$Y = 1 - 0.9z^{-1}Y - 0.2z^{-2}Y \qquad \ldots (2.6)$$

Solving for Y gives:

$$Y(z) = \frac{1}{1 + 0.9z^{-1} + 0.2z^{-2}} \qquad \ldots (2.7)$$

Multiplying top and bottom by z^2 gives:

$$Y(z) = \frac{z^2}{z^2 + 0.9z + 0.2} \qquad \ldots (2.8)$$

This equation can be decomposed into a sum of linear factors, by using partial fraction decomposition. By consulting a table of z-transform relations, it is found:

time sequence z-transform

$$r^n \qquad\qquad z/(z-r) \qquad \ldots (2.9)$$

To get equation (2.8) into this form, the whole equation is first divided by z. Then the denominator is factorized to give:

$$\frac{Y(z)}{z} = \frac{z}{(z + 0.5)(z + 0.4)}$$

A partial fraction expansion yields:

$$\frac{Y(z)}{z} = \frac{5}{z + 0.5} - \frac{4}{z + 0.4} \qquad \ldots (2.10)$$

The z transform is linear — the transform of a sum equals the sum of the individual transforms. Multiplying both sides of equation (2.10) by z and then using relation (2.9) gives:

$$y(k) = 5(-0.5)^k - 4(-0.4)^k \qquad \ldots (2.11)$$

Both equations (2.5) and (2.11) compute the impulse response y as $\{1 - 0.9000 \ 0.6100 \ -0.3690 \ ....\}$, showing that the method works.

This result is generalized as follows. The pure IIR difference equation, driven by a digital impulse, is:

$$y(k) = \delta(k) - b_1 y(k-1) - \ .... \ - b_q y(k-q) \qquad ... (2.12)$$

$$Y(z) = \frac{z^q}{z^q + b_1 z^{q-1} + b_2 z^{q-2} + \ .... \ + b_q} \qquad ... (2.13)$$

The roots r_i of the denominator are found, and then:

$$\frac{Y(z)}{z} = \frac{z^{q-1}}{(z-r_1)(z-r_2)...(z-r_q)} \qquad ... (2.14)$$

The partial fraction expansion and relation (2.9) show that, for $y(k)$ to be stable, all roots r_i of the denominator of equation (2.14) must lie inside the unit circle of the z-plane. Otherwise the sequence would diverge. This is a fundamental test for stability.

2.2.2   Poles and zeros

The original fixed filter equation (2.1) can be transformed to:

$$\frac{Y(z)}{X(z)} = \frac{a_0 z^p + a_1 z^{p-1} + \ .... \ + a_p}{z^q + b_1 z^{q-1} + b_2 z^{q-2} + \ .... \ + b_q} = \frac{N(z)}{D(z)} \qquad ... (2.15)$$

$D(z)$ has positive coefficients, which explains why the original equation had negative b values. In equation (2.15), the numerator $N(z)$ represents the FIR portion of the filter. The denominator $D(z)$ is responsible for the IIR behaviour, and hence controls stability. The roots of $N(z)$ are called zeros, and the roots of $D(z)$ are called the poles of the transfer function. For stability, the poles must lie inside the unit circle $|z| = 1$, as has just been seen.

Equation (2.15) has an interesting geometric interpretation via its poles and zeros. The magnitude response $|Y(z)/X(z)|$ in the complex z-plane can be thought of as a stretched rubber sheet. Zeros pin the rubber sheet down to zero value. Poles poke the rubber sheet up to infinite value. Figure 2.3(a) shows two complex poles at $0.9 \pm j0.1$, and two complex zeros at $0 \pm j1$. The log magnitude response of $1 + |Y/X|$ in the z-plane is shown in Fig. 2.3(b).

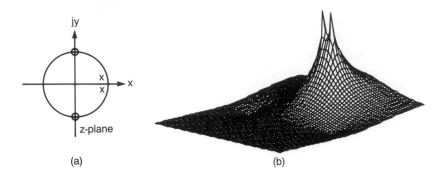

Fig. 2.3   Rubber sheet analogy to the magnitude response in the complex plane, showing (a) pole and zero locations, and (b) the rubber sheet. Poles stretch the sheet to infinity, zeros pin the sheet to the surface.

2.2.3   Frequency response

The z-transform operator z^{-1} was used above to represent a delay of T sec. Laplace transform theory gives the same delay as e^{-sT}, where s is the Laplace transform variable $s = \sigma + j\omega$. Equating these two gives $z = e^{sT}$. The frequency axis $s = j\omega$ is mapped to:

$$z = e^{j\omega T} \qquad \qquad \text{... (2.16)}$$

which is a circle of unit radius in the z-plane. The unit circle also seems to be the correct frequency axis boundary from the above discussion, for it is the boundary of stability.

The mapping of equation (2.16) is many to one; horizontal strips in the s-plane each map to the entire z-plane. A strip has the dimensions:

$$-\infty \leq \sigma \leq \infty, \ N\pi/T \leq \omega \leq (N+2)\pi/T \qquad \text{... (2.17)}$$

where N is a negative or positive integer.

Equation (2.16) shows that as ω goes from 0 to $2\pi/T$, z goes around the unit circle once in an anticlockwise direction, starting and finishing at $1 + j0$. Then as ω goes from $2\pi/T$ to $4\pi/T$, this performance is repeated, and so on.

This leads to the fact that a digital filter has a repetitive frequency response. Its response on $-\pi/T \leq \omega \leq \pi/T$ is duplicated on other strips, which are centred at $\pm 2\pi/T$, $\pm 4\pi/T$, and so on. The multiple passbands can be embarrassing; usually the input signal is conditioned by an anti-alias pre-filter to remove these higher frequencies before the signal is processed by the digital filter.

The frequency response magnitude can be inferred from rubber sheet analogy given above for poles and zeros. It is simply the height of the sheet above the unit circle, after it has been stretched by the poles and zeros. The dB magnitude response $10\log_{10}[|(Y(z)/X(z)| + 1]$ (the vertical bars denoting absolute value) is calculated on the unit circle for two complex poles at $0.9 \pm j0.1$, and two zeros at $0 \pm j1$, and displayed in Fig. 2.4.

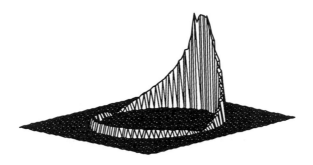

Fig. 2.4   Height of the rubber sheet above the unit circle equals the magnitude response of the filter.

The high response due to the two poles, and the two transmission zeros at $\pm j1$ are clearly seen in Fig. 2.4.

2.3   IIR DIGITAL FILTER DESIGN

Generally an IIR filter is much more selective than an FIR design, because a few poles near the unit circle can give the sharp selectivity achieved only by tens or hundreds of zeros.

Although there are many approaches to designing IIR filters, just two methods are considered here, the bilinear transformation and the digital all-pole response. When these two are compared below, a surprise will emerge.

2.3.1   Bilinear transformation

This method takes advantage of analogue filter designs, which are mapped into the digital domain. Let an analogue filter be based on the fictitious Laplace transform variable $w = u + jv$, and let $z = x + jy$ define the digital z-plane. A fictitious Laplace variable is used, because its frequency axis does not correspond to real frequencies.

The bilinear mapping is:

$$z = \frac{(1 + w)}{(1 - w)}, \text{ or } w = \frac{(z - 1)}{(z + 1)} \qquad \ldots (2.18)$$

There are at least two properties of interest with this mapping. The fictitious jv axis maps to the unit circle in the z-plane, as seen in Fig. 2.5.

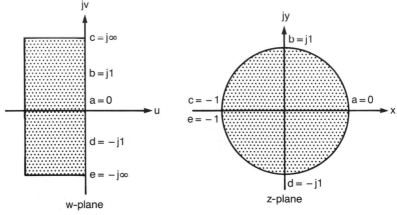

Fig. 2.5   Bilinear mapping links between the w-plane and z-plane.

Secondly, the left-half of the w-plane maps to the interior of the unit circle. For stability, all poles should lie within the shaded regions.

To see the bilinear method in action, a three-pole Chebyshev filter is designed in the w-plane. Let its cut-off frequency be at $v = 1$. The transfer function can be written:

$$H_3(w) = \frac{K}{(w - p_1)(w - p_2)(w - p_3)} \qquad \ldots (2.19)$$

To map this response to the z-plane, $w \rightarrow (z-1)/(z+1)$:

$$H_3(z) = \frac{K}{\left(\dfrac{z-1}{z+1} - p_1\right)\left(\dfrac{z-1}{z+1} - p_2\right)\left(\dfrac{z-1}{z+1} - p_3\right)} \qquad \ldots (2.20)$$

What is known about $H_3(z)$? Its magnitude squared response on the unit circle must be equiripple, because of the frequency axis mapping. The cut-off frequency is at:

$$z_c = \frac{1+j1}{1-j1} = e^{j\pi/2}$$

$H_3(z)$ has a numerator polynomial $K\,(z+1)^3$, giving three transmission zeros at $z = -1$, which correspond to the transmission zeros at infinity in the w-plane. Finally, the z-plane poles can be found by using the w to z mapping:

$$p_i(z) = \frac{1+p_i(w)}{1-p_i(w)} \qquad\qquad \text{... (2.21)}$$

If the w to z mapping does not produce the correct cut-off frequency, the w plane poles and zeros can be scaled appropriately. An example will make the above procedure clear.

Example — design a Chebyshev 3 pole digital filter, with a 1 dB passband ripple level. The cut-off frequency is 3 kHz, and the sample rate is 18 kHz.

● Find the w-plane prototype filter, with cut-off frequency $v = 1$. Standard formulae for the Chebyshev pole locations give the w-plane poles:

$$p_1 = -0.2471 + j0.9660, \; p_2 = -0.4942, \; p_3 = -0.2471 - j0.9660$$

● Scale this filter, so that it has the correct cut off frequency. The z-plane cut-off frequency has unit radius and the radian angle:

$$\phi_c = 2\pi f_c/f_s = \pi(3000)/9000 = \pi/3 \qquad\qquad \text{... (2.22)}$$

When $z = e^{j\phi_c}$, the z to w mapping gives:

$$w_c = \frac{z_c - 1}{z_c + 1} = j\tan\left(\frac{\phi_c}{2}\right) \qquad\qquad \text{... (2.23)}$$

which in this case gives $w_c = j0.5774$. For the prototype filter to have this cut-off point, all finite poles and zeros in the w-plane are scaled by 0.5774. The result is:

$$p_1 = -0.1427 + j0.5577, \; p_2 = -0.2853, \; p_3 = -0.1427 - j0.5577$$
$$\text{... (2.24)}$$

- Map poles and zeros to the z-plane. The mapping $z = (1 + w)/(1 - w)$ gives:

$$p_1 = 0.4136 + j0.6899, \quad p_2 = 0.5560, \quad p_3 = 0.4136 - j0.6899$$

Since the z-plane poles and zeros will be in complex conjugate pairs, there is no need to map all of them. It is only necessary to map poles and zeros from the upper half of the w-plane (and the real axis) to z. The other z-plane roots will be the complex conjugates of the mapped ones. This will reduce the computational work by roughly a factor of two. There are also three transmission zeros at $w = \infty$, which map to $z = -1$.

- Produce the difference equation. When the pole and zero factors are multiplied together, then:

$$H(z) = \frac{z^3 + 3z^2 + 3z + 1}{z^3 - 1.3822z^2 + 1.1070z - 0.3598} \qquad \ldots (2.25)$$

Then top and bottom are multiplied by z^{-3} to give:

$$H(z) = \frac{1 + 3z^{-1} + 3z^{-2} + z^{-3}}{1 - 1.3822z^{-1} + 1.1070z^{-2} - 0.3598z^{-3}} \qquad \ldots (2.26)$$

It is important to keep the first denominator coefficient at 1. However, $H(z)$ can be scaled by a constant to create, for example, unity transfer at d.c. $(z = 1)$. $H(z)$ then becomes:

$$H(z) = \frac{0.0455 + 0.1365z^{-1} + 0.1365z^{-2} + 0.0455z^{-3}}{1 - 1.3822z^{-1} + 1.1070z^{-2} - 0.3598z^{-3}} \qquad \ldots (2.27)$$

From this the difference equation can be written:

$$y(k) = 0.0455x(k) + 0.1365x(k-1) + 0.1365x(k-2) + 0.0455x(k-3) + 1.3832y(k-1) - 1.1070y(k-2) + 0.3598y(k-3) \qquad \ldots (2.28)$$

Although a Chebyshev filter has been used in this example, the bilinear transformation works for any magnitude squared response.

The method is easily extended to bandpass, highpass and bandstop designs. Cut-off frequencies on the unit circle of the z-plane are transferred via equation (2.23) to the jv axis of the w-plane. Then an analogue filter is designed, whose magnitude response meets the mapped specification. Finally, the poles and zeros are mapped to the z-plane, and the difference equation formed as above.

2.3.2   All-pole digital filters

To see the nature of an all-pole digital filter, consider a one-pole transfer function:

$$H_1(z) = \frac{1}{1 + a_1 z^{-1}} \qquad \ldots (2.29)$$

where a_1 can be complex.

Its magnitude squared frequency response is given by, letting $z = e^{j\phi}$:

$$
\begin{aligned}
H_1(\phi)H_1{}^*(\phi) &= \frac{1}{(1 + a_1 e^{-j\phi})(1 + a_1{}^* e^{j\phi})} \\
&= \frac{1}{1 + |a_1|^2 + (a_1 + a_1{}^*)\cos(\phi)} \qquad \ldots (2.30)
\end{aligned}
$$

where $|a_1|$ is the magnitude of a_1. An n-pole filter will consist of n products of the form of equation (2.30), and will therefore have an nth degree denominator polynomial in $\cos(\phi)$. In analogue filter work, however, the magnitude squared function is of degree $2n$. This result can be achieved with the trignometric substitution:

$$\cos(\phi) = 1 - 2\sin^2(\phi/2)$$

which then gives:

$$|H_n(\phi)|^2 = 1/P_n(\sin^2(\phi/2)) \qquad \ldots (2.31)$$

with the condition:

$$P_n(\sin^2(\phi/2)) \geq 0 \text{ for all } \phi$$

since this is a magnitude squared function. An easy way to satisfy this condition is to let:

$$P_n(\sin^2(\phi/2)) = 1 + e^2 P_n{}^2(\sin(\phi/2))$$

where P_n is an even or odd polynomial. For example, if $n = 5$ a pure odd polynomial P_5 can be used to form:

$$P_5(\sin^2\phi/2) = 1 + e^2[c_1\sin(\phi/2) + c_3\sin^3(\phi/2) + c_5\sin^5(\phi/2)]^2$$

which will be greater than zero for all ϕ. One more refinement is added. The frequency variable $\sin(\phi/2)$ is normalized to the cut-off frequency $\sin(\phi_c/2)$. Then:

$$P_{2n}(\sin(\phi/2)) = 1 + e^2 P_n^2\left[\frac{\sin(\phi/2)}{\sin(\phi_c/2)}\right]$$

which still satisfies the amplitude condition $P_{2n} \geq 0$. The transfer function magnitude squared response is then:

$$|H_n(\phi)|^2 = 1/[1 + e^2 P_n^2(\sin(\phi/2)/\sin(\phi_c/2))] \qquad \ldots (2.32)$$

where the constant e^2 gives the passband ripple level. Equation (2.32) is similar to the form of an analogue lowpass filter:

$$|H_n(v)|^2 = \frac{1}{1 + e^2 P_n^2(v)} \qquad \ldots (2.33)$$

whose cut-off frequency is assumed to be $v = 1$.

It is natural to equate the frequency variables, as follows:

$$v = \sin(\phi/2)/\sin(\phi_c/2) \qquad \ldots (2.34)$$

These frequency variables can be extended to complex variables, by using:

$$v \rightarrow w/j \qquad \ldots (2.35)$$

and

$$\sin(\phi/2)/\sin(\phi_c/2) \rightarrow (e^{j\phi/2} - e^{-j\phi/2})/[2j\sin(\phi_c/2)]$$
$$= (z-1)/[2jz^{1/2}\sin(\phi_c/2)] \qquad \ldots (2.36)$$

Equating (2.35) to equation (2.36) gives the link:

$$w = (z-1)/[2z^{1/2}\sin(\phi_c/2)] \qquad \ldots (2.37)$$

which, when solved for z, yields:

$$z = b \pm \sqrt{(b^2-1)}, \text{ where } b = 1 + 2w^2\sin(\phi_c/2) \qquad \ldots (2.38)$$

If an all-pole filter with cut-off frequency $v = 1$ is specified in the w-plane (equation (2.33)), its poles can be mapped via equation (2.38) to the z-plane to give an all-pole filter there.

To illustrate the method, a 3-pole all-pole digital filter with a passband ripple of 1 dB, and a z-plane cut-off frequency of $\pi/3$ will be found. The w-plane has the same pole locations as given in the bilinear example:

$$p_1 = -0.2471 + j0.9660, \; p_2 = -0.4942, \; p_3 = -0.2471 - j0.9660... \; (2.39)$$

When each of these is mapped to the z-plane, the sign in front of the square root in equation (2.38) is chosen such that the pole appears inside the unit circle. The result is:

$$p_1 = 0.4098 - j0.6346, \; p_2 = 0.6131, \; p_3 = 0.4098 + j0.6346 \; ... \; (2.40)$$

The pole factors are then multiplied together to form the transfer function:

$$H_3(z) = \frac{0.2906}{1.0000 - 1.4327z^{-1} + 0.0732z^{-2} - 0.3499z^{-3}} \qquad ... \; (2.41)$$

There is a factor z^{-3} in the numerator that has been ignored, since it does not contribute to the magnitude squared response on the unit circle. Also, the scale factor 0.2906 has been included in the numerator, to make the response unity at d.c. ($z = 1$).

A graph comparing the bilinear 3-pole response to the all-pole three pole response is shown in Fig. 2.6.

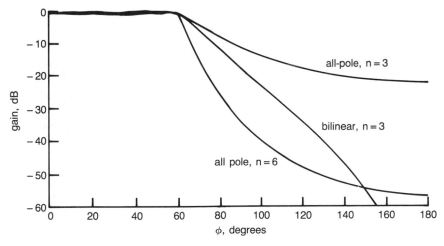

Fig. 2.6   Frequency responses of various filter designs.

It is seen that the bilinear response is more selective, but the comparison really isn't fair! The work done in computing the numerator of the bilinear response could be transferred to the denominator of an all-pole filter. The all-pole six pole response is also shown in Fig. 2.6. It is seen that this response is superior to the bilinear design, even though it has comparable numerical complexity.

In general, this result is true. It is better to compute denominator rather than numerator coefficients, when best selectivity is desired.

2.4   FIR FILTER DESIGN IN THE FREQUENCY DOMAIN

There are many methods available to find FIR filter coefficients which meet some frequency domain specification [1-4]. The one being looked at here has the features:

- it is fairly recent [5];

- it performs a joint real-part/imaginary-part optimization for arbitrary specification of these functions;

- it minimizes the integral squared error;

- it will handle any number of passbands and stopbands;

- it outperforms some other classic methods in approximation accuracy [5];

- it uses deterministic methods, resulting in quick run times compared to some other algorithms.

The filters that result from this method are called eigen-filters by Pei [5], because eigenvectors and eigenvalues are used in the computation process. The method is very general, and can also produce optimized Hilbert transformers, differentiators, and phase equalizers.

How does it work? The first step is to see how a special error function can be minimized. Let E have the quadratic form:

$$E = A^t Q A \qquad \qquad \text{... (2.42)}$$

In this equation A is a column vector of (unknown) FIR filter coefficients. The superscript t is used to denote transpose. Q is a known square matrix that is real, symmetric, and positive definite (which imply that all eigenvalues are greater than zero). It will also be assumed that the norm (squared length) of A is one, that is:

$$A^t A = 1 \qquad \qquad \text{... (2.43)}$$

If this were not the case, an obvious way to minimize E in equation (2.42) would be to set all elements of A equal to zero!

It is next shown that the best A is equal to the eigenvector of Q associated with its lowest eigenvalue. To see this, three properties about the eigenvalues λ_i and normalized eigenvectors u_i of Q are noted:

- decomposition : $Q = [u_1 \; u_2 \; ... \; u_n] \begin{bmatrix} \lambda_1 & 0 & ... & 0 \\ 0 & \lambda_2 & ... & 0 \\ & & ... & \\ 0 & 0 & ... & \lambda_n \end{bmatrix} \cdot \begin{bmatrix} u_1^t \\ u_2^t \\ ... \\ u_n^t \end{bmatrix}$

$$\text{... (2.44)}$$

- orthonormal eigenvectors: $u_i^t u_j = 0 \qquad i \neq j$

$$u_i^t u_i = 1 \qquad \qquad \text{... (2.45)}$$

- sorting: equation (2.44) can be arranged such that $\lambda_1 \leq \lambda_2 \leq ... \leq \lambda_n$

$$\text{... (2.46)}$$

With this in mind, let $A = u_1$, the eigenvector having the smallest eigenvalue λ_1. Then, from equations (2.42), (2.44), and (2.45):

$$E = \lambda_1$$

If then $A = k_1 u_1 + k_2 u_2$ (where $k_1 + k_2 = 1$) is tried, it is found that:

$$E = k_1^2 \lambda_1 + (1 - k_1^2)\lambda_2 \qquad 0 \leq k_1^2 \leq 1$$

The graph in Fig. 2.7 shows that E is smallest when $k_1^2 = 1$, that is, $A = u_1$. Similar arguments can be developed for trying to spread A amongst several eigenvectors — there is always minimal error when $A = u_1$.

This first step is summarized as follows. Given the symmetric, positive definite matrix Q, find the eigenvector u_1 associated with the smallest eigenvalue. Then the FIR coefficients $A = u_1$ give the smallest error. Fortunately, there are many efficient programmes around for finding the eigenvectors and eigenvalues of large matrices.

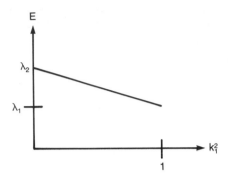

Fig. 2.7 Variation of error with $k_1{}^2$.

The second step is to knock the FIR approximation error into the form of equation (2.42). The method is illustrated with a three element filter, starting with a single frequency ω. The FIR transfer function is:

$$H(\omega) = a_0 + a_1 e^{-j\omega} + a_2 e^{-j2\omega} \text{ for } T = 1$$
$$= [a_0 + a_1 \cos(\omega) + a_2 \cos(2\omega)] - j[a_1 \sin(\omega) - a_2 \sin(2\omega)] \quad \text{... (2.47)}$$

Letting

$$A = [a_0 \ a_1 \ a_2]^t$$
$$C(\omega) = [1 \ \cos(\omega) \ \cos(2\omega)]^t, \text{ and } S(\omega) = [0 \ \sin(\omega) \ \sin(2\omega)]^t$$

equation (2.47) can be written as:

$$H(\omega) = A^{t*}C(\omega) - jA^{t*}S(\omega) = C^t(\omega)^*A - jS^t(\omega)^*A$$
$$= HR(\omega) - jHI(\omega) \quad \text{... (2.48)}$$

Also there will be a desired real and imaginary response in mind:

$$D(\omega) = DR(\omega) - jDI(\omega)$$

Looking at the squared error of the real parts at frequency ω gives:

$$ER(\omega) = [DR(\omega) - HR(\omega)]^2 \quad \text{... (2.49)}$$

In this form ER does not fit equation (2.42), since $DR(\omega)$ does not have A or A^t in it.

A reference frequency ω_0 is now defined where ER is exactly zero. Usually ω_0 will be placed in the middle of the positive frequency passband. The function $HR(\omega_0)$ does contain A, although the desired response $DR(\omega_0)$ of course does not. For zero error to exist at ω_0, the function is redefined to be:

$$ER(\omega) = \left[\frac{DR(\omega)HR(\omega_0)}{DR(\omega_0)} - HR(\omega) \right]^2 \qquad \ldots (2.50)$$

Since $HR(\omega 0)$ and $HR(\omega)$ contain A, this equation can be written as:

$$ER(\omega) = A^t \left[\frac{DR(\omega)C(\omega_0)}{DR(\omega_0)} - C(\omega) \right] \left[\frac{DR(\omega)C^t(\omega_0)}{DR(\omega_0)} - C^t(\omega) \right] A \ldots (2.51)$$

The product of the bracketed terms will be a positive definite symmetric square matrix.

The squared error of the imaginary parts gives a similar expression:

$$EI(\omega) = A^t \left[\frac{DI(\omega)S(\omega_0)}{DI(\omega_0)} - S(\omega) \right] \left[\frac{DI(\omega)S^t(\omega_0)}{DI(\omega_0)} - S^t(\omega) \right] A \quad \ldots (2.52)$$

For this method to work, it is seen that ω_0 should be chosen such that $DR(\omega_0)$ and $DI(\omega_0)$ are not close to zero. Usually one of these is zero at the origin, and so ω_0 is best chosen somewhere else in the passband, perhaps where $DR(\omega_0) = DI(\omega_0) \gg 0$.

Finally, the squared error over all passbands and stopbands is integrated:

$$E(\omega) = \int_R [ER(\omega) + EI(\omega)] d\omega = A^t Q_1 A \qquad \ldots (2.53)$$

where R includes all passband and stopband regions of interest on the frequency axis. This equation is in the form of equation (2.42). To be even more general, a weight factor $W(\omega)$ can be added to equation (2.53):

$$E(\omega) = \int_R [ER(\omega) + EI(\omega)] W(\omega) d\omega = A^t Q_1 A \qquad \ldots (2.54)$$

where $W(\omega)$ is normalized such that:

$$\int_R W(\omega) d\omega = 1 \qquad \ldots (2.55)$$

The above information may be collected together to form the final design equations.

The method for a 3-tap FIR filter whose passband is $0 \leq \omega \leq \pi/4$, and stopband is $\pi/2 \leq \omega \leq \pi$ is illustrated, with $W(\omega)$ being constant for this example. The desired response is a linear phase brick wall filter with ω_c at $\pi/4$:

$$D(\omega) = e^{-j\omega} = \cos(\omega) - j\sin(\omega) \qquad 0 \leq \omega \leq \pi/4$$

$$= 0 \qquad\qquad\qquad\qquad\qquad \pi/2 \leq \omega \leq \pi$$

where the group delay $t_g = 1$ s has been picked to suit a 3-tap filter. The zero error point ω_0 is placed at the middle of the positive passband, $\omega_0 = \pi/8$. The elements of Q_1 in equation (2.54) are [2]:

$$q(p,m) = \int_0^{\pi/4} \left\{ \left[\frac{\cos(t_g\omega)\cos(p\omega_0)}{\cos(t_g\omega_0)} - \cos(p\omega) \right] \right.$$

$$\left[\frac{\cos(t_g\omega)\cos(m\omega_0)}{\cos(t_g\omega_0)} - \cos(m\omega) \right]$$

$$+ \left[\frac{\sin(t_g\omega)\sin(p\omega_0)}{\sin(t_g\omega_0)} - \sin(p\omega) \right]$$

$$\left. \left[\frac{\sin(t_g\omega)\sin(m\omega_0)}{\sin(t_g\omega_0)} - \sin(m\omega) \right] \right\} d\omega$$

$$+ \int_{\pi/2}^{\pi} [\cos(p\omega)\cos(m\omega) + \sin(p\omega)\sin(m\omega)] \, d\omega \quad 0 \leq p, m \leq 2$$
$$\ldots \text{(2.56)}$$

Evaluation of equation (2.56) gives Q_1 to be:

$$Q = \begin{bmatrix} 1.5784 & -1.0000 & -0.0190 \\ -1.0000 & 1.5708 & -1.0000 \\ -0.0190 & -1.0000 & 1.6272 \end{bmatrix}$$

The eigenvalues of Q_1 are:

$$\lambda_1 = 0.1629, \quad \lambda_2 = 1.6218, \quad \lambda_3 = 2.9918$$

The eigenvector u_1 gives the FIR coefficients as:

$$A = u_1 = [0.5072 \ 0.7086 \ 0.4095]^t$$

It is interesting to note that although a linear phase response was specified, the A coefficients do not have symmetry about their midpoint. Thus the design attempts to balance the amplitude and phase errors. The performance of this simple filter is shown in Fig. 2.8.

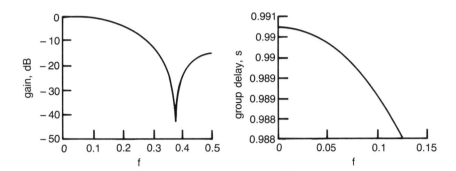

Fig. 2.8   The amplitude and group delay responses for the filter in the example. Note that only passband group delay variation is shown.

The responses illustrated in Fig. 2.8 are not bad, considering that the filter only has three taps!

As mentioned above, specifications for Hilbert transformers, differentiators, etc, can also be entered into $DR(\omega)$ and $DI(\omega)$ in equations (2.51) and (2.52). The FIR approximation is then computed. It is known that this design will always minimize the integral squared error of the real and imaginary parts, as shown by equations (2.51)-(2.53).

If the resulting error is larger than desired in some region of the passband or stopband, the weight factor $W(\omega)$ of equation (2.54) can be enlarged in this region, and be recomputed. The $W(\omega)$ function allows the error to be tailored to the designer's wishes. If for some reason this design does not fully meet the user's needs, at least it produces a result that will be close to the optimum. Then iterative procedures should quickly home in on the desired result.

Finally, it should be mentioned that linear phase filters form a special case for the method — either ER or EI is identically zero. For even symmetry, for example, a 5-tap filter would have the coefficients:

$$[a_0 \ a_1 \ a_2 \ a_1 \ a_0]$$

The frequency response is (disregarding the linear phase factor):

$$H(w) \ = \ a_2 + 2a_1\cos(\omega) + 2a_0\cos(2\omega) = b_0 + b_1\cos(\omega) + b_2\cos(2\omega)$$

The b_i can be found by setting EI to zero in the above design. It is then a simple matter to find the a_i. Thus the approach works for designs with built-in linear phase constraints.

2.5   MORE FIR FILTERS — THE FSK DISCRIMINATOR

In frequency shift keying systems, two tones are often used for sending isochronous information. For example, the V.23 modem specification uses the tones $f_1 = 1.3$ kHz, $f_2 = 2.1$ kHz, with a data rate of 1.2 kbit/s. The resulting data signal is defined to have continuous phase at element period boundaries — the waveshape is continuous.

One way of producing such a signal is by driving a VCO with the binary data waveform. It would be disastrous to form the signal by having two independent oscillators attached to an output switch (controlled by the data), for then the phase and amplitude would be discontinuous at the switching times.

Given the continuous phase modulated signal, the main problem is to detect accurately the two frequencies, so producing a clean binary output waveform. The word 'clean' needs some further discussion. If a switch from f_2 to f_1 is considered, the FSK and baseband (data) waveforms should have the appearance shown in Fig. 2.9.

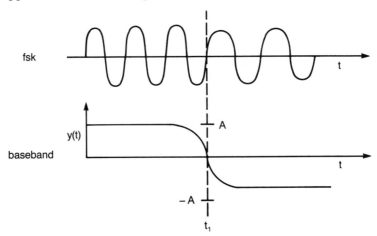

Fig. 2.9   Ideal demodulation waveform for an FSK signal.

It can be seen that the desired $y(t)$ has:

- a steady amplitude of $+A$ when f_2 is on;

- a steady amplitude of $-A$ when f_1 is on;
- a zero crossing at the element period boundary.

The first two imply that $y(t)$ is free from annoying high frequency ripple. The third condition is important for timing extraction purposes — if this is not met, $y(t)$ suffers from so-called telegraph distortion. In the design presented in Fig. 2.10, all three conditions are met; $y(t)$ then has a beautiful eye diagram.

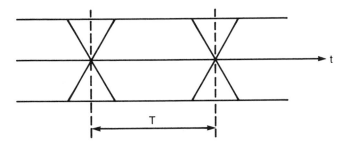

Fig. 2.10   Desired eye diagram for demodulated waveform.

It is completely open in the middle of the element period, and has precise zero crossings for timing extraction purposes. Of course an actual modem system will have impairments. The FSK input will not be ideal, due to the transmit and receive filters. There will also be system noise. However, these imperfections are ignored in designing our discriminator! The impairments (if not large) can then be mopped up by a final round of iterative optimization of the filter coefficients.

For various reasons FIR filters will be used in the design. They have a limited memory span, which frees them from pattern-dependent effects. The coefficients for different modem designs can be stored in a digital memory. To change modems, it is only necessary to redefine the coefficient addresses. Finally, the filters and most of the other processing in the modem can be implemented with a DSP device, needing only A/D and D/A converters to connect it to the outside world.

The structure of the discriminator is shown in Fig. 2.11 [6,7].

The upper two paths respond to the upper tone f_2. The *UE* FIR filter has even symmetry in its coefficients, that is:

$$ue(-k) = ue(k), \; k=1 \text{ to } n$$
$$ue(0) = 1$$

Likewise, the *UO* filter has odd coefficient symmetry:

$$uo(-k) = -uo(k), \; k=1 \text{ to } n$$
$$uo(0) = 0$$

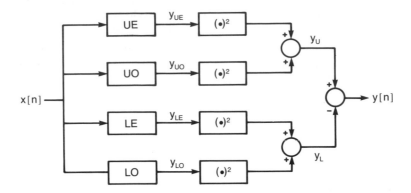

Fig. 2.11   FSK discriminator structure, incorporating four FIR filter sections.

The lower two paths respond to f_1. Again, even and odd symmetric filters are used. The reason for having even and odd symmetries is to use the fact:

$$\sin^2(x) + \cos^2(x) = 1 \qquad\qquad \dots (2.57)$$

This identity explains the reason for the squaring and addition stages in Fig. 2.11.

In the work that follows, a V.23 design will be described, having a sample frequency of 9.6 ksample/s, and using 9-tap FIR filters [8]. Because of coefficient symmetry, there are only 16 unknown filter coefficients to find.

The design is started by considering the response of the discriminator to the upper tone $\omega_2 = 2\pi f_2$. The UE filter has a (shifted) z-transform:

$$UE(z) = \sum_{k=1}^{4} ue(k)(z^k + z^{-k}) + 1$$

Using the identity $e^{jx} + e^{-jx} = 2\cos(x)$, the UE transfer function has the amplitude at ω_2:

$$UE(e^{j\omega_2}) = 2 \sum_{k=1}^{4} ue(k)\cos(k\omega_2) + 1 = a \qquad\qquad \dots (2.58)$$

where a is to be determined. Next the UO filter is constrained to have the same amplitude at ω_2:

$$UO(e^{j\omega_2}) = 2 \sum_{k=1}^{4} uo(k)\sin(k\omega_2) = a \qquad\qquad \dots (2.59)$$

Since *UE* responds to the even part of a sinusoid, and *UO* to the odd part, and since both have the same peak amplitude at frequency ω_2, the y_U summed output in Fig. 2.3 will be a^2, which is a constant level. If equation (2.59) did not hold, y_U would also have a $2\omega_2$ frequency component. The reasoning behind the structure of Fig. 2.11 is now seen more clearly.

The lower channel will not have much response at ω_2, but it is allowed to have some:

$$LE(e^{j\omega_2}) = 2 \sum_{k=1}^{4} le(k)\cos(k\omega_2) + 1 = b$$

$$LO(e^{j\omega_2}) = 2 \sum_{k=1}^{4} lo(k)\sin(k\omega_2) = b \qquad \qquad \text{... (2.60)}$$

Then $y_L = b^2$, another constant level, and the final output is:

$$y = a^2 - b^2 = A \qquad \qquad \text{... (2.61)}$$

These steps are repeated for the lower tone ω_1, but interchange a and b:

$$UE(e^{j\omega_1}) = 2 \sum_{k=1}^{4} ue(k)\cos(k\omega_1) + 1 = b$$

$$UO(e^{j\omega_1}) = 2 \sum_{k=1}^{4} uo(k)\sin(k\omega_1) = b$$

$$LE(e^{j\omega_1}) = 2 \sum_{k=1}^{4} le(k)\cos(k\omega_1) + 1 = a$$

$$LO(e^{j\omega_1}) = 2 \sum_{k=1}^{4} lo(k)\sin(k\omega_1) = a \qquad \qquad \text{... (2.62)}$$

Now the final output is the constant level:

$$y = -(a^2 + b^2) = -A$$

So far the first two objectives for $y(t)$ have been met. To satisfy the third condition (zero telegraph distortion), the situation is considered where an element boundary is in the middle of the FIR filters, and a frequency change is taking place. Specifically, let the phase angles for $T = 1$ be:

tap point = $\quad -4 \quad\quad -3 \quad\quad -2 \quad\quad -1 \quad 0 \quad 1 \quad\quad 2 \quad\quad 3 \quad\quad 4$

phase = $\phi - 4\omega_1 \quad \phi - 3\omega_1 \quad \phi - 2\omega_1 \quad \phi - \omega_1 \quad \phi \quad \phi + \omega_2 \quad \phi + 2\omega_2 \quad \phi + 3\omega_2 \quad \phi + 4\omega_2$

The final output y(0) needs to be zero at this time, regardless of ϕ. Some algebraic work shows that, if the following conditions are imposed:

$$\sum_{k=1}^{4} ue(k)[\sin(k\omega_1) - \sin(k\omega_2)] = c$$

$$\sum_{k=1}^{4} uo(k)[\cos(k\omega_2) - \cos(k\omega_1)] = c \qquad \dots (2.63)$$

where c is to be determined, then:

$$y_U = (a+b)^2/4 + c^2$$

which is independent of ϕ. Likewise, for the lower channel the following is forced:

$$\sum_{k=1}^{4} le(k)[\sin(k\omega_1) - \sin(k\omega_2)] = c$$

$$\sum_{k=1}^{4} lo(k)[\cos(k\omega_2) - \cos(k\omega_1)] = c \qquad \dots (2.64)$$

Then it can be shown that $y_L = y_U$, and so the final output $y(0) = 0$ at this time point.

Twelve constraint equations have now been accumulated for our 16 unknowns. The final equations come from a desire to minimize the noise power output of each FIR filter, subject to a white noise input. The noise power from *UE* for example is proportional to:

$$P_{ue} = \sum_{k=-4}^{4} ue^2(k)$$

The minimization of P_{ue} subject to constraints is solved by using the Lagrange multiplier method. A functional is formed:

$$L = \sum_{k=1}^{4} ue^2(k) + L_1 \left[\sum_{k=1}^{4} ue(k)\cos(k\omega_2) - \frac{a-1}{2} \right]$$

$$+ L_2 \left[\sum_{k=1}^{4} ue(k)\cos(k\omega_1) - \frac{b-1}{2} \right]$$

$$+ L_3 \left[\sum_{k=1}^{4} ue(k)\sin(k\omega_2) - \sin(k\omega_1) - c \right] \qquad \text{... (2.65)}$$

To minimize L, the following are set:

$$\partial L/\partial ue(k) = 0, \ k = 1 \text{ to } 4, \ \partial L/\partial L_1 = 0, \ \partial L/\partial L_2 = 0, \ \partial L/\partial L_3 = 0 \qquad \text{... (2.66)}$$

This gives seven equations with seven unknowns:

$$\begin{bmatrix} 1 & 0 & 0 & 0 & cu_1 & cl_1 & r_1 \\ 0 & 1 & 0 & 0 & cu_2 & cl_2 & r_2 \\ 0 & 0 & 1 & 0 & cu_3 & cl_3 & r_3 \\ 0 & 0 & 0 & 1 & cu_4 & cl_4 & r_4 \\ cu_1 & cu_2 & cu_3 & cu_4 & 0 & 0 & 0 \\ cl_1 & cl_2 & cl_3 & cl_4 & 0 & 0 & 0 \\ r_1 & r_2 & r_3 & r_4 & 0 & 0 & 0 \end{bmatrix} \begin{bmatrix} ue(1) \\ ue(2) \\ ue(3) \\ ue(4) \\ L_1 \\ L_2 \\ L_3 \end{bmatrix} = \begin{bmatrix} 0 \\ 0 \\ 0 \\ 0 \\ \frac{a-1}{2} \\ \frac{b-1}{2} \\ -c \end{bmatrix}$$

$$\text{... (2.67)}$$

where $cu_k = \cos(k\omega_2), cl_k = \cos(k\omega_1)$, and $r_k = \sin(k\omega_2) - \sin(k\omega_1)$.

This equation can be partitioned as:

$$\begin{bmatrix} I_4 & B \\ B^t & 0 \end{bmatrix} \begin{bmatrix} UE \\ L \end{bmatrix} \qquad \begin{bmatrix} 0 \\ K \end{bmatrix} \qquad \text{... (2.68)}$$

where I is the identity matrix. Let the inverse of the first matrix be:

$$\begin{bmatrix} E & F \\ G & H \end{bmatrix}$$

Then the following can be written:

$$\begin{bmatrix} E & F \\ G & H \end{bmatrix}\begin{bmatrix} I_4 & B \\ B^t & 0_3 \end{bmatrix} = \begin{bmatrix} 1_4 & 0 \\ 0 & 1_3 \end{bmatrix}$$

The F matrix is of particular interest. The above equation yields:

$$E + FB^t = 1, \quad EB = 0$$

Eliminating E gives:

$$(1 - FB^t)B = 0, \text{ or } F = B(B^tB)^{-1}$$

Returning to equation (2.68), both sides are pre-multiplied by:

$$\begin{bmatrix} E & F \\ G & H \end{bmatrix}$$

to give:

$$\begin{bmatrix} UE \\ L \end{bmatrix} = \begin{bmatrix} \cdots & F \\ \cdots & \cdots \end{bmatrix}\begin{bmatrix} 0 \\ K \end{bmatrix}$$

or:

$$UE = B(B^tB)^{-1}K \qquad\qquad \cdots (2.69)$$

where only a 3×3 matrix inversion is required.

The complete design process is summarized as follows. First, values of a, b and c are guessed. Second, equations similar to equation (2.69) are solved for each FIR filter. Third, the total signal-to-noise ratio $(a^2 - b^2)/P_{ue} + P_{uo} + P_{le} + P_{lo})$ is computed. The values of a, b and c are adjusted iteratively until this ratio is maximized.

Because of the small amount of work in equation (2.69), the complete design does not take long. Coefficient values resulting from this design are given in Rahim et al [8]. They produce the beautiful eye diagram displayed in Fig. 2.12 [8] (see also Fig. 2.10).

The conclusion is that fixed FIR filters with even and odd symmetry can work together to produce an excellent FSK demodulator.

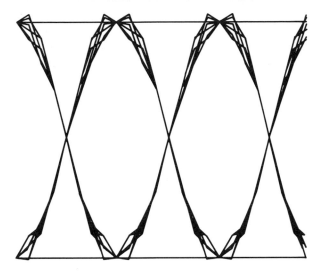

Fig. 2.12   Eye diagram for the example. Note the perfect crossing point for frequency transitions, and the horizontal pathways between transitions.

2.6   CONCLUSIONS

In this chapter the notion of poles and zeros in the z-plane was developed for fixed filters. In particular, it was shown that for stability, all poles must lie inside the unit circle.

The poles and zeros also have a direct effect on the magnitude of the frequency response, as seen by the rubber sheet analogy — in particular, by its height above the unit circle.

A method for the design of all-pole digital filters was then explained, and illustrated by example. The magnitude response of this filter was seen to be superior to that of a filter designed by using the bilinear transform, for a given amount of processing.

Next, a design method for FIR networks was presented. The key features were generality, a deterministic algorithm (no optimization necessary) and quick run times. It can be used for a variety of telecommunications applications, including group delay equalizers, joint amplitude and phase filters, and Hilbert transformers.

Finally, an FSK discriminator was designed, using fixed FIR filters. Both time and frequency constraints were placed on the design, in order to provide a good eye diagram. The eye has flat paths between sample points, and zero telegraph distortion at the zero crossings.

Hopefully, this chapter shows that there is still a lot of life in finding good fixed digital filters which satisfy telecommunications specifications. There are also times when it is best to allow a filter to become adaptive, to cope with a time variable system. Adaptive filters are the subject of Chapter 3.

REFERENCES

1.   Gold B and Rader C M: 'Digital processing of signals', McGraw-Hill (1969).

2.   Oppenheim A V and Schafer R W: 'Digital signal processing', Prentice-Hall (1975).

3.   Terrell T J: 'Introduction to digital filters', Macmillan (1980).

4.   Hamming R W: 'Digital filters (third edition)', Prentice-Hall (1989).

5.   Pei S and Shyu J: 'Eigen-approach for designing filters and all-pass phase equalizers with prescribed magnitude and phase response', IEEE Trans CAS II: Analog and Digital Signal Processing, 39, No 3, pp 137-146 (March 1992).

6.   Hughes P M, Hall M C and Lind L F: 'FSK discriminator', US Patent 5,065,409 (November 1991).

7.   Hughes P M, Hall M C and Lind L F: 'FSK discriminator', European Patent 0 305 125 B1 (June 1992).

8.   Rahim M, Goodyear C C and Hughes P M: 'Design of digital discriminator filters for voiceband FSK data modems', IEE Proceedings, 136, Pt G, No 4, pp 217-220 (August 1989).

3

ADAPTIVE FILTERS — A REVIEW OF TECHNIQUES

P J Hughes, S F A Ip and J Cook

3.1 INTRODUCTION

Practical adaptive filtering techniques have been available to the communications engineer since the mid-1960s when Lucky [1] proposed a successful technique for adaptive channel equalization. This was followed by a number of suggestions for other applications — for example, echo cancellation [2] — and it has since matured to the state where it has been usefully employed in many areas of telecommunications. A comprehensive review of current adaptive filter applications is given in Chapter 4, but, for example, adaptive filters used today enable:

- data transmission rates to be greatly increased over all types of media,

- echoes to be controlled in the telephone network,

- noise, echo and interference reduction for loudspeaking telephones and telephone kiosks,

- speech and image coding for reduced bit rate transmission systems.

This chapter firstly reviews, in section 3.2, three common system configurations using adaptive filters — direct system modelling, inverse system modelling and linear prediction. Section 3.3 studies the properties of the finite impulse response (FIR) adaptive filter (the most frequently used filter structure) in each configuration. Section 3.4 reviews the theory, analysis and performance of the adaptation techniques used with these filters, including the least mean squares (LMS) algorithm. The shortcomings of the LMS algorithm are explained and alternative adaptation schemes are discussed. The developing area of infinite impulse response (IIR) or pole-zero adaptive filters is then discussed in section 3.5. The timed-lattice filtering structure and algorithm is considered in section 3.6. Section 3.7 summarizes these linear algorithms and illustrates their behaviour using a simulation of an echo cancellation system.

Section 3.8 briefly reviews the structures and algorithms available for nonlinear adaptive filtering. Finally, the application of some of these techniques to jitter compensation is studied in section 3.9.

Adaptive filter analysis is inherently a mathematical subject and this chapter does not purport to provide a rigorous treatment — readers requiring such information are referred to the numerous text books on the subject [3-7]. The approach in this chapter is to present some important results, backed up by a mathematical analysis where appropriate, and to provide references to allow further reading. Some knowledge of signal theory, z-transforms and matrix theory is assumed.

3.2   CONFIGURATIONS FOR ADAPTIVE FILTERING APPLICATIONS

3.2.1   Open and closed loop application

There are two basic system configurations for adaptive filters — open loop and closed loop systems. In an open loop system shown in Fig. 3.1, the input signal, $x(t)$, to the adaptive filter is analysed and the filter characteristics are set up directly such that its output meets some performance criteria. There is no analysis of the success of this operation.

On the other hand, the closed loop system shown in Fig. 3.2 is a much more powerful technique, since it involves examination of the adaptive filter output signal. The input signal $x(t)$ is assumed to be in some way related to a desired signal $d(t)$, and the adaptive filter characteristics may be modified such that, according to some suitable criteria, its output, $\hat{d}(t)$, resembles $d(t)$ as closely as possible. The difference between $\hat{d}(t)$ and $d(t)$ is called the error, $e(t)$, defined as:

$$e(t) = d(t) - \hat{d}(t), \qquad\qquad \ldots (3.1)$$

The error is fed into an adaptation algorithm and used to modify the adaptive filter if required.

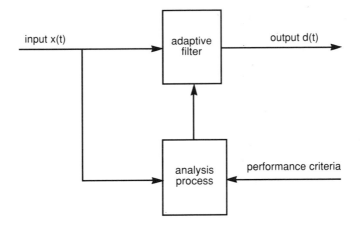

Fig. 3.1   Open loop adaptation system.

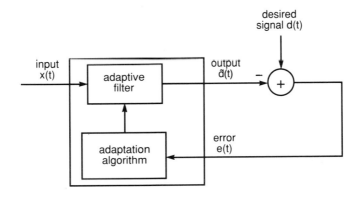

Fig. 3.2   Closed loop adaptive filter.

In a practical system some of the signals are likely to be contaminated by noise, represented in this chapter by $w(t)$. As is shown later, this can affect the performance of both the optimizing criterion and the adaptation algorithms.

The adaptation process is generally one of driving $e(t)$ towards zero. However, this is not always a realistic objective; so in practice some form of optimizing criteria is employed. Early work on adaptive equalization [1] focused on minimizing the peak distortion due to intersymbol interference.

This led to the zero forcing algorithm which gave successful operation, but the algorithm was susceptible to noise amplification, and became unstable if the peak intersymbol interference was greater than unity.

An alternative technique is to minimize the mean square error, leading to a discrete time equivalent to the Wiener-Hopf equation [8]. This is a very powerful technique since it leads to explicit definitions for the optimum filter shaping in terms of the correlation properties of the signal.

The many applications of the closed loop system may be further divided into three main configurations — direct system modelling, inverse system modelling and linear prediction. These are described next.

3.2.2 Direct system modelling

In the direct system modelling configuration, shown in Fig. 3.3, an adaptive filter is used to directly model the transfer function of an unknown system. The same signal is applied to the input of both the unknown system and the adaptive filter. The desired response is the output of the unknown system. Attempts to minimize some function of the error signal are made by modifying the adaptive filter such that it models the unknown system. In principle, it should always be possible to find a filter to exactly model the unknown system.

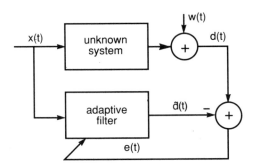

Fig. 3.3   Direct system modelling configuration.

The major practical use of this system in telecommunications is for echo and interference cancellation, where the unknown system may be either the echo response of a transmission line or the reverberation characteristic of an acoustic environment. Typically, the input signal, $x(t)$, will be either speech, data or noise.

3.2.3   Inverse system modelling

The primary use of inverse system modelling is for reducing intersymbol interference in digital receivers by equalizing the forward characteristics of a transmission line as shown in Fig. 3.4. Here the objective is to eliminate the distortion caused by the unknown system such that $\hat{d}(t) = s(t - \tau)$, where τ is the time delay through the unknown system and the adaptive filter. The adaptive filter will then ideally take the form of the inverse of the unknown system.

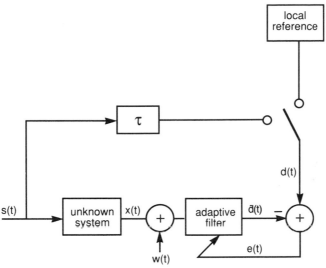

Fig. 3.4   Inverse system modelling configuration.

There are two complications with this configuration. Firstly, if the amplitude response of the unknown system were to approach zero at any frequency, the adaptive filter would require a correspondingly high gain to compensate, amplifying any noise present to unacceptable levels. In an extreme case, the required response may be unstable.

Secondly, in most cases, it is not known what was originally transmitted. There is, therefore, no desired response to facilitate adaptation. The solution is to initially adapt the system using either a local reference which is synchronized to the transmitter, or to restrict $s(t)$ to a quantized range of signals so that the receiver may make sufficient reliable estimates of $s(t)$ to allow adaptation to take place. Analysis of the system generally assumes that $d(t) = s(t - \tau)$. A very good summary of adaptive equalization techniques is given by Qureshi [9].

3.2.4   Linear predictive system

In the linear predictor case the intention is to model the characteristics of a signal so that its future values can be predicted from its previous values. For this the configuration shown in Fig. 3.5 is used where the desired signal is the instantaneous value and the input to the adaptive filter is a delayed version of the same signal.

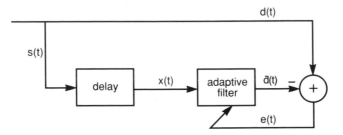

Fig. 3.5   Linear predictor configuration.

The success of the prediction is dependent on the correlative properties of the input signal — for example random noise is, by its nature, unpredictable, whereas sinusoids may be easily predicted. This system therefore is used for extracting signals from noise, where they are referred to as line enhancers, and in speech encoding schemes where the error signal rather than the input signal is transmitted. This latter application allows a dynamic range reduction, and hence a reduction in the quantity of data transmitted for a given speech quality. For more detail on linear prediction applications see Chapters 10 and 11, Makhoul [10] or Goodwin and Sin [11].

3.3   LINEAR FINITE IMPULSE RESPONSE FILTERING

3.3.1   The tapped delay line

By far the most common adaptive filter structure is the finite impulse response (FIT) filter, also referred to as a tapped delay line (TDL), non-recursive or feed-forward transversal filter. This consists of a series of delays, multipliers and accumulators as shown in Fig. 3.6. An important feature of this structure is that there are no poles in its transfer function. Consequently it is unconditionally stable, but it is rarely capable of exactly providing the required model (which would often contain poles as well). It has the major advantage of being simple to adapt. Provided a sufficient number of taps are employed — up to 250 is not uncommon — it can provide very good results.

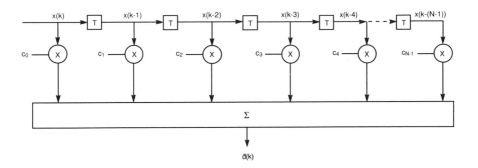

Fig. 3.6   Finite impulse response filter.

The input signal is assumed to be sampled at a Nyquist interval of T s such that the sample at a time instant kT is designated as $x(k)$. The qualities $\{c_n\}$ are the tap coefficients or multiplication weights applied to each sample on the TDL. The FIR filter output is given by:

$$\hat{d}(k) = \sum_{n=0}^{N-1} x(k-n) \cdot c_n \qquad \ldots (3.2)$$

In the analysis, it is convenient to use matrix notation. The tap coefficient and signal vectors at time kT are defined as:

$$\boldsymbol{C}^T = [c_0 \, c_1 \, c_2 \, c_3 \ldots c_{N-1}] \qquad \ldots (3.3)$$

$$\boldsymbol{X}^T(k) = [x(k) \, x(k-1) \, x(k-2) \ldots x(k-(N-1))] \qquad \ldots (3.4)$$

where bold type denotes a vector quantity and $[.]^T$ denotes a transpose. Using equation (3.1), the error at time kT is then expressed as:

$$e(k) = d(k) - \boldsymbol{C}^T.\boldsymbol{X}(k) \qquad \ldots (3.5)$$

3.3.2   Mean square error analysis

The case where the optimization is based on the mean square error (MSE) criterion is now considered. The effect of additive noise is included, since in the practical situation at least one of the signals will be derived from a noisy environment. This gives slightly different results for each configuration. Those for the direct modelling system are developed in some detail, and for the other systems the important results are simply stated.

3.3.2.1   Direct modelling systems

In this case the primary source of noise is likely to be the unknown system, resulting in a noise term, $w(k)$ added to $d(k)$. The mean square error is then given by:

$$E[e^2(k)] = E[((d(k)+w(k)) - C^T X(k))^2] \qquad \text{... (3.6)}$$

where $E[.]$ is the expectation operator. Multiplying, expanding and taking transposes where necessary gives:

$$E[e^2(k)] = E[(d(k)+w(k))^2 - 2.C^T.X(k).(d(k)+w(k))$$
$$+ \; C^T.X(k).X^T(k).C] \qquad \text{... (3.7)}$$

The expectation may be taken inside the brackets to give:

$$E[e^2(k)] = E[(d(k)+w(k))^2]$$
$$- \; 2.C^T.E[X(k).(d(k)+w(k))]$$
$$+ \; C^T.E[X(k).X^T(k)].C \qquad \text{... (3.8)}$$

Assuming that $w(k)$ is uncorrelated with either $x(k)$ or $d(k)$, all the cross-correlated noise terms become zero. Equation (3.8) therefore becomes:

$$E[e^2(k)] = E[d^2(k)] + E[w^2(k)] - 2.C^T.P + C^T.R_x.C \qquad \text{... (3.9)}$$

where:
$$R_x = E[X(k).X^T(k)]$$
is the $N \times N$ autocorrelation matrix of the input signal,   ... (3.10)

$$P = E[d(k).X(k)]$$
is the $N \times 1$ cross-correlation vector of the input and desired signals.
                                                                    ... (3.11)

Equation (3.9) is an N-dimensional quadratic equation in the tap coefficients, C, space which, if plotted in $N+1$ dimensional space against the mean square error, would give a bowl shaped 'surface', the precise shape of which would be determined by the correlation properties of the input and desired signals. This is illustrated for a 2-tap filter in Fig. 3.7. It is of course, impossible to draw this for a greater number of dimensions!

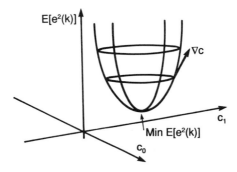

Fig. 3.7 Mean squared error surface of 2-tap FIR filter.

It is important to note that there is a global minimum and no local minima. This property enables the optimum value of C to be determined by obtaining the gradient of $E[e^2(k)]$ with respect to C and finding the position of zero slope. This gradient is defined as:

$$\nabla_c = \frac{d}{dC} E[e^2(k)] = 2.[R_x.C - P] \qquad \text{... (3.12)}$$

Equating this to zero gives the optimum value for C as:

$$C_{opt} = R_x^{-1}.P \qquad \text{... (3.13)}$$

This is a discrete time equivalent of the Wiener-Hopf equation [8] mentioned earlier in the text. So computing the optimum tap coefficient vector in a mean square sense requires knowledge of the autocorrelation properties of the input signal, and the cross-correlation properties of the input signal and the desired signal. Though the noise term, $E[w(k)^2]$ in equation (3.9) is clearly going to affect the observed MSE, its effect is simply to give an offset to the error surface, and does not affect the value of the optimum coefficient vector. However, as explained in section 3.4, it does affect the performance of some of the adaptation algorithms.

3.3.2.2   Inverse modelling system

As in the direct modelling case, the primary source of noise is likely to be the unknown network. However, in this case the noise term $w(k)$ is added to the input signal $x(k)$. The mean square error is then given by:

$$E[e^2(k)] = E[\{d(k) - C^T.(X(k) + W(k))\}^2] \qquad \ldots (3.14)$$

where $W(k)$ is the vector of noise terms. Applying the same expansion as before yields an error surface:

$$E[e^2(k)] = E[d^2(k)] - 2.C^T.P + C^T.[R_x + R_w].C \qquad \ldots (3.15)$$

where:

$$R_w = E[W(k).W^T(k)]$$
is the $N \times N$ autocorrelation matrix of the noise. $\qquad \ldots (3.16)$

The optimum coefficient vector becomes:

$$C_{opt} = [R_x + R_w]^{-1}.P \qquad \ldots (3.17)$$

So in this case the noise term has actually corrupted the shape of the error surface and the value of the optimum coefficient vector, rather than give a simple offset as in the direct modelling case. This is not necessarily a major practical problem because in a typical communications application the noise power would be significantly less than the signal power. However, this would not always be the case; so in a high noise environment the simple mean square optimizing criteria may not be the best choice.

3.3.2.3   Linear predictive system

The predictor error surface is slightly different, since the desired signal, $d(k)$, is an earlier sample of the input signal, $x(k)$ in Fig. 3.5. If the input signal comes from a noisy environment then both $x(k)$ and $d(k)$ will contain a noise term, $w(k)$. Replacing $d(k)$ in equation (3.2) with $x(k)$, and including the noise term gives:

$$E[e^2(k)] = E[\{x(k+j) + w(k+j) - C^T.(X(k) + W(k))\}^2] \qquad \ldots (3.18)$$

where j is the number of sample delays between the desired signal and the adaptive filter input.

Applying the same process as before yields a mean square error surface:

$$E[e^2(k)] = E[x^2(k)] + E[w^2(k)]$$
$$- 2.C^T.[Q_x + Q_w] + C^T[R_x + R_w]C \qquad \text{... (3.19)}$$

where:

$$Q_x = E[x(k+j).X(k)]$$
is the N-dimensional autocorrelation vector of the input signal,
$$\text{...(3.20)}$$

$$Q_w = E[w(k+j).W(k)]$$
is the N-dimensional autocorrelation vector of the noise,   ... (3.21)

and an optimum coefficient vector:

$$C_{opt} = [R_x + R_w]^{-1}.[Q_x + Q_w] \qquad \text{... (3.22)}$$

It can then be seen that the optimum coefficient vector has been seriously modified by the noise. It is now dependent on the sum of the noise and desired signal autocorrelation functions. This is what might intuitively be expected since the predictor has no noise-free reference and so treats the noise as part of the input signal.

3.3.2.4   Extension to complex signals

In some situations — for example, equalizers and echo cancellers in quadrature amplitude modulation (QAM) transmission systems — the signals involved are 2-dimensional or complex. The analysis given in this sector may be readily extended to a complex system where very similar results are obtained.

3.4   LINEAR FIR FILTER ADAPTATION TECHNIQUES

In section 3.3 it was shown that the optimum tap coefficient vector for the adaptive FIR filter could be defined by the statistical properties of the input and desired signals. This implies that if these properties were known then the optimum tap coefficients could be obtained directly. However, there is unlikely to be an accurate measurement of them available, they may be varying with time and the matrix inversion would require considerable computing power — especially if there were a large number of taps. Practical adaptation algorithms usually involve finding the optimum tap coefficient vector using iterative techniques. The following gives some algorithms suitable for practical real time applications. The analysis given here relates to the direct modelling system.

3.4.1   Gradient search techniques

A practical method is to take advantage of the shape of the error surface. Since this has a global minimum and no local minima, the tap coefficient vector, C, could be optimized by an iterative process. This approach gives rise to the well-known least mean squares (LMS) or stochastic gradient technique [12].

In this approach the value of C is successively modified using the negative direction of the gradient of the mean square error vector. After a sufficient number of iterations, the minimum point of the error surface is reached, at which the tap coefficient vector will be at its optimum, C_{opt}. This can be likened to releasing a ball at any point on the error surface. It will always fall in the direction of the steepest slope. If its motion is arrested at regular intervals of time, it will eventually reach the lowest point. This can be seen graphically from Fig. 3.6 and from the contour map of a 2-tap error surface in Fig. 3.8.

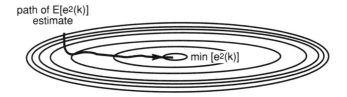

Fig. 3.8   Plan view of error surface illustrating the iterative technique for finding the minimum.

Let $\nabla_c(k)$ represent the value of the gradient vector of the error surface and $C(k)$ the tap coefficient vector at time kT. The update equation is then given by:

$$C(k+1) = C(k) + \beta.(-\nabla_c(k)) \qquad\qquad \dots (3.23)$$

where β is a non-negative constant which controls the step size at each iteration. Substituting the value of $\nabla_c(k)$ given by equation (3.12) into equation (3.23) gives:

$$C(k+1) = C(k) + 2.\beta.[P - R_x.C(k)] \qquad \ldots (3.24)$$

which is an iterative technique of obtaining C_{opt} called the steepest descent algorithm. This algorithm is really of academic interest only, since it still requires the expectation operation to obtain the values of P and R_x. However, the matrix inversion operation has now been removed and so, if some suitable approximation to P and R_x can be made, a practical technique will evolve. The simplest approximation is to use the instantaneous values of $x(k)$ and $d(k)$, i.e. the expectation in equations (3.10) and (3.11) are deleted to give:

$$R_x \approx X(k).X^T(k) \qquad \ldots (3.25)$$

$$P \approx d(k).X(k) \qquad \ldots (3.26)$$

Substituting these new values for P and R_x into equation (3.24) gives the well-known least mean squares (LMS) algorithm, first described by Widrow and Hoff [12]:

$$C(k+1) = C(k) + 2.\beta.X(k).e(k) \qquad \ldots (3.27)$$

$$e(k) = d(k) - X^T(k).C(k) \qquad \ldots (3.28)$$

Despite the crudeness of the approximation involved, the LMS algorithm will converge to the optimum coefficient vector in a wide variety of situations — a feature attributable to the averaging nature of the algorithm itself. It has become very popular since it may be implemented using only two multiply accumulate operations per tap coefficient in each sample interval. This is referred to as having a processing order of '2N'. Discussion now covers the adaptation features of the algorithm.

3.4.2 LMS adaptation performance

Since the optimum coefficient vector is obtained by following the contours of the mean square error surface, the rate of convergence will be dependent on their steepness in each of the N dimensions. For example, the convergence of the mean square error (MSE) in Fig. 3.8 will be rapid where the contour is steep, but will slow down during the final approach to the optimum coefficient vector.

The adaptation rate is usually analysed in terms of the error in the tap coefficient vector, given by:

$$\epsilon(k) = C(k) - C_{opt} \qquad \qquad \text{... (3.29)}$$

If equations (3.27) and (3.28) are written in terms of this error vector, it can be shown [3] that the mean square of this coefficient error vector adapts according to:

$$E[\epsilon(k+1)] = [I - 2.\beta.R_x].E[\epsilon(k)] \qquad \qquad \text{... (3.30)}$$

where I is the identity matrix.

It follows that $\epsilon(k)$ adapts towards zero following a path which is the sum of N 'trajectories', of the form:

$$(1 - 2.\beta.\lambda_0)^k + (1 - 2.\beta.\lambda_1)^k + ... + (1 - 2.\beta.\lambda_{N-1})^k \qquad \text{... (3.31)}$$

where $\lambda_0...\lambda_{N-1}$ are the eigenvalues of the autocorrelation matrix, R_x.

In general these eigenvalues will not be equal, so the largest is referred to as λ_{max} and the smallest as λ_{min}. The observed effect on the error, $e(k)$, will be such that the error slope has several different values as $e(k)$ reduces. This is illustrated for a 3-tap system in Fig. 3.9.

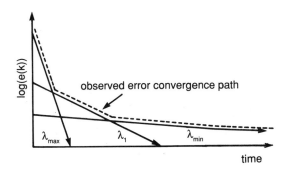

Fig. 3.9   Observed effect of eigenvalue spread on the error $e(k)$.

For stable convergence each term in equation (3.31) must be less than unity, so for convergence the following must exist:

$$0 < \beta < \frac{1}{\lambda_{max}} \qquad \qquad \text{... (3.32)}$$

This sets the maximum value of β, though this is not a sufficient condition for stability under all signal conditions. The final convergence rate of the algorithm is determined by the value of the smallest eigenvalue. An important characteristic of the input signal is therefore the eigenvalue spread or disparity, defined as:

$$\frac{\lambda_{max}}{\lambda_{min}} \qquad \qquad \text{... (3.33)}$$

So, from the point of view of convergence speed, the ideal value of the eigenvalue spread is unity; the larger the value, the slower will be the final convergence. It can be shown [3] that the eigenvalues of the autocorrelation matrix are bounded by the maximum and minimum values of the power spectral density of the input. Furthermore, as the order of the matrix $N \to \infty$:

$$\lambda_{min} \to \min \, [S(\omega)] \qquad \qquad \text{... (3.34)}$$

$$\lambda_{max} \to \max \, [S(\omega)] \qquad \qquad \text{... (3.35)}$$

where $S(\omega)$ is the power spectral density.

It is therefore concluded that the optimum signal for fastest convergence of the LMS algorithm is white noise, and that any form of colouring in the signal will increase the convergence time.

This dependence on convergence of the spectral characteristics of the input signal is a major problem with the LMS algorithm.

3.4.3 LMS echo canceller adaptation performance — uncorrelated inputs

Whilst the assumption of white noise input is not valid in the case of speech, it is close to the situations often encountered in data transmission applications and provides an indication of the behaviour of the system under ideal conditions. It has been shown by Messerschmitt [13] that the mean square tap coefficient error vector, equation (3.29), of an LMS adapted echo canceller is given by:

$$E[\epsilon^2(k)] = E[\epsilon(0)]^2.L^k + 4.\beta^2.N.E[x^2(k)].E[w^2(k)].\frac{1-L^k}{1-L} \quad \text{... (3.36)}$$

where:

$$L = 1 - 4.\beta.E[x^2(k)] + 4.\beta^2.N.E[x^2(k)]^2,$$
$$N = \text{number of taps,}$$

$E[w^2(k)]$ = interference noise power on the desired signal,
k = iteration number,
β = step size,
all other symbols having their previous meanings.

Similar expressions for a complex system have been derived by Weinstein [14]. This error in the coefficient vector may be related to the mean square error in the system by:

$$E[e^2(k)] = E[\epsilon(k)]^2.E[x^2(k)] + E[w^2(k)] \qquad \text{... (3.37)}$$

For binary data and $E[w^2(t)] = 0$, the convergence rate is:

$$10.\log (1 - 4.\beta.E[x^2(k)] + 4.\beta^2.N.E.[x^2(k)]^2) \text{ dB/iteration} \quad \text{... (3.38)}$$

The fastest convergence rate is obtained when $\dfrac{dL}{d\beta}$ equals zero, giving the optimum value for β as:

$$\beta_{opt} = \frac{1}{2.N.E[x^2(t)]} \qquad \text{... (3.39)}$$

The adaptation becomes unstable if $L > 1$. This sets the maximum value of β at:

$$\beta_{max} = \frac{1}{N.E[x^2(t)]} \qquad \text{... (3.40)}$$

For non-zero $E[w^2(t)]$ the mean steady-state error is obtained by setting $k = \infty$ in equation (3.36):

$$E[\epsilon(\infty)]^2 = \frac{\beta.N}{1 - \beta.N.E[x^2(t)]} \cdot E[w^2(t)] \qquad \text{... (3.41)}$$

3.4.4   Other gradient search techniques

The LMS algorithm arises from using instantaneous values to estimate the error surface gradient. Several algorithms have been proposed which attempt to gain a performance advantage by obtaining an improved estimate.

In the block LMS (BLMS) algorithm [15,16] the data is divided into blocks and the update to the filter is based on the complete data in each block. This approach lends itself to adaptive filtering in the frequency domain. The convolutions in equation (3.2) then become multiplications, which in some applications can lead to a decrease in the processing requirements.

Another technique is the momentum LMS (MLMS) algorithm, where the previous two coefficient vectors are used in the update algorithm, resulting in the 2nd order update algorithm:

$$C(k+1) = C(k) + 2.\beta.X(k).e(k) + \gamma[C(k) - C(k-1)] \qquad \dots (3.42)$$

where γ is a scalar < 1.

This has been analysed by Shink and Roy [17,18] who reported smoother, but not necessarily faster, adaptation.

In some practical situations the adaptation process may only be performed after a fixed delay, giving rise to the delayed LMS (DLMS) algorithm. The LMS equations (3.27) and (3.28) become:

$$C(k+1) = C(k) + 2.\beta.X(k-\phi).e(k-\phi)e(k) = d(k) - X^T(k).C(k)$$

where ϕ is the number of sample delays in the update operation. Provided ϕ is within a certain bound the behaviour is similar to the LMS algorithm, the differences being slightly slower adaptation and a reduced maximum value for β. The algorithm has been studied by Long et al [19] and Haimi-Cohen et al [20].

3.4.5 Fast adaptation algorithms

At the beginning of this section it was observed that direct solving of equation (3.13) to obtain C_{opt} was impractical because the auto- and cross-correlation functions were not known and the matrix inversion was a heavy computational burden. However, it is possible to solve for C_{opt} using an iterative technique known as the recursive least squares (RLS), or Kalman algorithm[1]. This has a variety of forms, which are given in detail in Haykin [3] and Grant et al [5]. For this, rather than the mean square error, the following summation is minimized:

$$\sum_{n=0}^{k} (d(n) - \hat{d}(n))^2.\alpha^{n-k} \qquad \dots (3.43)$$

where α is a constant, $0 < \alpha < 1$, and the other symbols have their previous meanings.

It should be noted that equation (3.43) is an algebraic rather than a statistical quantity as used in the gradient search techniques. The constant, α, is used to reduce the influence of the older samples to enable the algorithm to be used with non-stationary signals.

---

[1] Not to be confused with Kalman filters.

The data samples moving through the TDL are used to provide estimates of the correlation data, and the matrix inversion lemma — an iterative technique of finding the inverse of a matrix — may be used. The update sequence is:

$$C(k) = C(k-1) + r_x^{-1}(k).X^T(k).e(k) \qquad \text{... (3.44)}$$

where

$$r_x(k) = \sum_{n=0}^{k} X(n).X^T(n) \qquad \text{... (3.45)}$$

$$e(k) = d(k) - C^T(k-1).X(k) \qquad \text{... (3.46)}$$

The matrix inversion iteration is:

$$r_x^{-1}(k) = \frac{1}{\alpha}.\left(r_x^{-1}(k-1) - \frac{r_x^{-1}(k-1).x(k).x^T(k).r_x^{-1}(k-1)}{\alpha + x^T(k).r_x^{-1}(k-1).x(k)}\right) \text{... (3.47)}$$

This algorithm gives very fast convergence times — in the order of N iterations. It should be noted that all the available data is used in forming $\hat{d}(t)$, rather than the instantaneous data in the LMS algorithm. An added benefit is that the convergence behaviour is not dependent on the eigenvalues of the input signal. The drawback is that the processing requirements are of the order of N^2 — significantly greater than the order of $2N$ for the LMS algorithm.

A recent development is the fast Kalman algorithm [21]. This takes advantage of the structure of the autocorrelation matrix to obtain significant reductions in processing requirements — the order of $15N$. However, it is a less regular algorithm and can become unstable under some circumstances.

3.5   A COMBINED LINEAR INFINITE AND FINITE IMPULSE RESPONSE ADAPTIVE FILTERING STRUCTURE

The desired physical responses will often contain both poles and zeros and so the all-zero filters described in sections 3.3 and 3.4 can only approximate the required response. A strong pole may require several hundred zeros for satisfactory performance, resulting in a long delay line and leading to high cost and computational overhead. There is therefore a big attraction in using adaptive infinite impulse response (IIR) filters. These have the ability to exactly match any linear physical system. The drawback is that the adaptation process is significantly more complicated, the system is potentially unstable and more susceptible to noise. There have been a number of different approaches to adaptive IIR filtering [22]. The structures generally fall into two distinct categories — output error and equation error.

3.5.1   Output error adaptation

Output error adaptation is also known as the parallel technique since it uses the system error to adapt an all-poles structure directly. There are many versions of this, for example, the recursive prediction error (RPE) algorithm [23], and SHARF [24], based on the concept of hyperstability. A good summary is given by Shink [22].

The major problem with output error techniques is that the error surface is not quadratic and contains local minima [25]. This means that gradient search techniques are not guaranteed to find the optimum coefficient vectors. Furthermore, it is necessary to perform stability checks after every update iteration, so processing requirements are high.

3.5.2   Equation error structures

These structures overcome some of the problems with the output error adaptation by creating an auxiliary all-zero section in which to perform the adaptation using FIR techniques. This is also known as the series-parallel technique, shown in Fig. 3.10. Sections $A(z)$ and $B(z)$ are adaptive FIR filters which are adapted to minimize the equation error, $e(k)$. $A(z)$ models the zeros,

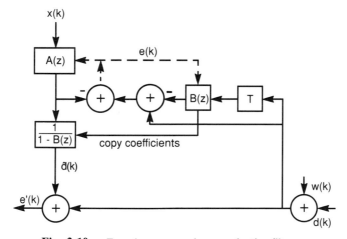

Fig. 3.10   Equation error pole-zero adaptive filter.

$B(z)$ the poles. $B(z)$ is then used to create a true all-poles filter which is used to form the output $\hat{d}(k)$, and hence the output error, $e'(k)$. The close similarities between the system surrounding $B(z)$ and the linear predictor system in Fig. 3.5 should be noted.

This structure has been investigated by several workers — Long, Schwed and Falconer [26] and Hughes [27]. The stability problem is reduced since the potentially unstable pole section is outside the adaptation loop. These have good adaptation properties (i.e. no local minima in the error surface), but have poor performance in the presence of noise and minimizing $e(k)$ does not necessarily minimize $e'(k)$. It is easy to show that the z-transforms of the two errors are related by:

$$E'(z) = E(z) \cdot \frac{1}{1 - B(z)} \qquad \qquad \text{... (3.48)}$$

An alternative implementation has been proposed by Jenkins [28] and Fan [29], in which the order of the pole and zero sections is reversed. This gives some improvements in the adaptation properties, but the stability problem is increased because the pole section is then inside the adaptation loop.

3.5.3   Mean squared error analysis of pole-zero structure

The direct modelling case is considered with the noise term initially considered to be zero. Using matrix notation as before the mean square error at time kT may be expressed as:

$$E[e^2(k)] = E[(d(k) - B^T.D(k-1) - A^T.X(k))^2] \qquad \text{... (3.49)}$$

where D is the vector of samples of the desired signal, and A and B are the vectors of zero and pole tap coefficients respectively.

Expanding as before gives the mean square error surface in terms of the correlation properties:

$$\begin{aligned} E[e^2(k)] &= E[d^2(k)] + B^T.R_d.B \\ &\quad + A^T.R_x.A - 2B^T.Q - 2A^T.P \\ &\quad + 2B^T.R_{dx}.A \end{aligned} \qquad \text{... (3.50)}$$

where

$$R_d = E[D(k).D^T(k)] \qquad \qquad \text{... (3.51)}$$

is the $M \times M$ autocorrelation matrix of the desired signal,

$$R_x = E[X(k).X^T(k)] \qquad \ldots (3.52)$$

is the $N \times N$ autocorrelation matrix of the input signal,

$$R_{dx} = E[D(k-1).X^T(k)] \qquad \ldots (3.53)$$

is the $M \times N$ cross-correlation matrix of desired and input signals,

$$Q = E[d(k).D(k-1)] \qquad \ldots (3.54)$$

is the $M \times 1$ autocorrelation vector of the desired signal,

$$P = E[d(k).x(k)] \qquad \ldots (3.55)$$

is the $N \times 1$ cross-correlation vector of the desired and input signal.

The error surface given by equation (3.50) is an $M + N + 1$ dimensional surface. Expressions for the optimum coefficient may be obtained in the same way as before. Taking partial derivatives of equation (3.50) with respect to A and B and equating to zero gives:

$$A_{opt} = R_x^{-1}.[P - R_{dx}^T.B] \qquad \ldots (3.56)$$

$$B_{opt} = R_d^{-1}.[Q - R_{dx}.A] \qquad \ldots (3.57)$$

which are the same as for the all-zero case but with an extra term consisting of the cross-correlation matrix multiplied by the other coefficient vector. Thus the optimum setting for each tap coefficient vector is dependent on the value of the other vector.

3.5.4 The effect of noise on the pole-zero system

If $d(k)$ is expanded to include the noise component $w(k)$, and the noise is assumed to be uncorrelated with $d(k)$ or $x(k)$, the optimal coefficient vectors become:

$$A_{opt} = R_x^{-1}.[E[y(k).X(k)] - E[X(k).D^T(k-1)].B] \qquad \ldots (3.58)$$

$$B_{opt} = [R_d + R_w]^{-1}.(E[d(k).D(k-1)] \\ - E[w(k).W(k-1)] - E[D(k-1).X^T(k)].A) \qquad \ldots (3.59)$$

It can therefore be seen that the optimum tap coefficient vectors are strongly affected by noise. This effect is referred to as coefficient bias, and represents a major problem with the equation error approach.

3.5.5   LMS adaptation for the pole-zero system

This is obtained using the same process as for the all-zeros system, but the gradient vector with respect to both A and B must be taken. This gives the steepest descent equations as:

$$A(k+1) = A(k) + 2.\beta_a.[P - R_x.A(k) - R_{dx}^T.B(k)] \qquad \ldots (3.60)$$

$$B(k+1) = B(k) + 2.\beta_b.[Q - R_d.B(k) - R_{dx}.A(k)] \qquad \ldots (3.61)$$

where β_a and β_b are the step size for the zero and pole sections.

The LMS algorithm is obtained by replacing the expected value with the instantaneous values to give:

$$A(k+1) = A(k) + 2.\beta_a.X(k).e(k). \qquad \ldots (3.62)$$

$$B(k+1) = B(k) + 2.\beta_b.D(k-1).e(k) \qquad \ldots (3.63)$$

$$e(k) = d(k) - B^T(k).D(k-1) - A^T(k).X(k) \qquad \ldots (3.64)$$

This has exactly the same form as the all-zeros case and so may be implemented using approximately $2(N+M)$ multiply accumulate instructions.

Moving on to the LMS adaptation properties of the pole-zero system, the coefficient error vectors are defined as:

$$\epsilon_a(k) = A(k) - A_{opt} \qquad \ldots (3.65)$$

$$\epsilon_b(k) = B(k) - B_{opt} \qquad \ldots (3.66)$$

It may be shown [27] that the coefficient error vectors after the kth iteration are given by:

$$E[\epsilon_a(k+1)] = [I - 2.\beta_a.R_x] .E[\epsilon_a(k)] \qquad \ldots (3.67)$$

$$E[\epsilon_b(k+1)] = [I - 2.\beta_b.R_d] .E[\epsilon_b(k)] \qquad \ldots (3.68)$$

which might appear to show that each set of tap coefficients converges to its current optimum value at a rate dependent on the autocorrelation matrices of their input signals. However, there is a different optimum value of A for each value of B, and vice versa, so the adaptation of the pole and zero sections to their mutual optimum values is not truly independent. Nevertheless, this is a useful result because it enables the gain values, β_a and β_b, to be chosen based on the power in the appropriate signals.

An interesting feature not revealed by this analysis is that, for good adaptation performance, the adaptive filter must be such that there is an equal number of spare poles and zeros above that required to exactly match the unknown system. This is described in Long et al [26] and Hughes [27].

3.6   A LINEAR COMBINED LATTICE AND FIR ADAPTIVE FILTERING STRUCTURE

In section 3.4 it was shown that the adaptation rate of an FIR filter adapted using the LMS algorithm was dependent on the eigenvalue spread of the autocorrelation matrix of the input signal, and that this could be improved by employing the computationally more expensive RLS algorithm. In the adaptive lattice filter the improvement in adapatation performance is achieved by modifying the actual structure of the filter. A new chain of adaptive reflection coefficients, $\{K(k,m)\}$, is added, as shown in Fig. 3.11.

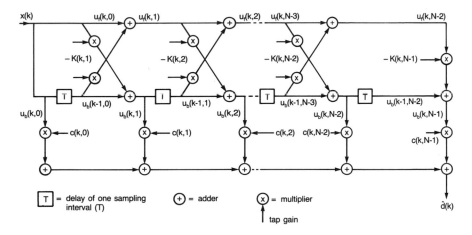

Fig. 3.11   Adaptive lattice structure.

The filter maintains an FIR character, but it now has the property that each lattice stage has a decorrelating effect on the signal which speeds up the rate of convergence under many signal conditions [30-34]. This is because the lattice structure has the ability to orthogonalize correlated input signals before being fed into the transversal filter in order to reduce the amount of interaction between the transversal filter taps. The degree of orthogonalizing is data-dependent — for example, an uncorrelated signal cannot be further orthogonalized.

The lattice filter structure handles correlated signals well but suffers from an effect called time-spread [34] which eventually requires more taps in the transversal filter. It also suffers from a higher level of algorithm noise [34], resulting in a higher steady state error despite the improvement in convergence speed. In channel equalization, Rutter and Grant [34,35] suggested the freezing of the values of the reflection coefficients after a given number of iterations, which is known as the timed lattice algorithm. However the effectiveness of this technique requires the signal to be stationary. The method reduces the level of algorithm noise and hence improves the value of the steady state error. It also overcomes the poorer performance of the lattice filter with white noise input. The effect of the timed lattice filter will be studied later in section 3.7 in the simulation of an echo canceller.

3.6.1   Timed lattice filter adaptation algorithm

The lattice filter is employed to model an unknown system and minimize the mean square value of $e(k)$ in Fig. 3.2. The lattice filter comprises two parts, an adaptive lattice filter with N_s stages and a transversal filter with N taps. Figure 3.11 depicts the complete lattice filter with the number of stages in the lattice filter, N_s, equal to $N-1$ (the value of N_s is not necessarily required to be equal to $N-1$).

3.6.1.1   The lattice section

The data in the new nodes are referred to as the forward and backward residuals, $\{u_f(k,m)\}$ and $\{u_b(k,m)\}$ respectively. These are generated from the input signal using the lattice coefficients $\{K(k,m)\}$. The timed lattice algorithm differs from the standard lattice filter in that these coefficients are only allowed to adapt for short timed periods.

On the receipt of the new input sample $x(k)$, at time kT, the lattice residuals are generated by:

$$u_f(k,0) = x(k) \qquad\qquad \ldots (3.69)$$

$$u_f(k,m) = u_f(k,m-1) - K(k,m).u_b(k-1,m-1) \text{ for } m = 1,2,\ldots\ldots.N_s$$
$$\ldots (3.70)$$

$$u_b(k,0) = x(k) \qquad\qquad \ldots (3.71)$$

$$u_b(k,m) = u_b(k-1,m-1) - K(k,m).u_f(k,m-1) \text{ for } m = 1,2,\ldots\ldots.N$$
$$\ldots (3.72)$$

If there are less stages in the lattice than the transversal section, i.e. $N_s < N - 1$, then:

$$u_b(k, N_s + m) = u_b(k - m, N_s) \text{ for } m = 1, \ldots \ldots N - 1 - N_s \qquad \ldots (3.73)$$

and

$$u_f(k, N_s + m) = 0 \text{ for } m = 1, \ldots \ldots N - 1 - N_s \qquad \ldots (3.74)$$

The lattice reflection coefficients, $\{K(k,m)\}$, are initialized to zero and then updated as follows:

$$K(k+1, m) =$$

$$K(k,m) + \gamma. \left(\frac{u_f(k,m).u_b(k-1,m-1) + u_f(k,m-1).u_b(k,m)}{\delta(k,m)} \right)$$

$$\text{for } m = 1, 2, \ldots \ldots N_s \qquad \ldots (3.75)$$

where

μ and λ = lattice gain constants

m = the mth stage of the lattice, and

$$\delta(k,m) = (1 - \mu).\delta(k-1,m) + \gamma.(|u_f(k,m-1)|^2 + |u_b(k-1,m-1)|^2)$$

for $m = 0 \ldots . . N$ $\ldots (3.76)$

is a power normalization factor for each lattice stage.

The timed nature of this lattice structure arises when the update of the lattice section is fixed after a number of iterations, i.e. after a period of $n_f T$ seconds from initialization ($k = 0$), where the parameter n_f is the timed-factor, the values of $\{K(k,m)\}$ and $\{\delta(k,m)\}$ remain unchanged.

3.6.1.2   The transversal section

The backward residuals $\{u_b(k,m)\}$ are then fed into the feedforward transversal filter, where the output $\hat{d}(k)$ is given by the convolution:

$$\hat{d}(k) = \sum_{m=0}^{N-1} c(k,m).u_b(k,m) \qquad \ldots (3.77)$$

where $c(k,m)$ = tap number 'm' at time kT.

The feedforward transversal filter tap gains (initialized to zero) are updated adaptively using the conventional power-normalized LMS algorithm:

$$c(k+1,m) = c(k,m) + \frac{\sigma.e(k)u_b(k,m)}{\xi.\delta(k,m+1)}$$

for $m = 0,1.......N-1$ ... (3.78)

where

$\sigma =$ the stepsize for the transversal section
$e(k) = d(k) - \hat{d}(k)$

and the sum of the squares of the backward residuals is given by:

$$\xi = \sum_{m=0}^{N-1} e_b^{2}(k,m) \qquad \text{... (3.79)}$$

In order to reduce processing, the value for ξ may be estimated using the techniques described in Satorius and Pack [32] and Rutter and Grant [34].

3.6.2   Performance

The convergence behaviour of the combined lattice and transversal filter structure is difficult to analyse because the convergence of transversal tap weights cannot start until the reflection coefficients of the lattice filter have converged. The higher-order reflection coefficients are also affected by the convergence of the lower-order reflection coefficients. However, as the results in the next section show, the adaptation performance is superior to the LMS algorithm and the performance in the presence of noise is very good. However, whilst the timed lattice filter works well with stationary signal, the filter requires careful control on the timing to freeze the lattice section, in particular when working with non-stationary input signals.

Assuming initially only the lattice section is adapting (equations (3.63)-(3.76)), and the operations in the transversal section are ignored (equations (3.77)-(3.79)), the processing requirements are of the order of 8N. This reduces to 5N with the lattice frozen, so the processing requirements are 3 to 4 times as much as the LMS algorithm, but considerably less than the RLS algorithm.

3.7   SUMMARY OF LINEAR TECHNIQUES

3.7.1   Simulated results

Before the more advanced topics of adaptive nonlinear systems and time invariant systems are discussed it would be useful to illustrate the performance of the various linear algorithms discussed in the previous sections. The configuration chosen for this is the direct modelling, or echo cancellation circuit, shown in Fig. 3.12. The simulated results are obtained using the signal processing worksystem (SPW) simulation tool produced by COMDISCO Systems and discussed in detail in Chapter 5. The model includes a Gaussian noise source, an optional raised cosine filter to correlate the noise source, an unknown system and an adaptive filtering structure which forward-models the unknown system. All the digital signal samples are assumed to be separated by an interval of T seconds.

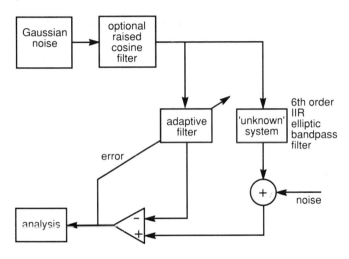

Fig. 3.12   Schematic diagram of test circuit.

The 'unknown' system is a 6th order IIR elliptic bandpass filter having an impulse response in excess of 60 sample intervals. The input signal consists of a Gaussian noise source, providing an eigenvalue spread of unity, which may be filtered by a raised-cosine filter, providing a wide range of eigenvalues with a ratio in excess of 120. The adaptive filter is one of the linear adaptive structures described earlier. Additive noise may be injected into the system as shown to simulate the effect of a noisy channel.

Figures 3.13 and 3.14 show the log mean square error adaptation curves of a 40-tap TDL, adapted using both the LMS and RLS algorithms, and a timed lattice structure. The results are the ensemble average of 100 identical runs with a different noise seed for each run. The adaptation is frozen during the first 100 samples. The performances with the Gaussian noise input are shown in Fig. 3.13 and filtered (correlated) Gaussian noise in Fig. 3.14. The gain in the LMS algorithm was set to 0.0125, the optimum gain for white noise input. The lattice filter chosen for this test had a 39-stage lattice and

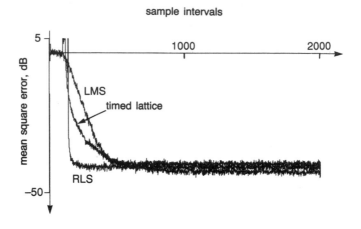

Fig. 3.13   Adaptive FIR filter convergence with unfiltered input.

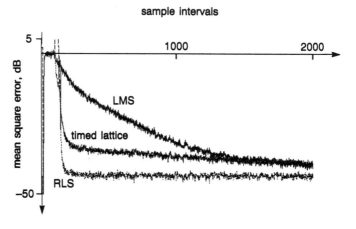

Fig 3.14   Adaptive FIR filter convergence with filtered input.

a 40-stage transversal section, i.e. $N_s = N - 1$. The lattice section, with the gains μ and γ set to 0.001, was adapted for the first eight iterations of adaptation and frozen thereafter. The transversal action gain, σ, was set to 0.002. The factors governing this choice are discussed in Rutter and Grant [34,35].

In both cases the RLS algorithm adapts very quickly and achieves the steady state error in about 50 samples. The LMS algorithm is very slow with the correlated signal, and even with the white noise signal reaches steady state only after about 400 samples. The LMS, RLS and the timed lattice filter eventually converged to the same level of MSE at the steady state. It should be noted that the level of the steady state MSE is due to the length of the TDL being insufficient to accurately model the impulse response of the unknown system.

Figure 3.20 shows the effect of noise on the adaptation of the filters. Noise with a power level approximately 10 dB below the input signal power is added to the output of the unknown system and subtracted from the error in the analysis section. The result in Fig. 3.15 shows the superiority of the time lattice algorithm over the LMS and RLS filters, both in terms of the convergence speed and steady state algorithm noise floor.

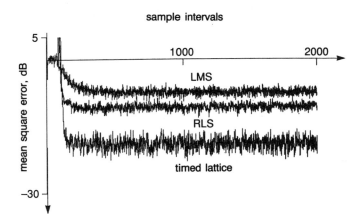

Fig. 3.15   Adaptive FIR filter convergence with interference noise.

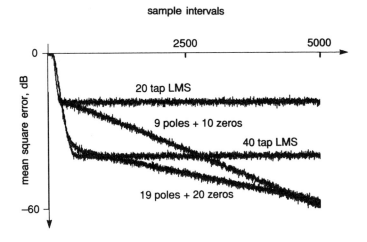

Fig. 3.16   Adaptive IIR and FIR filter convergence.

Figure 3.16 shows the benefits that may be obtained using the adaptive pole-zero filter. Adaptation curves for a 20- and 40-tap LMS adapted TDL filter are compared with two pole-zero systems having either 10 zeros and 9 poles or 20 zeros and 19 poles and after approximately 2500 samples a total of 19 active coefficients in the pole-zero achieves better performance than 40 coefficients in the conventional FIR adaptive filter. The two distinct slopes in the adaptation curves are attributed to the differing eigenvalue spreads of the input signal and the desired signal.

3.7.2   Choice of algorithm

So, given the differing performances, how is a suitable algorithm chosen? As has already been mentioned, in the past the available processing technology usually led to the LMS algorithm being the only practical choice. In many ways it remains the first choice, since its behaviour is well understood and working around its drawbacks is often preferable to the expense and risk of using one of the alternatives. Nevertheless, the other algorithms are being used in practical applications; for example the RLS algorithm has been found to be very useful for reducing training times in data modems, and speech coding schemes commonly employ recursive elements. Table 3.1 gives a brief summary of the linear algorithms.

Table 3.1   Summary of linear adaptive filter algorithms.

Structure/adaptation technique		Properties	Processing order
FIR	LMS	Adaptation rate slow and dependent on input signal properties. Stable, robust algorithm.	$2N$
	RLS	Fast adaptation for all.	$2N^2$
	Fast Kalman	Fast adaptation with reduced processing. Very intricate.	$7-15N$
Timed Lattice	Gradient	Faster adaptation than FIR LMS. Very good noise performance. Needs careful control.	$8N$ before freezing and $5N$ thereafter
IIR	Output Error	Possible convergence to local minima. Slow adaptation. Good noise immunity. Stability testing required.	Wide range of values depending on system employed.
	Equation Error	Gradient adaptation. No local minima. Poor noise immunity. Stability testing may not be required.	$3N$ (total no of co-efficients).

3.8   NONLINEAR ADAPTIVE FILTERING TECHNIQUES

In the preceding discussion it has been implicitly assumed that the systems being modelled by the adaptive filters are linear in the sense that they exhibit the superposition property. In most cases the systems being dealt with are sufficiently linear that this assumption does not give rise to problems but in some applications the degree of nonlinearity may become significant and in these circumstances nonlinear adaptive filters can be employed.

The best example application which commonly needs nonlinear techniques is that of echo cancellation in digital transmission systems. In such systems it is common to work with channel attenuations as high as 40 dB, while to obtain good performance the received signal-to-noise ratio may need to be of the order of 20 dB when all noise sources are taken into account. As nonlinear functions of the near-end echo contribute to noise these must apparently be maintained at a level greater than 60 dB below the transmit level, a degree of linearity which may not be practically achievable. The problem has commonly been resolved by the use of nonlinear echo cancellers in transceiver designs.

In the earlier sections linear echo cancellers have been classified according to the various models of linear systems, for example FIR, IIR or lattice filters. An analogous kind of categorization can be applied to nonlinear echo cancellers using techniques such as table look-up, Volterra-series expansion and others. The first two of these are perhaps the most commonly used and are looked at in outline in the following sections. Other techniques, including those of neural networks (see Chapter 16), are beyond the scope of this chapter.

3.8.1   Table look-up filters

The basic structure of the table look-up filter is that shown in Fig. 3.17. The input signal, x, enters a delay line and the contents of the delay line are used to specify an address for a memory, M. The content of M pointed to by the address is used as the output, y, of the filter. The filter can be made adaptive by allowing the contents of the memory to be modified by some algorithm.

The structure uses the existence of a unique address associated with each possible history, $\{x(k)\}$, to produce a unique response, $y(k)$, and so can model a wide range of nonlinearities subject only to the requirement that they produce a deterministic response to a stimulus of any length less than that of the delay line.

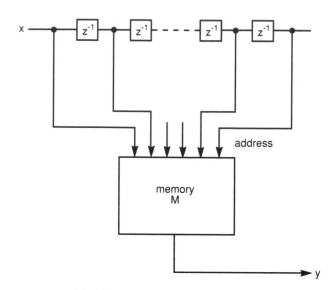

Fig. 3.17   Table look-up filter structure.

A suitable means of adaptation can be found by analogy with the LMS algorithm for an FIR filter:

$$y(k) = M(k)_{\text{add}\{x(k)\}}$$

$$e(k) = d(k) - y(k) \qquad\qquad \dots (3.80)$$

$$M(k+1)_{\text{add}\{x(k)\}} = M(k)_{\text{add}\{x(k)\}} + \beta.e(k) \qquad\qquad \dots (3.81)$$

where add{.} is a function which converts the history vector, $X(k)$, into an address for the memory, M.

A practical problem with the implementation of a table look-up filter is the size of memory which may be required. In the digital transmission system application, the length, N, of the echo can be tens or even hundreds of symbols, and if each symbol has one of L possible values the number of memory addresses required is L^N, which for any L rapidly becomes unmanageably large as N increases. It is worth noting too that the method is only applicable for small L; if x is a sampled analogue filter, table look-up is probably only possible for $N=1$ or 2.

The adaptation algorithm is one which has been studied in the literature [36,37] and is amenable to analysis of behaviour. The general conclusion is that although the filter can be relied upon to converge, the time taken to reach convergence is in proportion to the order of L^N iterations, i.e. it is proportional to the number of memory addresses which must be adapted. This means even if the application can afford the size and cost of a full table look-up filter the penalty will be very slow convergence.

These problems mean that a direct form table look-up is often avoided in favour of some hybrid technique. There are many possibilities but for example a table look-up filter can be split into a number of sections [38] as shown in Fig. 3.18(a) or a shorter table look-up filter could be combined with a linear FIR filter as shown in Fig. 3.18(b).

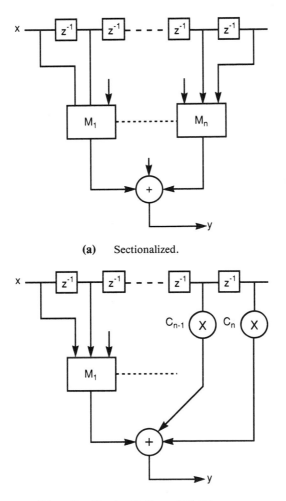

(a)   Sectionalized.

(b)   Combined with linear FIR filter.

Fig. 3.18   Hybrid table look-up filters.

In a split filter with M sections and total length N the number of coefficients has been reduced to $M.L^{N/M}$ which is, potentially, vastly smaller than for the full table look-up, and it has been shown [38] that the convergence time remains proportional to the coefficient count, so there is a corresponding speed increase as well as a complexity saving.

The combined table look-up and linear filter structure is designed to more closely represent the sources of nonlinearity in the real echo to be cancelled. It is usually found that the echo response in the time domain is greatest shortly after the stimulus and decays away slowly to progressively smaller values. It is also reasonable to assume that the majority of the nonlinearity will be associated with the largest parts of the echo, so the table look-up part of the hybrid filter covers the early part of the echo response and the tail is covered by the following linear FIR part. As the length of the table look-up part is only a small proportion of the whole the total coefficient count and hence complexity and convergence times are much reduced.

There are, of course, many other hybrid possibilities; for example, to increase coverage the sections of the sectionalized filter can be made to operate from overlapping ranges of the delay line or the look-up table filter could be combined with other linear structures such as IIR or lattice types.

The disadvantage of all these hybrid techniques is that the generality of the full table look-up has been lost, and although it is possible to assess the capabilities of a given structure, it can be very difficult to assess the usefulness of those capabilities in a given situation.

3.8.2   Volterra-series based filters

Another way of looking at the problem of modelling nonlinear systems is that of the Volterra-series [39], which can be regarded as a generalization of the power series approximation of functions of a single variable. In the case of the Volterra-series the input to the system being modelled is a time varying function so that there is an extra dimension involved resulting in an infinite number of infinite summations. The series is represented in the discrete causal case as shown below:

$$y(k) = h_0 + \sum_{i1=0}^{\infty} h_1(i1) + x(k - i1)$$

$$+ \sum_{i1=0}^{\infty} \sum_{i2=0}^{\infty} h_2(i1,i2).x(k - i1).x(k - i2)$$

$$+ \sum_{i1=0}^{\infty} \sum_{i2=0}^{\infty} \sum_{i3=0}^{\infty} \ldots + \ldots \qquad \ldots (3.82)$$

Each group of summations added to the approximation is said to increase the order of the series, while the limits on the individual summations increase the number of terms or span in each order. The quantities $h_n(.)$ are often called kernels and can be identified with coefficients in realizations of Volterra filters.

Clearly any practical realization has to have a finite number of coefficients which means that the equivalent series expansion must be truncated in both order and span. In particular the order is likely to be very severely truncated since the number of coefficients increases very rapidly with this parameter.

The zero order approximation consists of the dc component h_0 only, while the first order approximation corresponds with the usual linear model. It is only by extending the order to second or higher that any nonlinear behaviour is modelled. It is at this order that truncation is often placed for practical realizations [40], and it is worth looking further at the terms of this order. There are terms in various products of delayed version of the input signal x of the form $x(k-i1).x(k-i2)$, but with the given limits of summation for $i1 \neq i2$ each term is duplicated since the product $x(k-i2).x(k-i1)$ is indistinguishable from the former product. These duplicated terms can be eliminated without decreasing the generality of the expansion by moving the limits of the summation or equivalently by changing the indexing of the terms so that the second order terms become:

$$\sum_{i1=0}^{\infty} \sum_{i2=0}^{\infty} h'_2(i1,i2).x(k-i1).x(k-i2-i1) \qquad \ldots (3.83)$$

As the kernels have been rearranged in this notation, they have been marked with a prime. In this form both the input terms have been delayed by $i1$ samples so an alternative way of representing the terms can be stated using the delay operator z^{-i1}:

$$\sum_{i1=0}^{\infty} \sum_{i2=0}^{\infty} h'_2(i1,i2) \, z^{-i1} \{x(k).x(k-i1)\} \qquad \ldots (3.84)$$

In this form the number of products which has to be computed is reduced since many of those required can simply be recalled from products computed earlier for use by different kernels. A filter structure based on this form is shown in Fig. 3.19, where the span of $i1$ and $i2$ have both been limited to 2.

In the form shown in Fig. 3.19 the filter is suitable for use with any kind of input signal x, unlike the table look-up filter which is only suitable for use when the elements of x are chosen from a small symbol set such as found in echo cancellation of digital signals. If a Volterra-derived filter is used with a small symbol set, then the multipliers which form the second-order functions of the input signals can be very simple and the total hardware complexity

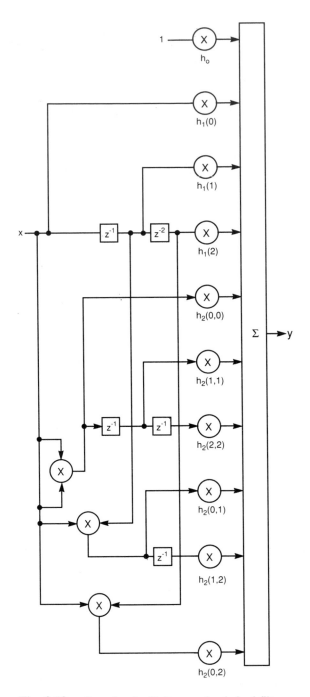

Fig. 3.19   Second-order Volterra-series derived filter.

of the filter of Fig. 3.19 becomes comparable with that of the 9-tap linear FIR canceller, so the structure is practical to implement, and the LMS algorithm can be used to adapt the coefficients in the same way.

It has been found that it is not necessary to use the nonlinear functions directly derived from the Volterra-series and, instead, any set of independent functions of the input can be used; for example, the use of simple logical functions has been described by Aggazzi et al [41]. This fact is the key which links the table look-up approach to its Volterra-series roots since by choosing the set of functions which are unity for a unique input sequence and zero elsewhere they effectively become address decoders and each kernel becomes one element of a memory.

The flexibility of the Volterra approach lies in the ability to choose kernels which are significant for the particular application in hand. There is no need to truncate the series cleanly as has been shown in the filter above, but, instead, particular higher-span terms could be included while some lower-span terms might be excluded if found insigificant. The same difficulty arises here as with the simplified table look-up structures in that a detailed knowledge of the nonlinear channel may be needed in order to decide which kernels are significant and which can be ignored.

3.9   JITTER COMPENSATION

A further assumption which has been made in the discussion of linear adaptive filters is that of time invariance of the system being modelled. In fact the LMS algorithm can cope well when the model must change slowly to track drift, for example, owing to thermal or ageing effects, but as time variation becomes more significant the lag in adaptation places limits on the accuracy of the modelling that can be achieved [42,43]. Often the effect of time variation is small enough to be safely ignored and in many cases there is no *a priori* information about the nature of the time variation which could be used to mitigate its effect, but there are exceptions.

For example, in the construction of digital transmission systems it is usual to have some means of changing the sampling point of the incoming signal for the purpose of maintaining synchronization and this can be regarded as causing time variation in both the transmission and the echo channels. The variation in the transmission channel is small enough to be ignored but, as was seen earlier, the fact that the echo can be so much larger than the received signal can make the variation in the echo channel significant. The magnitude of the effect can of course be reduced by using very narrowband timing systems but these may be difficult to implement and slow to operate. A number of techniques have been used to combat this problem using digital phase locked loops (DPLL). In a DPLL timing phase changes occur discretely at known times so that their effects can be confined to immaterial instants or cancelled out.

Examples of the former are combined echo cancellation and burst mode [44], in which timing adjustments are made in gaps between transmission bursts, and framed timing adjustment [45], in which adjustments are made during frame synchronization words. These techniques work well at the slave end of a link where changes in the echo channel are transient but are less effective at the master end where the echo canceller must re-adapt before the noise floor will return to its nominal value.

In the second category are fixed interpolation [46], multiple tap sets [47] and jitter compensation [48]. Fixed interpolation interpolates between adjacent echo canceller taps to modify tap values when a timing step is executed and works best where the echo canceller is oversampled. The multiple tap set technique provides an alternative set of echo canceller taps for each of two or more timing phases, but the most general form is that of jitter compensation for which estimates of the derivatives of the echo functions in the vicinity of the sample point are maintained.

For the purpose of explanation c'_n are the samples of the echo response $c(t)$ at times which deviate slightly from times nT as required because of perturbations either in the time at which previous samples of $x(k-n)$ were transmitted or in the time at which the response was sampled. In general these time perturbations are different for each sample and can be labelled $\partial T_n(k)$. This gives a set of Taylor-series expansions:

$$c'_n = c(nT + \partial T_n(k)) \approx c(nT) + \frac{dc}{dt}(nT).\partial T_n(k)$$

$$+ \frac{d^2 c}{dt^2}(nT).\{\partial T_n(k)\}^2 + \dots \qquad \dots (3.85)$$

It is possible to build higher-order jitter compensation but it is usually sufficient to consider only the first derivative of the echo. Making this assumption and allowing $c(nT) = c_n$ and $\frac{dc}{dt}(nT) = \nabla c_n$ gives:

$$c'_n \approx c_n + \nabla c_n.\partial T_n(k) \qquad \dots (3.86)$$

c_n are the nominal samples of the echo response and ∇c_n are the corresponding deviations due to timing variations of magnitude ∂T. The meaning of this equation can be seen more clearly with the aid of Fig. 3.20 which shows an example $c(t)$ with sample points nT and quantities c_n, ∇c_n and ∂T_n marked.

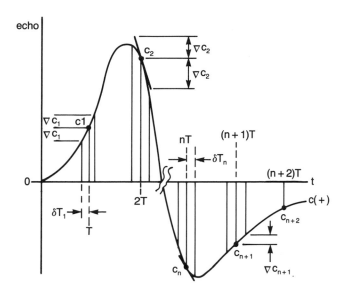

Fig. 3.20   Echo sampling point approximation for a first-order jitter-compensated
echo canceller.

The output, y, of the canceller is found by convolution and the error
signal, e, by subtraction of y from the desired signal:

$$y(k) = \sum_n c'_n.x(k-n)$$

$$e(k) = d(k) - y(k) \qquad \qquad ... (3.87)$$

The LMS algorithm can be used to maintain the required coefficients c_n
and ∇c_n; updated equations for normal operation then become:

$$c_n(k+1) = c_n(k) + \beta.e(k).x(k-n)$$

$$\nabla c_n(k+1) = \nabla c_n(k) + \beta.e(k).\partial T_n(k).x(k-n) \qquad ... (3.88)$$

For this to function correctly the sequences ∂T and x need to be zero mean
and uncorrelated. In practical applications this can be satisfied by the simple
training sequence $\partial T(k) = (-1)^k$ used for the majority of the time. The zero
mean property implies that there is no shift in the mean timing phase, which
is not very helpful for the main purpose which is to maintain synchronization.
When a shift in phase is required this can be obtained by a special modification
of the c_n:

$$c_n(k+1) \ = \ c_n(k) + \nabla c_n(k).\{\partial T_n(k) - \partial T_n(k-1)\} \qquad \ldots (3.89)$$

In other words the c_n are changed by an amount equal to the derivative coefficients in a direction determined by the change in the mean timing phase.

In order to describe this process in detail and to understand how a distinction is made between the mean and instantaneous timing phase it is necessary to distinguish between the various contributions to $\partial T_n(k)$:

$$\partial T_n(k) \ = \ \Psi_t(k) \ + \ \Psi_r(k) - \Phi(k-n) \qquad \ldots (3.90)$$

where $\Phi(k)$ represents the phase at which symbol $x(k)$ was transmitted, $\Psi_t(k)$ represents deliberate variations in the sampling phase for the purpose of training and $\Psi_r(k)$ represents the required variations in mean sample phase to maintain synchronization. Typically $\Psi_t(k) = (-1)^k$ while $\Phi(k)$ and $\Psi_r(k)$ take non-zero values only very sparsely. A typical signal flow chart for just one jitter-compensated echo canceller tap is shown in Fig. 3.21.

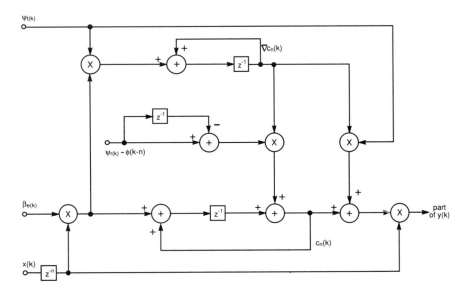

Fig. 3.21 Signal flow chart for one jitter-compensated echo canceller tap.

It should be noted that only the $\Psi_t(k)$ component of $\partial T_n(k)$ is used for adapting ∇c_n, the sparseness of the other components rendering them insignificant in this role. As $\Psi_t(k)$ has to be zero mean, it need not contribute to special updates of c_n, so only the other components, $\Psi_r(k)$ and $\Phi(k-n)$, are used for this application.

In this description no distinction has been drawn between the actual sampled channel responses c_n and ∇c_n and their approximations maintained by the canceller. The issue of convergence is dealt with in detail in Carpenter et al [48] where it is found that jitter compensator taps respond to the LMS algorithm as they do to ordinary adaptive filter taps but with a modified gain constant.

Jitter compensation may seem very complex, but this is deceptive. Only those echo canceller coefficients with the largest absolute values will have significant associated derivative terms, so only a small proportion of the echo canceller taps will need to be jitter compensated. A modest increase in complexity can therefore allow perhaps an order of magnitude increase in the tolerable DPLL step size, simplifying design and allowing faster system training and greater synchronization tracking capability.

3.10   CONCLUSIONS

The three decades since the first adaptive filter algorithm was published have seen adaptive filters grow from theoretical novelties into vital components in modern telecommunications systems. During this time the majority of applications have employed the LMS algorithm and have been restricted to linear modelling. In the short term, this trend is likely to continue, owing to the simplicity and robustness of the algorithm, and the fact that in many situations a linear model is adequate. However, alternative techniques, some of which have been discussed in this paper, are receiving greater attention as cost of processing decreases, and are already being used in real applications. In the future, greater use of adaptive filters in nonlinear environments and improved performance in the current linear applications can be expected.

Adaptive filtering remains a very active area of research, and the techniques discussed in this chapter represent only part of a mature but developing field. For reasons of space, numerous issues have not been discussed, for example, IIR lattice networks and further nonlinear techniques, such as multilayer perceptrons. Nevertheless, this chapter has presented an overview of the major adaptive filter issues in telecommunications and the authors hope that it has provided the reader with a useful overview of the subject.

REFERENCES

1.   Lucky R W: 'Automatic equalization for digital communication', Bell System Tech Journal, 44 , pp 547-588 (1965).

2. Sondhi M M: 'An adaptive echo canceller', Bell System Tech Journal, 46 , pp 497-511 (1967)

3. Haykin S: 'Adaptive filter theory', Prentice Hall Information and Sciences Series (1986).

4. Widrow B and Stearns S D: 'Adaptive signal processing', Prentice Hall Signal Processing Series (1985).

5. Grant P M, Cowan C F N, Mulgrew B and Drips J H: 'Analogue and digital signal processing and coding', Chartwell-Bratt Studentlitterliteratur (1989).

6. Hoing M and Messerschmitt D G: 'Adaptive filters structures, algorithms and applications', Kluwer Academic Publishers (1984).

7. Cowan C F N and Grant P M: 'Adaptive filters', Prentice Hall Signal Processing Series (1985).

8. Wiener N and Hopf E: 'On a class of singular integral equations', Proc Prussian Acad Maths and Physics (1931).

9. Qureshi S U H: 'Adaptive equalization', Proc IEEE, 73 , No 9, pp 1349-1387 (September 1985).

10. Makhoul J: 'Linear prediction: a tutorial review', Proc IEEE, 63 , No 4 (April 1975);

11. Goodwin G C and Sin K S: 'Adaptive filter prediction and control', Prentice-Hall (1984).

12. Widrow B and Hoff M E Jr: 'Adaptive switching circuits', IRE WESCON Convention Record, Part 4, pp 96-104 (1960).

13. Messerschmitt D G: 'Echo cancellation in speech and data transmission', IEEE Selected Areas in Comms, $SAC-2$, No 2, pp 283-297 (March 1984).

14. Weinstein S: 'A passband data-driven echo canceller for full duplex transmission on two wire circuits', IEEE Trans on Communications, $COM-25$, No 7, pp 645-666 (July 1977).

15. Clark G A, Mitra S K and Parker S R: 'Block implementation of adaptive digital filters', IEEE Trans Circuits and Systems, $CAS-28$, No 6, pp 584-592 (June 1981).

16. Dentino M, McCool J and Widrow B: 'Adaptive filtering in the frequency domain', Proc IEEE, 66 , pp 1658-1659 (December 1978).

17. Shink J J and Roy S: 'The LMS algorithm with momentum updating', Proc IEEE Int Sym on Circ and Sys, pp 2651-2654 (June 1988).

18. Roy S and Shink J J: 'Analysis of the momentum LMS algorithm', IEEE Trans Acou Speech and Sig Proc, 38 , No 12, pp 2088-2098 (1990).

19. Long G, Ling F and Prokais J G: 'The LMS algorithm with delayed coefficient adaptation', IEEE Trans Acou Speech and Sig Proc, 37 , No 9, pp 1397-1405 (December 1990).

20. Haimi-Cohen R, Herzberg H and Beéry Y: 'Delayed adaptive filtering, current issues', ICASSP 90, 3 , Albuquerque, USA, pp 1273-1276 (1990).

21. Coiffi J M and Kailaith T: 'Fast recursive least squares transversal filters for adaptive filtering', IEEE Trans Acoust, Speech and Signal Processing, ASSP32 , No 2, pp 304-338 (April 1984).

22. Shink J J: 'Adaptive IIR filtering', IEEE ASSP Magazine (April 1989).

23. White S A: 'An adaptive recursive digital filter', Proc 9th Asilomar Conf Circuits Systems, Computers, Pacific Grove California, pp 21-25 (1975).

24. Larimore M G, Treichler J R and Johnson C R: 'SHARF: An algorithm for adapting IIR digital filters', IEEE Trans Acou Speech and Sig Proc, 28 , No 4, pp 428-441 (August 1980).

25. Stearns S D: 'Error surfaces of recursive adaptive filters', IEEE Trans, CAS − 28 , No 6, pp 603-606 (June 1981).

26. Long G, Schwed D and Falconer D: 'Study of a pole-zero adaptive echo canceller', IEEE Trans on Circuits & Systems, CAS − 34 , No 7, pp 765-769 (July 1987).

27. Hughes P J: 'An adaptive error pole — zero echo canceller, MSc dissertation, Essex University (1990).

28. Jenkins W K and Nayeri M: 'Adaptive filters realized with second order sections', ICASSP, Tokyo, pp 2103-2106 (1986).

29. Fan H and Jenkins W K: 'A new adaptive IIR filter', IEEE Trans Circuits & Systems, CAS − 33 , No 10, pp 948-957 (October 1986).

30. Narayan S S, Peterson A M and Narasimha M J: 'Transform domain LMS algorithm', IEEE Trans Acoust, Speech and Signal Processing, ASSP − 31 , No 3, pp 609-619 (June 1983).

31. Friedlander B: 'Lattice filters for adaptive processing', Proc of the IEEE, 70 , No 8, pp 829-867 (August 1982).

32. Satorius E H and Pack J D: 'Application of least squares lattice algorithms to adaptive equalization', IEEE Trans on Commun, COM − 29 , No 2, pp 136-142 (February 1981).

33. Honig M L: 'Echo cancellation of voiceband data signals using recursive least squares and stochastic gradient algorithms', IEEE Trans on COM − 33 , No 1, pp 65-73 (January 1985).

34. Grant P M and Rutter M J: 'Application of gradient adaptive lattice filters to channel equalization', IEE Proc, 131 , Pt F, No 5, pp 473-479 (August 1984).

35. Rutter M J and Grant P M: 'Timed gradient adaptive lattice equalizer', IEEE Proceedings, 132 , Pt F, No 3, pp 181-186 (June 1985).

36. Justnes B O: 'A transmission module for the digital subscriber loop', Proc Communi '80, pp 73-76 (April 1980).

37. Cowan C F N and Adams P F: 'Nonlinear system modelling: concept and applications', Proc ICASSP84, $\underline{3}$, pp 45.6/1-4 (1984).

38. Ernst P G: 'Multistage RAM: an FIR filter for echo-cancellation in a 2-wire subscriber loop', IEEE Trans, $\underline{CAS-34}$, No 3, pp 225-233 (March 1987).

39. Schetzen M: 'Nonlinear system modelling based on the Weiner theory', Proc IEEE, $\underline{69}$, No 12, pp 557-1573 (December 1981).

40. Koh T and Powers E J: 'An adaptive nonlinear digital filter with lattice orthogonalization', Proceedings ICASSP83, pp 37-40 (1983).

41. Agazzi O, Messerchsmitt D G and Hodges D A: 'Nonlinear echo cancellation of data signals', IEEE Trans Com, $\underline{COM-30}$, No 11, pp 2421-2433 (November 1981).

42. Thomas E J: 'An adaptive echo canceller in a non-ideal environment (nonlinear and time variant)', BSTJ, $\underline{50}$, No 8, pp 2779-2795 (October 1971).

43. Falconer D D: 'Timing jitter effects on digital subscriber loop echo cancellers: part I — analysis of the effect', Trans IEEE, $\underline{COM-33}$, pp 826-832 (August 1985).

44. Brosio A, Mogavero C and Tofanelli A: 'Echo canceller burst mode: a new technique for digital subscriber lines', ICC85, $\underline{3}$, pp 1487-1491 (1985).

45. Sailer H, Schenk H and Schmid E: 'A VLSI transceiver for ISDN customer access', ICC85, $\underline{3}$, pp 1448-1451 (1985).

46. Lin N S, Hodges D A and Messerschmitt G D: 'Partial response coding in digital subscriber loops', IEEE Globecom, $\underline{3}$, pp 1322-1328 (December 1985).

47. Messerschmitt D G: 'Asynchronous and timing jitter insensitive data echo cancellation', IEEE Trans, $\underline{COM-34}$, No 12, pp 1209-1217 (1986).

48. Carpenter R B P, Cox S A and Adams P F: 'Jitter compensation in echo cancellers', IASTED85 (June 1985).

4

ADAPTIVE FILTERING — APPLICATIONS IN TELEPHONY

A Lewis

4.1   INTRODUCTION

Engineers have been designing adaptive systems, which change their behaviour depending on external circumstances, for many centuries. The temperature-compensated clock is a good candidate for the first consciously adaptive system design. All modern engineers are familiar with the concept of filtering, in the frequency-selective sense, but seldom consider any link with adaptivity, since almost all practical filters have been fixed in design.

There are two key reasons for this lack of awareness and exploitation of the powerful nature of adaptive filtering. First is the mathematical intractability of making conventional, analogue filters usefully adaptive. Second, and more serious, is the deleterious accumulation of drift and noise that can occur in analogue systems that have many adaptive elements.

The developments that have taken place in digital signal processing (DSP) theory and technology over the past two decades are changing this situation. The inherent precision, repeatability and programmability of digital computer technology ideally suits adaptive filter implementation. The mathematics of digital filtering, by considering signal manipulation in the time domain, has fostered the development of powerful, general and robust algorithms for adapting filters.

Telephony applications are in the vanguard of this minor revolution. The latest high-speed modems use as many as four adaptive filters, to transmit data over speech channels more than eight times faster than earlier designs without such filters. International telephone calls increasingly depend on adaptive filters to remove audible echo and improve conversational speech quality. Echo cancellation has also been applied in video-conference terminals to give duplex hands-free loud speech. Future applications include the mitigation of ambient noise transmitted from telephones in workshop or concourse environments. Many telecommunications services of the future, from mobile telephones to videophones and machines that listen and talk, will exploit filters that adapt.

This chapter reviews two applications of adaptive filters in telephony, for control of echoes and for hands-free speech. It also explains some architectural issues relevant to telephony and discusses a variety of hardware implementations.

4.2   ADAPTIVE SYSTEMS

Simple adaptive systems, where one parameter is designed to auto-regulate, are commonplace — as in the circuits used for many decades to provide automatic gain control (AGC) in radios, televisions and tape recorders. Adaptive filtering is essentially the generalization of such methods to multiple parameters, such as frequency and phase response variables. The architecture of the signal processing depends strongly on the application — algorithms are affected by the statistics of input signals and in practical systems appropriate control is very important. Large and complex adaptive systems can be constructed that involve the transfer of both filtered signals and filter parameters between adaptive filters to achieve otherwise intractable or seemingly impossible tasks [1].

In this chapter, the term 'adaption' is used specifically to distinguish the process of adapting the system from its final adapted state, for which the word 'adaptation' is employed in its common usage. Chapter 3 describes the principles of adaptive filtering and the mathematics of some adaption algorithms. The signal processing in artificial neural networks can be regarded as a special class of nonlinear adaptive filtering (see Chapter 16).

In telecommunications, almost all current applications use linear finite impulse response (FIR) architectures, where the filter output is a weighted sum of past inputs. The transfer function denominator is inherently scalar and this lack of recursive or output-dependent terms restricts the filter to an all-zero frequency response. Real-world transfer functions have both poles and zeros, requiring verbose description in FIR form. Nonetheless, FIR architectures are popular for several good reasons. They are inherently stable, unlike recursive filters which can oscillate with unsuitable coefficients. Adapting recursive filters requires an algorithm capable of navigating to the global optimum in a mathematical error space containing local optima — a problem currently without a practical general solution.

Some variation of the Widrow-Hoff least mean square (LMS) algorithm [2] is often used for adapting FIR filters. This well-established and much-studied algorithm has a computational complexity roughly equal to that of the filter itself. It is mathematically well-conditioned and remarkably rugged in the face of real-world problems, such as noise and nonlinearity. It achieves adaptation iteratively, by successive directed summations over many hundreds or thousands of sample cycles, as in equation (4.1):

$$c_i(j+1) = c_i(j) + \Delta c_i(j+1) \qquad\qquad \dots (4.1)$$

where c is the set of filter coefficients indexed by number i, j is the signal sample number or time index and Δc is the small, iterative adjustment factor which may be positive or negative. Calculating Δc is the ingenious part, to make the coefficients statistically converge close to their optimum values with a suitable input signal. Such stochastic convergence is a feature of many adaption algorithms. Other algorithms and architectures [3-5] can offer better performance, but at the cost of greater computational load, reduced robustness to signal properties or increased complexity of adaption control (see Chapter 3). All the applications described in this chapter use FIR and LMS techniques.

Neither superposition, nor the bulk of Fourier mathematics, strictly apply in linear adaptive filters, since though the signal processing may be linear to arbitrary precision, the coefficients are not time-invariant. In practice, for adaption that is sufficiently slow, these concepts remain useful. Much academic literature on the subject of adaptive filtering views it as a system identification task, where the signals involved are largely incidental and what matters are the filter coefficients after adaptation. Chapter 16 takes this view. However, in many telecommunications applications exactly the reverse is true. For this reason, care is needed to interpret academic conclusions and the balance of benefits against problems is not always what it may seem.

4.3   ECHOES IN TELEPHONY

4.3.1   Origin and perception

The phenomenon of echo arises when signals undergo delayed reflections. In telecommunications, signals can be reflected from one direction of transmission into the other, either at the electrical interface between 2-wire and 4-wire circuits or by acoustic coupling at the handset. The effects of telephony echo on customers are complex, depending on electric and acoustic factors as well as the motivation of both customers. Infrequent telephone users tolerate echo on international calls, especially if there is a strong sentimental or urgent factual content, but frequent users are often disturbed if there is audible echo.

Both talker and listener can suffer from echo problems. Talker echo can disturb the normal pattern of speech production, while listener echo reduces intelligibility. The severity of these effects is strongly dependent on reflection amplitude (echo loss) and the delay between the original signal and its echo. Conversation can become almost impossible if echo loss is low and the delay is more than about 100 ms — roughly the duration of a syllable. In some cases, objectionable echo can still be heard despite over 56 dB of echo loss [6].

Echo is not audible if the delay is less than about 20 ms. The reflected signals are then perceived as sidetone, although a metallic or hollow quality may be detectable. Echo becomes audible when the delay exceeds about 30 ms for low echo loss. CCITT Recommendations do not require echo control devices on circuits with less than 50 ms echo delay, although this limit is based on former, lossy, all-analogue networks and may be reduced in future. The UK is geographically a small country and echo control has not been necessary on national calls, unlike the situation in large countries like the USA.

4.3.2   Echo solutions

Echo suppressors conceal echo by detecting when the distant customer is speaking and the near customer is silent. The signal from the near customer is then heavily attenuated, so preventing echoes being heard. This alternate-simplex mode of operation is enhanced by detecting when both customers are talking (double talk) and inserting a small loss in the send path [7]. Satisfactory operation depends on accurate speech detection to avoid syllable clipping and utterance blocking. Ambient noise or excessive delay in the echo path (also called the end or tail circuit) can cause false operation of these speech detectors.

The alternative approach of echo cancellation was first proposed by Sondhi [8]. A replica of the echo signal is synthesized by an adaptive filter and subtracted from the send signal (Fig. 4.1). High synthesis accuracy is required for significant subtractive attenuation, so the adaptive filter must model the echo path in very fine detail. Cancellers offer somewhat better call quality than suppressors on long delay circuits, because they have greater signal transparency and fewer speech clipping effects.

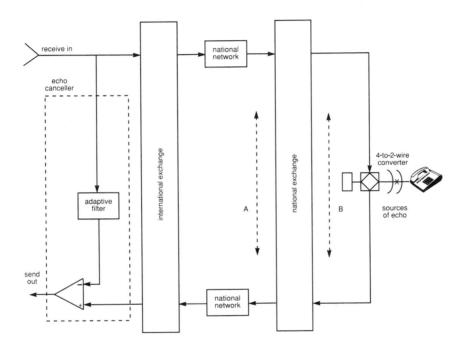

Fig. 4.1   Location of UK network echo cancellers.

The first single-chip custom VLSI implementation [9] was produced in 1969, with 128 FIR coefficients for an echo span of 16 ms. Later designs [10,11] were cascadable for greater delays. Figure 4.2 plots the modulus of the FIR coefficient values, measured with an experimental DSP echo measuring set (see Chapter 17) on a satellite circuit from London to Washington DC, USA. Figure 4.3 shows the frequency response of the distant echo signal on this circuit, with a slight ripple produced by interference between closely-spaced echo sources of different amplitudes. This can just be resolved in Fig. 4.2, which is essentially a picture of echo energy against time, as a small secondary peak following the main reflection.

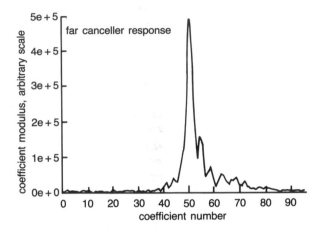

Fig. 4.2   Modulus of the FIR filter coefficients, modelling the circuit of Fig.4.3.

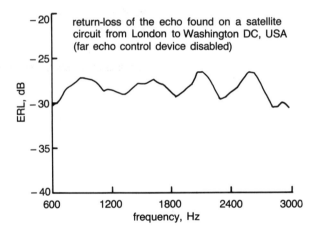

Fig. 4.3   Echo return loss data, from an experimental DSP echo measuring set.

Telephony signals are conventionally digitized using pseudo-logarithmic, instantaneous A-law or μ-law companding [12]. This method trades a reduction in a signal-to-noise ratio, due to signal-dependent quantization noise, for increased dynamic range within an 8-bit sample word. Linear adaptive filters cannot model this nonlinearity in the echo signal, so that cancellation (strictly, echo return loss enhancement [13]) is limited to the instantaneous signal-to-noise ratio of the echo path. In practice, the nature both of speech signals and of other effects (section 4.6) reduces the cancellation to 20 dB or less in many commercially available designs. Low bit rate speech encoding creates greater nonlinearity and further reduces the achievable cancellation if it is employed in the echo path.

4.3.3   Network echo canceller design

Commercially available network echo cancellers are intended for connection to 2048 and 1544 kbit/s PCM highways [13]. They therefore contain circuits for multiplexing, supervisory, remote-control and self-test functions as well as echo cancellation. Several configuration parameters can be selected by the user — the main choice being end delay. This is chosen to be greater than the largest echo delay expected, plus a margin of about 5 ms for group delay dispersion. In most European countries this is less than 32 ms [14], but echo cancellers with a longer delay span are sometimes necessary.

Echo cancellation of at least 38 dB is required on long-delay international circuits, such as satellite links, for telephones of usual sensitivity [15]. To achieve this, a centre-clipper or nonlinear processor (NLP) circuit is added in the send path (Fig. 4.4) of current designs. This blocks, or attenuates, the send output of the canceller when it is more than 20 dB or so below the receive signal level.

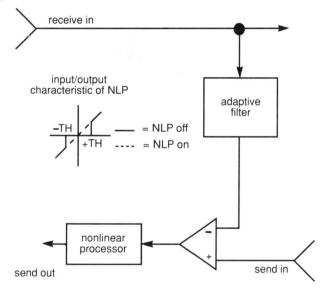

Fig. 4.4   A network echo canceller with a nonlinear processor.

These designs are in truth a combination of canceller and low-level suppressor, restricting duplex transmission to energetic speech signals, which erodes some of the advantages of using cancellers. The NLP circuit also causes ambient noise from the near talker to be modulated in level by the far speech bursts. Some designs inject a controlled amount of synthetic noise after the NLP to conceal this modulation. Control of adaption and NLP operation is by means of speech detectors, sensing the level of signals at the receive, send-in and canceller output points (Fig. 4.5).

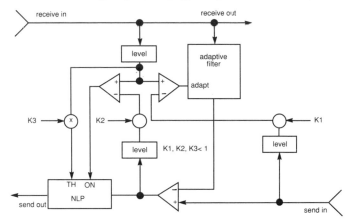

Fig 4.5   Simple level-based control of a network echo canceller and nonlinear processor.

Current designs of network cancellers use a similar kind of speech detector as suppressors, but employ the information differently — to enable and disable filter adaption rather than to open and close the send path. This method of canceller control gives satisfactory results under good conditions, i.e. with low-loss end circuits, telephones of matched sensitivity and customers of equal vocal level. However, when an unfortunate combination of asymmetrical circuit loss and asymmetrical sensitivities occurs, with customers of widely different vocal levels, then this type of control suffers false operation. Audible echo or speech clipping and mutilation can be the unpleasant, although occasional, result. Advances in DSP make more effective speech detectors feasible (section 4.6.2), although these have not yet been implemented commercially. Existing types of network cancellers are not compatible with high-speed modems and some facsimile protocols. They are often fitted with tone-detection circuits to bypass canceller operation on such calls.

4.3.4 Future trends

CCITT Recommendation G.165 for network cancellers has been little modified since its introduction in the early 1970s. The tests described are convenient to implement but not necessarily representative of the conditions of real use. DSP technology has changed beyond recognition since that time, but commercial inertia has so far prevented any major impact on echo canceller specifications, design or price. Current designs of cancellers can have reduced performance in tandem connections, in connections with widely-asymmetric speech levels and on speech-encoded echo paths. They require care in planning, positioning and network-signalling control, if optimum technical performance is to be achieved.

New network services are emerging, such as the groupe speciale mobile (GSM) cellular radio system, the digital European cordless telephony standard and the personal communications network [16]. Other new developments promise to increase network flexibility, such as the synchronous digital hierarchy, asynchronous transfer mode and frame-relay packetized speech systems. Taken together, these new services and systems are changing both the topology and the technical performance of the primary telephone network. Deregulation is accelerating the pace of this change. Optical fibre technology is providing almost an embarrassment of bandwidth, but market pressures are driving speech-channel bit rates ever lower. BT's primary network purpose is to transmit acceptable speech and the customer is the arbiter of quality. Levels of noise, degree of linearity, low values of transmission delay and bit transparency at 64 kbit/s cannot be taken for granted in future. Therefore, the long-term trend will be to move echo control devices away from their current position, near the centre of international networks, towards the periphery of national networks, even on to the line card (positions 'A' and 'B' in Fig. 4.1).

Cancelling echoes close to where they occur is the optimum technical solution. It maximizes both performance and network service flexibility, removing the need for the 'pseudo-duplex' NLP circuit. It could minimize the number of filter coefficients, the computational complexity, the cost and the risk of future interactions with new equipment. However, it would dramatically increase the total cost if every main exchange circuit or line card were to be equipped with the type of echo canceller designed for international access. The cost of DSP technology is rapidly declining and the familiar PCM combo-codec devices, fitted to today's line cards, may well be replaced by DSP-core ASIC devices to integrate channel filtering, codec and echo cancellation functions on one chip. Such devices would allow near-perfect 2-wire to 4-wire converters (autobalancing hybrids) to be economically designed.

The advent of the ISDN and universal, 4-wire, audio transmission might be thought of as the final solution to echo. Sadly, this is not true. Acoustic coupling, from earphone to microphone at the handset, can be sufficient to cause audible echo on long-delay connections [17]. Customers usually value aesthetic appearance and convenience in handset design more highly than acoustic performance, so that ISDN telephones will probably have internal echo control. ISDN telephones are likely to use adaptive filters for a different kind of echo cancellation — to give long-reach digital access over existing cables (see Chapter 6).

Network echo control strategy over the next decade will be influenced by a complicated mixture of technical, commercial and social factors. It is far from clear how this mixture will evolve, other than that it will be strongly market and customer driven. Assuming that telephony performance in the 21st century will be at least the equal of that of today, then telephones will surely become the most numerous single application of adaptive filters in the world.

4.4   DUPLEX HANDS-FREE SPEECH SYSTEMS

4.4.1   Introduction to hands-free speech

Telephone handsets attempt to give '4-wire' coupling to the human head. Hands-free systems often have the opposite arrangement, where the transducers are well coupled to each other and poorly coupled to the head. If subjective sensitivities are maintained, then extra send and receive gain is needed and oscillation can result.

The first hands-free or loudspeaking telephone (LST) designed by BT was the LST1, produced in 1961 (Fig. 4.6). It had a line-powered germanium-transistor amplifier, with separate boxes for the microphone and loudspeaker. It gave duplex (both-way simultaneous) operation, since its receive and send channels were active at low sensitivity simultaneously, although performance was acceptable only on short lines in quiet rooms. Speech detectors and voice-operated, channel-switching relays allowed gains to be raised and loud speech produced, but at the cost of alternate-simplex operation. BT's LST4, produced in the late 1960s, used these techniques in electronic form [18]. More complex speech detectors and continuously variable gain modulation, gave a modest performance improvement in the Bell '3A' and the Northern Electric 'Companion' set [19,20].

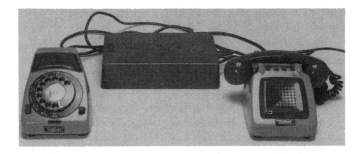

Fig. 4.6   BT's loudspeaking telephone No. 1 (circa 1960).

Loudspeaking telephones have had a chequered history. They have occasionally been used more as indicators of personal status than for operational need. Some customers strongly dislike being at the 'poor-relation' end of a handset-to-LST call, for reasons that probably involve both technical performance and perceived status. Hands-free telephony does have genuine uses, e.g. when holding a handset is inconvenient or dangerous because the user's hands are dirty or otherwise occupied. The truest application, in both business and personal communications, is when more than one person wishes to take part in the conversation, such as in audio and video conferences or in family telephone calls. When the purpose of telephony is to abolish the distance between two or more co-operative groups of people, then the single-user ethos of handset communication can be an unacceptable inhibition to group interaction.

4.4.2   Banes, boons and alternatives

Simple hands-free systems can suffer technical deficiencies in some circumstances. Alternate-simplex or voice-switched operation cannot transmit intelligible double talk and can produce syllable clipping and even speech mutilation during single talk, especially if the ambient noise is high in level or impulsive in nature. Ambient noise can also reduce the intelligibility of transmitted speech and reveal the gain-changing of alternate-simplex systems through unpleasantly distracting noise modulation. Speech quality and intelligibility can be significantly degraded if the acoustic environment of the LST is markedly reverberant. These effects can make conversation stilted, difficult or confused. Simple LSTs work best in quiet and acoustically-damped rooms, when all talkers adopt a conversational discipline that consciously avoids double talk. However, in conferences or family calls double talk is much more likely to occur than in handset telephony, because of the greater chance of excitement, interruption or argument.

Adaptive filters can solve the stability problem by echo cancellation, so that very little or no voice-operated gain changing is required [21]. As well as avoiding speech mutilation, duplex hands-free speech has subjective benefits — in greater conversational relaxation and in a naturalness which can be greater than that of handset telephony. Adaptive filters can also reduce the effects of reverberation on the transmitted speech by a variety of techniques including beam-steering microphone synthesis (section 4.5). Lastly, the lack of significant gain modulation makes ambient noise less obvious and its audibility can be reduced by a variety of adaptive signal processing techniques (see Chapter 14). The total complexity of these three signal processing tasks is high and beyond current economic application to domestic LSTs.

Simpler alternatives to complex hands-free signal processing are possible, if users are prepared to wear headsets or microphones (which could be cordless) or use push-to-talk buttons. However, most genuine users of hands-free systems prefer unencumbered naturalness. The decreasing cost of adaptive filtering with DSP technology will in future provide economic solutions to these requirements.

4.4.3   Duplex hands-free telephones

The feasibility of using adaptive filters for duplex hands-free telephony has been studied at BT Laboratories over the past decade. The first demonstration model, exhibited in 1980, processed signals with a mixture of discrete-component digital logic, 8080 microprocessor control and analogue speech-detector and gain-changing circuits. It required six, densely-packed 19-inch racks of experimental equipment and 40 A at 5 V to prove the idea technically feasible. In 1985 a second model had much improved performance, with all-digital signal processing and a low-cost, discrete-component, data-conversion system — all of which fitted on a single circuit board (Fig. 4.7). This design was further adapted to study a potential hands-free application in 1989.

The hardware architecture of this duplex hands-free telephone is shown in Fig. 4.8. Three custom large-scale integration (LSI) digital adaptive filter (DAF) chip-sets (section 4.7.2) provided cancellation of the transhybrid and loudspeaker-microphone coupling with 240 and 480 FIR coefficients respectively. The DAF chip-sets had a foreground/background architecture, as explained in section 4.6.3. A TMS32010 DSP device implemented all the other signal processing tasks, including speech detection, adaptive filter control, adaptive depth voice switching and stability control [22]. A refinement of mean level sensing speech detection was the basis for all control decisions.

Fig. 4.7 Duplex LST circuit board, showing the adaptive filter modules, TMS 32010 and data conversion circuits.

At the start of a call, conventionally deep voice-switching (>40 dB gain change) was used until the filters had adapted to sufficient precision. The DSP control system then reduced the depth of voice switching towards zero, giving duplex operation. In practice, voice-switching depths up to 12 dB proved to be subjectively equivalent to duplex operation. About five seconds of speech in each direction was required to reach this state. The control system continued to estimate loop stability, since the loudspeaker-microphone coupling was strongly dependent on the position of the user and nearby objects. A large stability margin was chosen, which meant that the depth of voice switching was often higher than necessary. Even so, it proved impossible to guarantee stability under extreme circumstances, such as intentionally placing large reflecting surfaces close to the loudspeaker.

This equipment gave an insight into the benefits of duplex hands-free operation. People accustomed to using conventional, alternate-simplex LSTs would change behaviour after about 15 s conversation on the duplex model. Relaxation was both visible and audible, with an increase in the occurrence of repartee and double-talk leading to a more informal and lively conversational style. Deep voice-switched operation could be forced by a user-operated switch, allowing easy comparisons. Duplex benefits were most

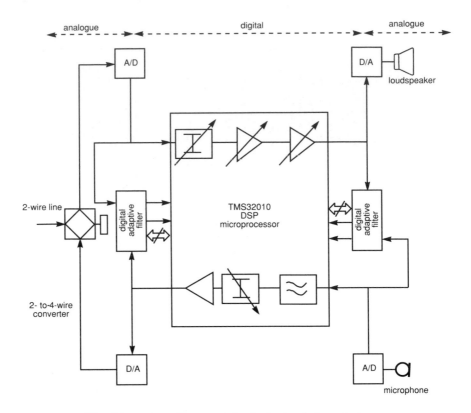

Fig. 4.8   Block diagram of the duplex hands-free telephone.

marked when the ambient noise level (Hoth spectrum) at the LST was moderately high, at around 65 dB(A) sound pressure level (SPL). The lack of noticeable modulation of the ambient noise allowed easy conversation, despite the degraded intelligibility of the send speech in this condition.

Several additional features were studied in the 1989 model, including multiple transducers and automatic volume-tracking with the background noise level (Fig. 4.9). This was evaluated in a reverberation chamber with ambient noise from a recording of traffic noise. This model showed that satisfactory hands-free performance, including the ability to make a 'mini-conference' call, was possible even at ambient noise levels as high as 85 dB(A) SPL.

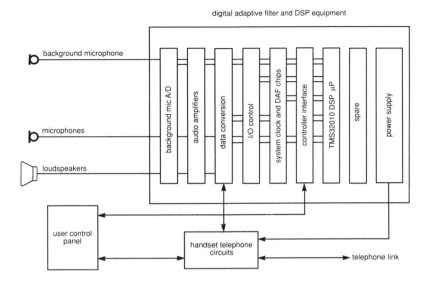

Fig. 4.9   Duplex hands-free demonstrator — hardware element.

4.4.4   Duplex videoconference terminals.

Unlike LST applications, the ambient noise and reverberation of video-conference rooms is well-controlled and low. Binaural loudspeaker listening reduces the psycho-acoustic audibility of echo compared with handset telephony. The 4-wire audio circuit has no 2-wire echo nor loss components and the microphones are placed out of the direct field of the loudspeakers. The overall audio system is usually specified as stable when open loop without cancellation. Providing good hands-free speech in this more favourable situation might seem easier. However, other aspects are considerably more demanding — the speech channel is often double the bandwidth of normal telephony, the video picture mercilessly reveals the disadvantages of alternate-simplex audio and talker echo is more audible to the local listeners than to the talker. The video encoding and decoding delay is typically 400 ms [23,24], so that the echo delay is at least 800 ms. High echo-control loss is therefore needed on all videoconference calls.

Acoustic echo cancellers for duplex conferencing [25] are an order of magnitude higher in computational complexity than network cancellers, typically having wide-band speech channels and over 4000 adaptive FIR coefficients to span room reflections up to 250 ms. Unlike network echo, the acoustic echo changes with time, as people move around during the conference. The adaption speed of conference cancellers should therefore be high, but the commonly-used LMS algorithm has an adaption speed that decreases as the number of coefficients is increased. Therefore current designs of conference canceller have reduced performance, or revert to echo-suppressor operation, if there is frequent or pronounced movement in the conference room.

In future, automatically-steered, super-directional microphones may provide significant benefits in conferencing. Their narrow beamwidth can reduce the audibility of ambient noise and reverberation, capturing speech at a distance with a quality close to that of a hand-held microphone. Such highly-directional microphones effectively increase the apparent loss of the loudspeaker-microphone echo path.

4.4.5   Beam-steering microphones

Super-directional microphones can be produced by acousto-mechanical design in gun or dish form and by electronic aperture-synthesis or phased-array techniques. Each signal from an array of discrete microphones can be filtered with a different frequency and phase response, prior to summing. By appropriate choice of these responses, a microphone with strong directional properties can be created [26] with the advantage that the beam direction can be electronically steered. Theoretical response calculation is possible, but in practice it is difficult to allow for all the mechanical and electrical tolerances or for the acoustic effects of near-field operation at low frequencies.

Adaptive filters solve this problem, either by a training period in the real environment or by self-adaption during use [27]. Two separate beams can be created from a single microphone array, so that the second beam can search for the next talker while the first points at the current one. The hardware complexity is high, typically needing one DSP device per microphone as well as one or more DSP device for system control. The number of microphones needed can be large, especially if a two-dimensional array is used to minimize the solid angle of the beam and thereby maximize the microphone's 'reach' in a large room. In future, such beam-steering microphones might look like a large piece of wall decoration, but contain many hundreds of individual microphones each attached to a single-chip, integral A/D DSP microprocessor.

4.5 OTHER APPLICATIONS

Adaptive filters can subtractively cancel interference or noise if a reference signal can be obtained that has a causal relationship with the interference at the point of interest [28,29]. The filter adapts to find this relationship, forming a prediction or replica of the interfering component. The simplest example is the removal of mains hum and its harmonics, even if not stable in frequency, without severe high-pass filtering. This technique is not generally useful in telephony, where interfering noise often comes from multiple and inaccessible sources. However, noise mitigation is possible without exact knowledge of the noise waveform, by exploiting differences in statistical properties between the wanted and unwanted signals. For example, adaptive nonlinear spectral manipulation methods, in single- or dual-channel form [21,30] can separate semi-periodic signals, such as speech, from uncorrelated noise. There are many other interesting applications of adaptive filters, such as line equalizers and rotating echo cancellers in high speed modems (see Chapter 7), one- and two-dimensional linear predictors in low bit rate speech and video encoding [31,32], high sampling rate applications in ISDN local loop transceivers (see Chapter 6), broadcast TV 'ghost' cancellers, acoustic noise cancellation in engine exhausts or fan ducts and multipath correction of troposcatter signals [33].

4.6 ARCHITECTURE AND PERFORMANCE

4.6.1 Adaptive filters and speech signals

The LMS adaption algorithm gives optimum performance with a white noise input signal, where individual samples are uncorrelated and the power spectrum is flat. In both network echo cancellation and duplex hands-free telephony the input signal is unavoidably speech. There is significant correlation within voiced phonemes if the pitch is moderately stable, as well as a wide variation in short-term talker level. The power spectrum of speech is not flat, but has short-term fluctuations around a long-term mean value that falls at both low and high frequencies. In the language of adaptive filtering, speech signals are said to be non-stationary and to have a marked spread of eigenvalues (see Chapter 3). The practical result is that LMS adaptive filters with speech input have coefficients that do not converge to optimum values, but dither with the fluctuations in speech [34]. The magnitude of this dither can be reduced by slowing the adaption speed (step size) as adaption proceeds, but at the cost of more complexity in system control. Other algorithms (see Chapter 3) are less sensitive to eigenvalue spread, although more computationally complex.

4.6.2 Appropriate control of adaption

In telecommunications, speech is almost always part of a dialogue or conversation. Double talk occurs to some extent in all human conversation, depending on social and ethnic factors. It is particularly noticeable during arguments and interruptions. It also occurs briefly during periods of extended single talk, as confirmatory affirmations, and when the listener and talker exchange roles.

In echo cancellation and duplex hands-free telephony, adaption must stop when the source of stronger signal lies inside the path modelled by the filter, otherwise the coefficients rapidly diverge from their optimum values. Appropriate control of adaption is therefore vital for best performance. This not only requires accurate speech detection but good estimation of the signal-to-interference ratio, in order to optimally set the speed of adaption. Control based on measurement of short-term mean level can achieve this satisfactorily, with one serious exception — on the basis of level alone, changes in the echo path are inherently indistinguishable from double talk.

Commercial systems often use level-based control with a simple 'time-out' declaration of path change, assuming that double talk does not persist for long periods. More subtle speech detection and control techniques are becoming economically feasible as VLSI technology develops. These can effectively discriminate speech from noise, based on linear predictive coding or other level-independent methods derived from speech encoding [35]. In conversational speech applications, good control of a simple adaption algorithm can be more cost-effective than employing a complex algorithm.

4.6.3 Stability and instability

In duplex hands-free systems, the send and receive channels are active simultaneously. Without the subtractive loss provided by the converged adaptive filters, objectionable oscillation would occur. Therefore, it is essential to avoid adaption during periods of unfavourable signals, such as double talk, so that the coefficients do not diverge from nearly-optimum values. A parsimonious control strategy, that is reluctant to enable adaption except under the most favourable signal conditions, is imperative.

The loudspeaker-microphone coupling path in duplex hands-free systems is a dynamic one, and not stable as a function of time, due to the small but inevitable movements of the user and other objects close to the LST. These movements can occur during extended periods of send speech or double talk, when the acoustic canceller cannot be usefully adapted. But fast tracking of adaptation is essential, to recover subtractive loss after possible movement and minimize the loss of stability margin. A generous control strategy, that is eager to enable adaption except under the most unfavourable signal conditions, is imperative.

This paradox is impossible to resolve if a conventional adaptive filter, with a single array of coefficients and a simple, level-based control strategy, is used. Some degree of compromise, in either double-talk immunity or tracking speed, must be accepted. However, Ochiai et al [36] showed how two arrays of coefficients could be used to provide 'virtually complete double-talking protection...' in a foreground/background architecture. The background array can be allowed to adapt generously, with parsimonious transfer of coefficient data into the foreground array, which is used to calculate the echo replica (Fig. 4.10). The DAF chip-set (section 4.7.2) used this architecture, with the further refinement of two-way coefficient transfer between arrays. This allowed faster tracking of dynamic echo paths, after a period of double-talk divergence, by resetting the background coefficients to the last-known good set, stored in the foreground array.

A foreground/background architecture can remove several of the disadvantages of simple, level-based control. Equation (4.2) shows the condition for background-to-foreground transfer (update), where L is the short-term mean level, expressed in dB, of the signals shown in Fig. 4.10 and ΔL is a small threshold value.

$$L_{BE} \leq L_{FE} - \Delta L_{U} \qquad \qquad \text{... (4.2)}$$

If the background coefficients diverge, L_{BE} tends to exceed L_{FE}. When interfering noise, or double talk, or periods of silence occur and adaption is disabled, the two levels in equation (4.2) inherently tend towards equality. In both cases update does not occur. Equation (4.3) shows the condition for foreground-to-background transfer (downdate).

$$L_{FE} \leq L_{BE} - \Delta L_{D} \qquad \qquad \text{... (4.3)}$$

For this condition to be met, not only must the coefficients of the foreground array be a better model of the echo path than the background (adaptive) array, but also the signal conditions must be propitious for simple, level-based measurement of this difference.

The practical benefits of the foreground/background architecture cost an increase in computational load of at least 50%. A conventional adaptive filter requires two 'multiply and add' operations per coefficient in a sample period, but a foreground/background filter requires three such operations since a second echo replica must be formed. For this reason, it has not been widely employed in commercial systems. It is however an extremely robust and successful technique for acoustic echo canceller control. In the duplex hands-free loudspeaking telephone (section 4.4.3), the level-based voice detectors occasionally made incorrect decisions, but erroneous foreground-to-background transfers were never observed.

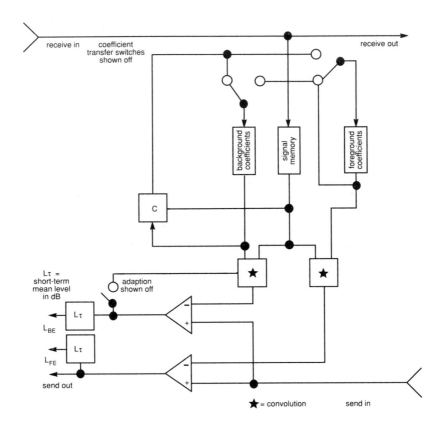

Fig. 4.10   Foreground/background architecture in an adaptive filter.

4.7 HARDWARE IMPLEMENTATIONS

4.7.1 Overview

Chapter 1 provided a comprehensive overview of the history and current status of DSP hardware and tools for software development. The rate of change in the past decade has been breathtaking. In 1986 DSP was described as a 'technology in search of applications' [37] but in 1990 the 'maturing of DSP' was reported [38]. Digital adaptive filters have been implemented in all possible forms — with custom, ASIC, component and programmable DSP devices.

4.7.2 Custom and ASIC VLSI devices

Network echo cancellers, produced in large numbers, justify the design of custom or ASIC implementations, with self-contained control and serial PCM interface circuits, as in the AT&T and Tellabs devices. Acoustic cancellers for teleconferencing are sold in small numbers and their greater complexity requires cascaded, multichip hardware. Designs have been produced with custom [39], component and mixed ASIC/DSP microprocessor implementations [40], often incorporating other signal processing functions.

The custom LSI digital adaptive filter (DAF) chip-set mentioned earlier was designed at BT Laboratories during 1982—84 [41], with 240 FIR coefficients and a foreground/background architecture optimized for duplex hands-free applications. The design, in 3 μm NMOS technology, operated at a cycle time of 244 ns and was partitioned into three parts, with a 2 μm CMOS ASIC 16.384 MHz clock generator [42]. A triple shift-register with gated recirculation logic stored the coefficient and signal data, a 12×16 bit multiplier/triple-accumulator performed the convolution and signal power calculations, while a convergence chip implemented the power-normalized LMS adaption algorithm and serial interface functions [43,44]. The multiplexed parallel and serial interconnections were chosen to allow additional functions, such as cascading up to four chip-sets for 120 ms impulse response, transferring coefficients to other systems and extending coefficient precision for adaptation in noise [45]. Figure 4.11 shows this chip-set, which was not commercially released, mounted on a multilayer circuit board.

Fig. 4.11   BT DAF adaptive filter chip-set, mounted on a multilayer PCB.

4.7.3   Component DSP devices

Component DSP devices vary from partly programmable types to dedicated, single-function parts. Some devices retain a programmable structure, but with a very limited instruction set, highly optimized for specific tasks. The A66 family from Array, capable of two-dimensional frequency-domain adaptive filtering, falls into this category along with the Zoran DFP and VSP. Other components offer a half-way house between a fully soft- and a fully hard-wired architecture. Such programmable bit-slice components, designed for 'DIY CPU' applications in DSP, include the AMD 29300 and Texas 74AS888 families. Very-high-speed adaptive filters in video and radar signal processing demand hard-wired, elementary logic components, such as register files, address sequencers, adders and multipliers, in bipolar and ECL technology. These are available from a wide range of specialist manufacturers, such as Analog Devices, TRW, Weitek, IDT, Cypress, BIT and others. Finally, algorithm-specific VLSI devices are available for FIR filtering, such as the Plessey PDSP 16256 and the LSI logic L64260/1. Paradoxically included in this last category of single-function components are devices, derived from programmable DSP microcomputers[1] by mask customization or ROM programming, such as the Motorola DSP56200 cascadable-adaptive FIR digital filter and the OKI MSM7520 echo canceller.

4.7.4   DSP microprocessor devices

Signals in telecommunications seldom require more than 16 bits of precision, but many applications demand more than 16-bit representation of filter coefficients. Double-precision software implies a substantial speed penalty,

---

[1] The term 'microcomputer' means a self-contained microprocessor with all the necessary memory and support resources on one chip.

while floating-point DSP microprocessors currently carry a significant cost premium. In practice, 24-bit precision is an acceptable compromise, retaining the cost advantages of fixed-point DSP microprocessors.

The implementation efficiency of adaptive filters in general-purpose DSP microprocessors has improved markedly in recent devices. The key to this improvement is the increased number of parallel operations, so that instruction and data fetching, multiplication, ALU functions, address generation and data storage can occur in a smaller number of machine cycles. Multiple-bused data spaces, wider and deeper on-chip memory and hardware support for looping and modulo addressing have also improved performance. For example, the Texas TMS320C25 requires seven cycles per LMS FIR coefficient, while the more recent TMS320C30 is more than twice as efficient at only three cycles per coefficient [46]. For simplicity, initialization, input/output, ancillary and system-control cycles are not shown in the examples below, although a fixed margin of 100 cycles per sample period is allowed.

A conventional LMS FIR adaptive filter, with a single array of real coefficients, could be implemented in the original Motorola DSP56001 with three cycles per coefficient. One of the architectural modifications in Revision C of this device allows coefficient data to be moved out of and into an accumulator in the same cycle. If coefficients are processed in pairs, as with the code kernel shown in Table 4.1 [47], only two cycles per LMS FIR coefficient are then required, although one sample delay is introduced in the adaption process (see Chapter 3).

In a single-chip DSP56001, the internal RAM size limits the number of coefficients to about 120. If external memory is added, extra cycles are needed for access so three cycles per coefficient are required. This means that, using the above code, a 33 MHz DSP56001 and MCM56824 RAM could process about 650 coefficients at a sampling rate of 8 kHz.

A foreground/background LMS FIR filter requires further cycles for transferring the coefficients between arrays, as well as additional memory space. If adaption is abandoned for one signal sample and substituted by coefficient transfer, then six cycles per coefficient are required by the DSP56001 with external memory. At 33 MHz this configuration could process about 325 foreground/background coefficients with a sampling rate of 8 kHz.

By comparison, the DAF custom design provided 240 coefficients per chip-set at the same sampling rate. It is a tribute to the design of today's DSP microprocessors that software can now equal a level of performance that demanded custom VLSI just less than a decade ago. It is also a sobering thought that silicon technology has advanced by three or four generations, from 3 μm NMOS to 1 μm CMOS, in that time.

Table 4.1   DSP56001 (Rev. C) code kernel for a single-real-array LMS adaptive filter.

Opcode	Operands	Data move	Data move	Comments — see equation (4.1)
mac	x0,y0,a	y0,b	b,y:(r5) +	$a = c_0(j)*x(n), b = c_0(j)$
macr	x1,y1,b	x:(r0) + ,x0	y:(r4) + ,y0	$b = c_0(j) + e*x(n-1) \rightarrow c_0(j+1)$
mac	x1,y0,a	y0,b	b,y:(r5) +	$a = a + c_1(j)*x(n-1), b = c_1(j)$
macr	x0,y1,b	x:(r0) + ,x1	y:(r4) + ,y0	$b = c_1(j) + e*x(n-2) \rightarrow c_1(j+1)$

It is clear that the implementation efficiency of adaptive filters in DSP microprocessors is profoundly influenced by relatively minor changes in device architecture, as shown by the example in Table 4.1. The extent of this influence is strongly dependent on the mathematics of the adaption algorithm. Most DSP microprocessors produced to date have chosen to concentrate chip resources on solving these architectural and algorithmic interaction issues, rather than expand other areas of on-chip support such as memory, I/O or data conversion functions. As a result, telecommunications applications of adaptive filtering tend at present to require multi-chip implementations in DSP microprocessor form.

4.7.5   Future trends

Chapter 1 described the key trends in both silicon technology and tools for hardware and software DSP development. Large and complex applications will still need multichip solutions in future. The optimum design and partitioning of such multiprocessor systems is a major challenge to DSP developers and to the researchers and vendors of DSP software tools alike. However, the scope and capacity of single-chip solutions is rapidly expanding. The integration of programmable DSP microprocessor cores with data conversion [48], semi-custom DSP and interfacing hardware, all in single-chip ASIC form, will significantly reduce both costs and development timescales. The growing computational power and declining cost of DSP devices will propel adaptive filtering to ever wider applications in telecommunications networks, terminals and services.

4.8   CONCLUSIONS

High-speed modems would not be possible without adaptive filters. Low bit rate encoding of speech and video uses several kinds of adaptive filtering techniques. The conversational quality of echo control and the naturalness of hands-free speech is enhanced by such methods. Long-reach ISDN local loop transceivers depend on high-speed adaptive filters.

The ability of adaptive filters, in the right circumstances, to learn the response required is uniquely powerful, though not widely understood. The subject has immense potential application to all sorts of engineering disciplines, from avionic, industrial and automotive control to consumer entertainment and bio-medical signal processing. It is particularly relevant when the required response changes erratically or unpredictably because of external factors like human movement or nonlinear interactions. In such cases, conventional wisdom declares filtering to be impossible or inappropriate. Professor Widrow, widely regarded as the 'father' of the LMS algorithm, has wryly reflected [49] on why it has taken over 25 years for the world to understand what he and his collaborators said.

The use of adaptive filters does require a different mind-set on the part of the designer, who has to forego the familiar search for a precise, *a priori* specification and instead design suitable adaptive behaviour. Adaptivity is a paramount feature of biological systems and the key to their widespread success. Designing with adaptive filters is rather like listing good things to do when bringing up children, instead of trying to define all the qualities of good adults. The system will need a learning period, and (just like people) be subject to some controlled degree of statistical uncertainty in behaviour after learning.

The future holds a dramatic flowering of digital signal processing. Adaptive filtering will play a major part in this quiet, yet far-reaching, revolution, profoundly affecting both telecommunications and everyday lives.

REFERENCES

1.   Widrow B and Walach E: 'Adaptive signal processing for adaptive control', Proc IEEE ICASSP, 2, pp 2.1.1/1-4, San Diego (March 1984).

2.   Widrow B and Hoff M E: 'Adaptive switching circuits', WESCON, Pt 4, p 96 (1960).

3.   Alexander S T and Ardalan S H: 'Adaptive telephony echo cancellation using fast Kalman pole-zero modelling', Proc IEEE ICCOM, 3, pp 1477-1481, Chicago (June 1985).

4.   Gritton C W K and Lin D W: 'Echo cancellation algorithms', IEEE ASSP magazine, pp 30-38 (April 1984).

5.   Treichler J R, Johnson C R Jr and Larimore M G: 'Theory and design of adaptive filters', Wiley-Interscience (1987).

6.   Richards D L: 'Telecommunication by speech', Ch 4.4.2, Butterworths (1973).

7. CCITT Recommendation G.164, 'Echo suppressors', Melbourne (November 1988).

8. Sondhi M M and Presti A J: 'A self-adaptive echo canceller', Bell Sys Tech J, 45 , pp 1851-1854 (December 1966).

9. Duttweiler D L and Chen Y S: 'A single-chip VLSI echo canceller', Bell Sys Tech J, 59 , No 2, pp 149-160 (February 1980).

10. Tao Y G et al: 'A cascadable VLSI echo canceller', IEEE J SAC, 2 ,No 2, pp 297-303 (March 1984).

11. Furuya N et al: 'High performance custom VLSI echo canceller', Proc IEEE ICCOM, 3 , pp 1492-1497 Chicago (June 1985).

12. CCITT Recommendation G.711, 'Pulse code modulation (PCM) of voice frequencies', Melbourne (November 1988).

13. Tellabs Inc, 2541/A echo canceller technical manual (March 1990).

14. Wehermann W and Koch W: 'Transmission characteristics of echo paths', EUROCON 71 (October 1971).

15. CCITT Recommendation G.131, 'Stability and echo', Melbourne (November 1988).

16. Groves I S: 'Personal mobile communications — a vision of the future', BT Technol J, 8 , No 1, pp 7-11 (January 1990).

17. Coleman A E and Stevens A E: 'Measurement and effect of echo caused by the acoustic loss path of telephone sets', Internal BT memorandum (1987).

18. 'Development of a voice-switched loudspeaking telephone (PO No 4) — circuit design', Internal BT memorandum (May 1963).

19. Clemency W F and Goodale W D Jr: 'Functional design of a voice-switched speakerphone', Bell Sys Tech J, 15 , No 3, pp 649 (May 1963).

20. Clarke W and Gale J: 'A new look at loudspeaking telephones', Telesis, 3 , No 3, pp 79-84 (1973).

21. South C R, Hoppitt C E and Lewis A V: 'Adaptive filters to improve loudspeaker telephone', Electronics Letts, 15 , No 21, pp 673-674 (October 1979).

22. South C R: 'Signal processing in a loudspeaking telephone', PhD thesis, Aston University (1985).

23. CCITT Recommendation H261: 'Video codec for audio-visual services at $P \times 64$ kbit/s', Melbourne (November 1988).

24. Whybray M W et al: 'Videophony', BT Technol J, 8 , No 3, pp 43-54 (July 1990).

25. NEC AEC-700, Acoustic echo canceller operating manual (1989).

26. Flanagan J L et al: 'Computer-steered microphone arrays for sound transduction in large rooms', J Acoust Soc Am, 78 , No 5, pp 1508-1518 (November 1989).

27. Widrow B and Stearns S D: 'Adaptive signal processing', Prentice-Hall Inc, New Jersey (1985).

28. Widrow B et al: 'Adaptive noise cancelling: principles and applications', Proc IEEE, $\underline{63}$, No 12, pp 1692-1716 (December 1975).

29. Harrison W A et al: 'A new application of adaptive noise cancellation', IEEE Trans ASSP, $\underline{34}$, No 1, pp 21-27 (February 1989).

30. Munday E: 'Noise reduction using frequency-domain nonlinear processing for the enhancement of speech', BT Technol J, $\underline{6}$, No 2, pp 71-83 (April 1988).

31. Hanes R B: 'The application of speech coding to telecommunications networks', BT Technol J, $\underline{1}$, No 2, pp 57-67 (October 1983).

32. Jayant N S and Noll P: 'Digital coding of waveforms', Prentice-Hall, New Jersey (1984).

33. Treichler J R, Johnson C R Jr and Larimore M G: 'Theory and design of adaptive filters', Chapter 8.4, Wiley-Interscience (1987).

34. South C R: 'Estimation of random spectral misadjustment of an adaptive filter', Proc IEEE ICASSP, pp 2995-2998, Tokyo (1986).

35. Freeman D K et al: 'The voice activity detector for the pan-European digital cellular mobile telephone service', Proc IEEE ICASSP, pp 369, Glasgow (May 1989).

36. Ochiai K, Araseki T and Ogihara T: 'Echo canceller with two echo path models', IEEE Trans Comm, $\underline{25}$, No 6, pp 589-595 (June 1977).

37. Marrin K: 'DSP: a technology in search of applications', Computer Design, pp 59-61 (November 1986).

38. Dyer S A and Higgins R J: 'The maturing of DSP', IEEE Micro, pp 11 (October 1990).

39. Itoh Y et al: 'An acoustic echo canceller for teleconference', Proc IEEE ICCOM, $\underline{3}$, pp 1498-1502 (1985).

40. Coherent, AC-1200/1 acoustic echo canceller operating manual (1989).

41. South C R: 'An adaptive filter in LSI', BT Technol J, $\underline{3}$, No 1, pp 30-46 (January 1985).

42. Reger J D: 'The development of a gate array for a loudspeaking telephone application', Internal BT memorandum (1984).

43. Harbridge J R: 'The design of an integrated circuit for use in loudspeaking telephones', Internal BT memorandum (1984).

44. Denyer Walmsley Microelectronics Ltd: 'Development of multiplier and shift register memory integrated circuits for an adaptive filter system', BT contract 600105 (1982).

45. South C R and Lewis A V: 'Extension facilities and performance of an LSI adaptive filter', Proc IEEE ICASSP, $\underline{1}$, pp 3.4/1-4, San Diego (March 1984).

46. Kuo S and Chen C: 'Implementation of adaptive filters with the TMS320C25 or the TMS320C30', Digital signal processing with the TMS320 family, 3 , Texas Instruments Inc (July 1990).

47. Motorola DSP56001 User's manual, Rev 1, Appendix B-4.

48. Davis H, Fine R and Regimbal D: 'Merging data converters and DSPs for mixed-signal processors', IEEE Micro, 10 , No 5, pp 17-27 (October 1990).

49. Widrow B, private communication (June 1986).

5

A SIGNAL PROCESSING APPLICATION DEVELOPMENT ENVIRONMENT

R J Knowles, P R Benyon and D K Freeman

5.1.   INTRODUCTION

Over the last decade, digital signal processing (DSP) devices (also known as digital signal processors) have increasingly been used in information technology products. A DSP device is a class of exceptionally high-speed microprocessor, targeted at applications requiring fast multiply-add operations. Whilst the life span of each generation of DSP devices is reducing, the complexity of both their architecture and their applications is increasing. There is also growing pressure to reduce the time from design conception to implementation and manufacture of a product[1].

These trends place more pressure on the algorithm developer. Historically, the design and development route for a system utilizing a programmable DSP device has been to program directly at the assembler level. This is then tested by execution on either target hardware or a commercial development system. The developer is often faced with debugging the algorithm, hardware and software implementations simultaneously. Often there is a real time constraint because the software is interrupt-driven [2]. The result is often an algorithm which is targeted for implementation on a specific device. It lacks portability and is also difficult to maintain or enhance. The final documentation does not necessarily describe accurately the resulting implementation. The design environment developed within BT Laboratories to address these issues is known as SPADE (signal processing application development environment) [3]. SPADE is designed to support the development cycle from conception, through specification, simulation and analysis to prototyping. Wherever possible, commercially available hardware development platforms and software tools are utilized. Due to this modular approach, the tools and platforms can be updated to match the user demand. Recent literature [4-6] provides details of some other design environments developed elsewhere as research projects over the last few years. The trend of these environments is to provide some form of graphical description of a DSP algorithm coupled with simulation and analysis capability. SPADE differs from these other environments in the level of integration of hardware development platforms and software development tools.

Parallelism is often utilized within DSP devices to increase their computational power. The most common use of parallelism within DSP devices are instruction pre-fetching, and arithmetic pipelining. When the parallelism is implicit, the programmer must be aware of its implications. This may compromise the ease of programming the device.

The computational power of DSP hardware can also be increased by exploiting parallelism in a more course-grained fashion. This is typically achieved by employing many processing elements which communicate either via shared memory or by message passing. DSP devices have recently become available which have been designed specifically for this purpose. However, DSP algorithms are conventionally developed using notations which are inherently sequential. This is a problem that future methodologies must overcome — object oriented or applicative/data-flow notations may be appropriate.

This chapter provides an overview of the SPADE environment, and, as parallel DSP capabilities are currently not supported by SPADE, it goes on to discuss some of the problems involved in providing support for parallel processing architectures.

5.2   OVERVIEW

The purpose of SPADE is to support the user with a variety of design techniques and tools to describe, simulate and implement DSP applications. Considerable emphasis has been placed upon tool integration, whilst accepting that design methods will often vary depending on the particular application. The user may need to use only a subset of the tools available. Thus SPADE provides the user with the choice of using as many of the tools as necessary to perform a particular task. In the development of SPADE, a human factor issue — the users' perception of the system, particularly of how easy it is to use — has been carefully considered [7]. The essential features of SPADE are:

- automatic software generation from a high-level specification;
- a notation suitable for DSP algorithm specification;
- functional decomposition of the specification;
- standard library routines which are user-extensible;
- signal analysis of results;
- data acquisition systems to provide real world data for simulations;
- simulation capability for specification verification;
- DSP device development boards for verification of prototype implementation.

Figure 5.1 illustrates the major components of SPADE. The terms and abbreviations used in Fig. 5.1 are described in the following section.

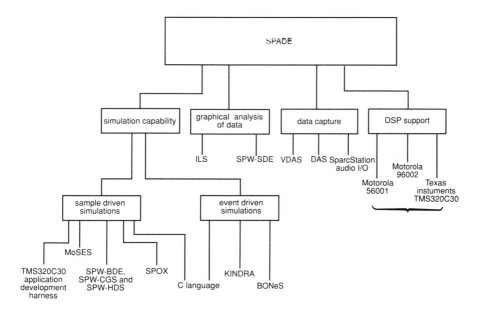

Fig. 5.1   Block diagram of the components of SPADE.

5.3   SPADE COMPONENTS

The following sub-sections provide a brief overview of each component within SPADE. Since SPADE is modular in design, tools may be added or removed as required [3].

5.3.1   Simulation tools

5.3.1.1   Signal Processing WorkSystem

Signal Processing WorkSystem^{TM} (SPW) is a commercial tool developed by Comdisco Systems. It is a graphical environment for the simulation of systems which are essentially synchronous in nature such as most signal processing systems. In these sample-driven simulations, all functions are executed in a source-to-destination fashion synchronously with a simulation clock. Multirate systems may be implemented, but, to accomplish this, functions must be selectively enabled or disabled by user-provided control logic.

SPW consists of the following components:

- BDE — block diagram editor;
- SDE — signal display editor;
- CGS — code generation system;
- HDS — hardware design system;
- FDS — filter design system.

The SPW-BDE is a graphical block diagram editor which allows the building of a simulation from standard library or custom user modules. It supports hierarchical representations to encourage the use of subsystems. Block diagrams are a natural way of describing many DSP algorithms and they have the additional benefit of providing a level of documentation above the actual program code. The hierarchy allows the level of detail to be hidden as appropriate.

The block diagram may be simulated from within the SPW-BDE to verify the functionality of the design. No programming is required unless custom-coded blocks are needed. If the desired functionality is not in the standard libraries, then the user may use C to implement the function — this is referred to as generating a custom-coded block. SPW will generate a code template for any custom-coded block from the graphical description.

An example of the SPW-BDE display is shown in Fig. 5.2. Initial use of SPW in the specification and design of an adaptive lattice echo canceller algorithm produced the first comparisons between the execution times of programs generated by Kindra (see section 5.3.1.6), SPW and hand-coded C. The conclusions of these comparisons were that although the algorithm execution speed was faster using hand-coded C, the design tools provided a useful method for prototyping. Thus the graphical libraries developed were more likely to be reused than a hand-coded C program, since they were more accessible, more amenable to manipulation and also more readily understood.

The SPW-SDE may be used to display, process and analyse signals generated from simulations constructed within the SPW-BDE. Signals may also be generated or edited within the SPW-SDE for use in subsequent simulations. Both scalar and vector operations are possible.

The SPW-CGS is divided into three variants, producing executable code to run on a platform which supports:

- generic C code;

- Texas Instruments TMS320C30 code;

- Motorola 96002 code.

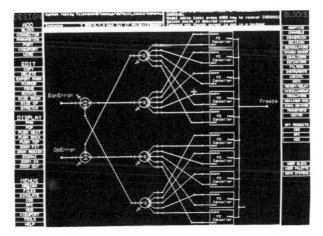

Fig. 5.2   An example of the SPW-BDE display.

The generic C variant will translate a block diagram into generic C for faster simulation than from within the SPW-BDE itself, although in practice there is no hard and fast rule as to how much quicker this will be. The C code may also be transferred to a different platform for execution.

The other two variants, the TMS320C30 and the Motorola 96002, allow the system developed within the SPW-BDE to be executed on a DSP device development board. The code generation in this case is in two phases. The first is performed by the SPW-CGS on the Sun workstation, the second by a cross-compiler on a PC. The first phase converts the block diagram into a series of library function calls. Any code within custom-coded blocks will be passed straight to the PC cross-compiler. The cross-compiler links the SPW library modules with any compilation of custom-coded blocks. The resulting program can be executed on the DSP device development board. All these operations may be invoked from the Sun workstation. Various tests have been performed to assess the efficiency of code generated from C source, especially for the TMS320C30. Although the assembly code produced can be far less efficient than hand-coded assembler, simple rules for code optimization have been developed [8].

In many cases, the reduction in algorithm development time together with the advantages of using a high-level language can outweigh the inefficiencies introduced through the use of a cross-compiler. Results obtained from an exercise carried out whilst SPADE was under development have been encouraging, leading to the execution of a simple adaptive differential pulse code modulation codec in real time [9].

The SPW-HDS is designed to allow the constraints of fixed precision processing to be applied to simulations developed within the SPW-BDE. This can simulate the effect of truncation and rounding errors prior to implementation on either a fixed-point DSP device or application specific integrated circuit (ASIC). At present, only certain blocks of the standard libraries allow fixed precision representation. Custom-coded blocks may be created; however they are more complex than those for the SPW-CGS because of the necessity to handle fixed precision arithmetic. In addition, the execution is slower because of this overhead. At present, there is no way of translating a floating-point simulation to fixed precision automatically.

SPW-FDS allows the specification and design of various filter types. Filters designed this way may be used within simulations using the SPW-BDE or for processing data within the SPW-SDE.

5.3.1.2   TMS320C30 application development harness

The TMS320C30 applications development harness was designed to provide a low-cost method for developing algorithms on a PC equipped with a TMS320C30 development board. The primary motivation for this work was a requirement to port several existing algorithms coded in C from a VAX to the TMS320C30. This enabled the comparison of simulation results produced from two different floating-point processors. This study concluded that it was viable to use the TMS320C30 as a target device for prototyping and simulation.

The harness incorporates several software modules, which have been written to be both reusable and extensible. It attempts to present the user with a consistent interface, whilst encouraging the reuse of command line options and code. The interface between the PC and DSP device is regarded as a simple memory model, whilst hiding details of how the data is actually exchanged.

During the software development cycle, the harness provides the means for handling:

- command line interpretation;
- interactive user input;
- help and usage message generation;
- data file open/closing operations;
- communications with the DSP device.

New algorithms may be added to the harness with the minimum of modification to the existing code. The harness assumes that algorithm data input and output will be via data files and that all code will be written in the C language. Although developed primarily for use within a PC and DSP device platform, a limited facility is provided for debugging algorithms on other hardware such as a Sun or DEC workstation.

The target system used to develop the harness consists of a PC running MS-DOS, equipped with a TMS320C30 development board. The harness provides a basic framework in which DSP algorithm development can take place with the minimum of hardware and software overhead. Within the SPADE environment, many of the facilities provided by the harness have been superseded by the more powerful tools and techniques embodied within the SPW toolset.

5.3.1.3 Signal processing operating executive

Signal processing operating executive (SPOX)^{TM} is a commercial package developed by SPECTRON MicroSystems Inc. This emerged in response to the demand for efficient DSP device software generation from a high-level language. At present, SPOX implements a set of library functions, specifically for the Texas Instruments' TMS320C30 and Motorola's MC56001 DSP devices, which may be called from a C program [10].

SPOX actually consists of three discrete elements:

- SPOX-LINK — providing functions to enable communication between processors (either between different DSP devices, or host and DSP devices);

- SPOX-DSP — providing functions for stream I/O, memory management and optimized DSP device mathematics;

- SPOX-RTK (real time kernel) — providing functions to create, prioritize and run tasks on a target system.

The efficiency of the library routines is difficult to ascertain since they are in object format. An evaluation of the SPOX-DSP environment was carried out to determine the merits of using SPOX [11]. This evaluation was an example of the level of tool integration within SPADE. Initial and final algorithm processing was performed using SPW, whilst a time-critical, nonlinear transform function was implemented using SPOX routines.

5.3.1.4   Modem simulation evaluation system

The modem simulation evaluation system (MoSES) was developed within BT. It was developed to provide a user-friendly interface with a modem simulation program.

MoSES was required to provide a form-based menu system for variable initialization, with tracing facilities for a number of user-defined variables. MoSES would provide graphical output with additional analysis if required.

A major application of MoSES was during the modification of a simulated V.32 modem receiver to test train-up on real training sequences. These training sequences were recorded using a VAX and digital audio tape based data acquisition system, which is a variant of that originally developed by the speech applications area at BT Laboratories. The modifications to the modem simulation program allowed the simulation to more accurately represent the actual hardware implementation. This system now provides a vehicle for simulation of further modem receiver enhancements.

5.3.1.5   Block-oriented network simulator

The block-oriented network simulator (BONeS™) is a commercial tool developed by Comdisco Systems. It is a graphical environment for the simulation of systems that are essentially asynchronous in nature. As such, it differs from the SPW tools since it provides event-driven rather than sample-driven modelling. Thus BONeS is more applicable to the simulation of protocols and communications networks.

As with SPW, BONeS provides standard libraries of functions and these may be extended with functions custom-coded in C. BONeS consists of the following components:

- BDE — block diagram editor;
- DSE — data structure editor;
- SIM — simulation manager;
- PP — post-processor.

The BONeS-BDE is a block diagram editor which is used to create, edit, document and store the network diagrams. The diagram may be hierarchical in nature. As with SPW, custom-coded blocks may be created using C to implement a function not contained in the standard libraries. When a network diagram is saved as complete, the BONeS-BDE generates and compiles the corresponding C code automatically.

Figure 5.3 provides an example of a BONeS network diagram, within the BONeS-BDE.

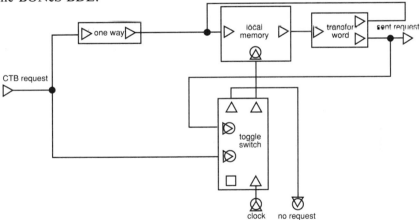

Fig. 5.3   An example of a BONeS block diagram.

The BONeS-DSE is a data structure editor which performs similar editing functions to the BONeS-BDE but applied to the data structures. These data structures may have an arbitrary number of fields and they are defined hierarchically. Each field may be a simple type such as integer, or another data structure. Data structures inherit the properties of any parent.

The BONeS-SIM generates a main program for the simulation. It also controls the submission and the progress of the simulation. Parameters which require initialization at run-time are specified and probes may be attached to extract data from the simulation, in a form usable by the post-processor. All simulation results are stored in a database for subsequent analysis and display by the post-processor.

The BONeS-PP is used to analyse and display simulation results. Statistical information may be calculated from the simulation data for different simulation runs under different conditions.

BONeS was used to model some important features of the CCITT Recommendation G.764, packetized voice protocol, with a G.727 ADPCM codec being modelled in SPW. With these two models, it was possible to simulate the effects of network congestion on ADPCM coded speech transmitted over a frame relay network (see Chapter 17). This example represented the marriage of event and sample-driven simulations within one design environment.

5.3.1.6   Kindra

Kindra allows the specification of algorithms or software, firstly at a high level and then subsequently, at lower, more detailed levels using the CCITT specification description language (SDL) notation. Kindra was developed internally at BT Laboratories [1].

The Kindra implementation is based on a series of charts. The top-level chart contains the complete specification of the software, with further lower-level charts containing the detail. Charts may also be referenced as procedures or macros, the former allows the generation of function calls, the latter in-line code. The Kindra coder combines all the separate charts to form a complete, compilable program.

The use of Kindra demonstrated that a structured tool could be applied to the DSP design process although this particular tool was not ideal. The SPW tools are far more powerful and hence they form the main backbone of software tools within SPADE.

Kindra was the first tool to be introduced into SPADE and it has now been superseded to a greater extent by the SPW tools. However, SDL-based tools are useful for the description of algorithms that require coding for simulation, since they may be readily implemented by other software engineers. There are now other commercially produced SDL-based tools marketed and one of these may be integrated into SPADE instead of Kindra. SDL-based tools also permit the specification of event-driven rather than sample-driven modelling.

5.3.2   Data capture

The acquisition of real-world data for use in simulation programs ensures there is a greater degree of confidence in the simulation results obtained. There are three forms of data acquisition within SPADE:

- DAS — data acquisition system;

- VDAS — VAX data acquisition system;

- SparcStation input/output.

The DAS system is a commercial board housed inside a PC. Sample rates of up to 16 kHz, dual channel with 16-bit precision, are supported. The DAS produces files of a particular format that have been adopted as the standard for data recordings. Essentially a header is appended to the data recording which provides information concerning the file. This is in two parts — optional user comments and system information (such as sample rate). The DAS format files may be used directly within an SPW simulation as conversion blocks have been developed.

The playback utilities on Sun SparcStations are via the built-in loudspeaker and digital to analogue converter. Playback may be controlled from the SPW block diagram editor using the SparcStation audio output.

The VDAS system is based on a digital audio tape (DAT) recorder system, originally developed by the speech applications area at BT Laboratories. After recording the data on the DAT recorder, the data is transferred to a MicroVAX using BT-developed software. The transferred data is by default 16-bit precision, dual channel, at the 48 kHz sample rate of the DAT recorder, although options exist for downsampling the data to 8, 16 or 24 kHz. VDAS produces data files on the MicroVAX in the standard DAS format.

5.3.3   Digital signal processing boards

SPADE supports the commercial PC-based DSP development boards which have the following devices:

- Motorola 56001;

- Motorola 96002;

- Texas Instruments TMS320C30.

A simulation system was developed to allow algorithms to be run on the Motorola 56001, with results from the processing captured for later analysis. The system was not real time, but operated only two to three times slower than the effective real time rate of 8 kHz sampling. This was significantly faster than some other simulation systems, being a factor of ten faster than the Motorola application development system and over a thousand times faster than the Motorola simulator [12].

The fact that the system does not run in real time can be beneficial in many ways:

- it removes some development problems encountered when there is insufficient time in the interrupt cycle to perform all the processing;

- it allows a 'what if' approach where, for example, filter lengths may be extended beyond the real time processing limit;

- data may be applied repetitively, so that the effects of algorithm modifications may be analysed in isolation from data dependent effects;

- algorithm output may be analysed using graphical analysis tools or user coded routines.

A similar system was developed for the Texas Instruments TMS320C30. However, the main use of the Texas Instruments TMS320C30 and the Motorola 96002 has been as target processors for code developed using the SPW-CGS. Due to the postponed release of the Motorola 96002 processor, it is not as yet fully functional within SPADE.

5.4   PARALLEL PROCESSING REQUIREMENTS

Many real time applications require more than one DSP device to achieve the required processing throughput. These multiprocessor solutions are not necessarily utilizing processing tasks in parallel, rather they are performing sequential partitioning to distribute the computational effort. Parallel processing requires the synchronization of processors, parallel task partitioning and accessing shared resources. Currently, none of the design tools within SPADE provide support for parallel processing simulations. Some of the inherent problems in providing support for multiprocessor simulations are that:

- the partitioning scheme requires knowledge of the target hardware configuration (this configuration information should be easy to change for different implementations);

- the task scheduling must take account of the required interprocessor communication overhead (this will depend on both the processors and the target hardware configuration);

- some means of monitoring the scheduling performance is required, this being necessary to determine whether the partitioning is an efficient use of the processing resources;

- the software specification must contain sufficient information for a scheduler to determine which tasks may be executed in parallel.

Parallelism is often utilized within DSP devices to increase their computational power. The most common uses of parallelism within DSP devices are instruction pre-fetching, and arithmetic pipelining. When the parallelism is implicit, the programmer must be aware of its implications. This may compromise the ease of programming the device. Explicit parallelism may allow the programmer to perform pointer modification in parallel to an arithmetic operation, or even two simultaneous arithmetic operations.

5.4.1   Multiprocessor topologies

When parallelism is exploited in a multiprocessor fashion, this is usually accomplished by message passing. The connections between processors are usually message-passing channels for practical systems. These connections can be fixed or dynamically reconfigurable. A fixed configuration may be appropriate to the specific task that the hardware is to perform, like the DSP-computer interconnections. The DSP-Computer (Network Emulator) (see Chapter 12) is a general purpose development engine. Therefore it allows these connections to be manually configured (via jump leads) to suit the requirements of the desired system.

The Texas Instruments TMS320C40 (C40) has been designed specifically for DSP multiprocessing applications. There are six 20 Mbyte/s parallel ports on a C40 (asynchronous bi-directional). The communication can be accomplished via DMA so there is no computational penalty. On power-up, three of the communications ports are initially in output mode, and the other three as inputs. Communications can change direction at the request of the sending DSP device, which transmits a special token message to the receiving DSP device. The DSP devices then change roles. The C40 has no serial communications ports, and so interfacing it to A/D—D/A chips requires a little glue logic.

Some general-purpose systems reconfigure the interconnections by software. This can be achieved in one of two ways. Either the communication links are routed via a switching network, or the communication links are fixed and the processors decide which ones are to be used and which are not. For the fixed link scenario, the ideal situation is to have fully interconnected processors. The fully interconnected architectures shown in Fig. 5.4 are suitable for a TMS320C40, the tetrahedron configuration being interconnected with two links between each processor, hence avoiding token passing overheads to change direction of communication. However, a totally interconnected architecture is often prohibited due to hardware constraints on the number of communications ports per processor. In this case, they would be interconnected in some regular structure such as a tree or a grid.

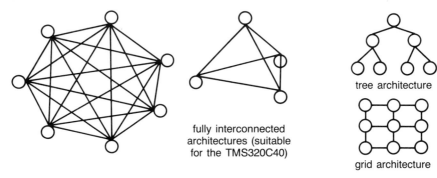

tree architecture

fully interconnected architectures (suitable for the TMS320C40)

grid architecture

Fig. 5.4   Multiprocessor architectures.

The tree structure suits algorithms that have been written using a subroutine methodology. The main program running on the root node parent processor, passes parameters to subroutines running on its child node processors (which in turn may pass parameters to subroutines running on its child node processors). Because this sort of architecture may waste processors, virtual tree structures are more common (this can be anything that 'looks the same' from a software point of view).

The grid structure is very popular because it is simple. Each processing element only requires four interconnection channels (or six for a two-dimensional grid). Hardware placement of processing elements is therefore simple, as is the message-passing software.

However, the ideal interconnection scheme minimizes not only the number of interconnection channels required per processor (hardware cost),

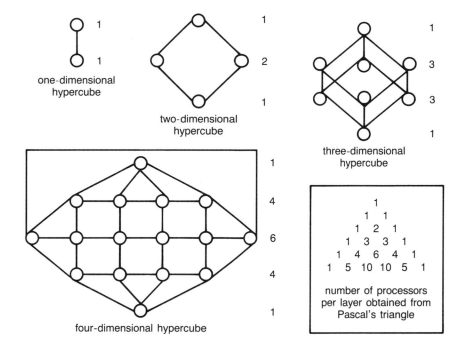

Fig. 5.5   Hypercube structures.

but also the average distance between processors (performance cost). The Hypercube is claimed to be the best compromise between these two (see Fig. 5.5).

5.4.2   Dynamic scheduling

The allocation of resources to a dynamic process must be performed as they are needed at run time. This is known as dynamic scheduling. The task of dynamic scheduling is normally performed by an operating system. An operating system for a multiprocessor system will have a kernel resident in each processor. This kernel will handle hardware dependent calls, message passing and (in co-operation with the other processor kernels) processor resource allocation. Such operating systems for DSP devices are a relatively new idea, but it is likely that the increasing demand for parallel processing will encourage further development in this area.

A 'good' process to processor allocation for a dynamically scheduled algorithm can be obtained from a simulation profile. A profile is an analysis of details such as the utilization of each processor and interprocessor communications overheads. Such a profile will give an indication of where the bottlenecks of the configuration are. Processes may be moved around to compensate for these bottlenecks. This is traditionally done by hand but there is no reason why a rule-based automated system could not perform this task. Profiling would typically be performed in non-real time on a single processor emulating a virtual parallel system. In this way the profiling itself can be easily excluded from the measurements.

5.4.3   Static scheduling

The allocation of resources to a static process can be performed prior to execution (provided these resources are not shared). This is known as static scheduling. Implicit static scheduling is where the allocation of tasks to processors is not specified by the programmer. A 'good' allocation must be worked out using a scheduler program. An optimum allocation can be found by looking at every possible allocation and selecting the best. However, there may be 'many' possible allocations and so there exist algorithms which perform a less exhaustive search and yield an allocation which is 'good'.

The factors that should be considered by such an algorithm are the task execution times (which may be different for each processor in a heterogeneous system) and the interprocessor communication delays. Scheduling may be performed in a pre-emptive or non-pre-emptive fashion.

5.4.4   Communications issues

There are possible dangers that can be caused by communication between dynamically scheduled processes. The operation of the concurrent processes must be synchronized. A particular process may suspend its operation while it waits for another processor to pass data, on which it depends. In this situation it is possible for deadlock to occur. This is where a circular wait condition prevents any further processing. When the processors share memory to communicate, it is important to enforce restrictions or impose a protocol on the way in which processors can access this memory. Otherwise it is possible to misinterpret partially-completed communications, or lose processor synchronization.

Processes in concurrent programs can communicate either by competition or by co-operation. Competition strategies are usually employed when the processors communicate via shared memory, whereas co-operation strategies are more suited to communications channels. When processes compete for a shared resource (such as memory) this is known as mutual exclusion. The sequence of statements within a process, which refer to a shared resource, is known as a critical region. Languages that support communication in this way provide a mechanism to ensure that only one process can execute its critical region for a particular resource at any one time. Semaphores and monitors are two ways of achieving this.

There are few useful high-level notations for expression of parallelism within DSP algorithms and communication between DSP tasks. Available tools may provide a means of scheduling and simulating an algorithm, but there is no automated path to efficient DSP code. The DSP tasks are typically hand-coded in the low-level DSP assembly language, and the message passing routines tend to be constructed in the same way. The standard of DSP development software lags very far behind the impressive capabilities of DSP hardware. This is what prevents multi-DSP from achieving its full potential.

5.4.5   Object oriented concurrent programming

Object oriented programming (OOP) has been hailed as one of the most important building blocks of next generation software. The ideas behind OOP actually originated in the late 1960s, but have only recently come into vogue. Several conventional languages (such as C + + and ADA) have been updated to support OOP. However, in such languages, the programmer has the freedom to choose whether to use a procedural methodology, or an object oriented one. There is, therefore, a danger that such programs may be written using both! Eiffel [13] is an example of a 'pure' object oriented language that supports concurrency.

The OOP philosophy is simply to encapsulate the data structures for components of the software with the operations or methods that are performed upon it, into an entity known as an object. This encapsulation ensures that objects are reusable and can be easily manipulated during execution. An object may inherit features from other objects. The interface of this object to its environment (the rest of the program) is tightly specified. In Eiffel, a consumer/supplier model is used, in that the object is a supplier to various consumers where it is invoked. Various clauses may be constructed within an object as to what each party expects to receive from the other. Failure to meet such a clause at run time, may trigger an exception. This feature of Eiffel can be exploited to ensure software reliability. The static definition of an object as defined within a program is known as the class of that object, whereas the dynamically allocated run time entity is known as an instance.

An object has similar properties to a process within a multiprocess environment, in the way in which it communicates with its environment, and retains its state between invocations. This property has been exploited in Eiffel by allowing concurrency by the same mechanism that is used to envoke an object (with only minor semantic differences). A 'lazy wait' mechanism is utilized to allow the program execution to effectively fork when this occurs. Lazy waiting involves the object returning a reference to a result which is still in the process of being evaluated. Meanwhile, the consumer is able to continue, using this reference in place of the result. A reference can be used in activities such as assignments or parameter passing, without need for its associated result. There are activities such as arithmetic, for which the consumer may have to wait for the supplying object to complete its evaluation of the result. Also the clause mechanism may also cause waiting so in order to adhere to data dependencies.

The combination of object oriented and concurrent principles in this way is very elegant because, in using an object oriented methodology, the programmer has very few other considerations to write programs which are concurrent.

5.4.6   Applicative languages and data-flow

A function in mathematics is a rule of correspondence which associates each member of its domain to member(s) in the range. This mapping from the domain to the range is static. Calls of the function using the same arguments (values within the domain) will always produce the same result (a value in the range). This is a very important property of functions, and is known as referential transparency. Conventional programming languages are imperative. An imperative language achieves its effect by changing the state of variables by assignment. An applicative language, however, is a language which achieves its effect by applying functions either by composition or recursively. Because there is never any redefinition of the state of variables in an applicative program, it can always be represented graphically as a data-flow graph (provided iteration and recursion are treated sensibly). Also, a data-flow graph can be viewed as an applicative description.

A DSP algorithm designer will sometimes prefer to graphically manipulate blocks in hierarchical data-flow diagrams, and sometimes it is more natural to describe things textually (e.g. equations). In the future there may be development environments that allow the user to switch between the two modes of interaction (see Fig. 5.6). These notations are far more natural for describing DSP algorithms than conventional (imperative) programming methods. It is also far more descriptive than simple block diagram simulation tools because high-level abstractions always boil down to the language axioms. The parallelism within the algorithm is implicit and so the algorithm designer does not waste effort bothering about concurrent tasks and communications. An applicative program is always deterministic, which is an important consideration for parallel systems which are usually notoriously difficult to

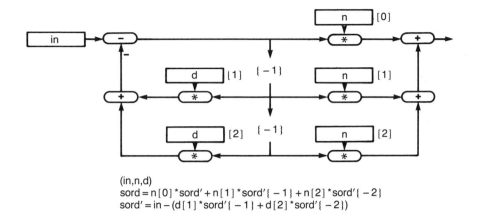

(in,n,d)
sord = n[0]*sord' + n[1]*sord'{-1} + n[2]*sord'{-2}
sord' = in - (d[1]*sord'{-1} + d[2]*sord'{-2})

Fig. 5.6 User entered data-flow description, and automatically generated functional description for a second order filter.

debug. Applicative languages can be used, not only to describe the algorithm, but also to implement formal program verification methods.

An applicative description is a description of what needs to be done. An imperative description describes how you want the processor to go about it. How the processor will go about a task should depend on the architecture of that processor. For a non-specific environment, such details should be abstracted from the programmer. The higher the level of the description of an algorithm, the more freedom a compiler would have to perform optimization, and the fewer implementation dependent distractions the algorithm designer has to contend with.

The data dependencies and fine-grain parallelism are easily apparent from a data-flow graph. A compiler analysing an internal representation of a data-flow graph should be able to alleviate the fine grain parallelism and pipelining delay problems, for compilation on to a DSP device which exploits such properties. Also, the analysis of the best way to split up an algorithm on to different processors (for efficient parallelism) can also be derived from a data-flow description. The granularity of such parallelism determination is not constrained by the syntax of the language.

The graphical and textual applicative notations are ideal for describing DSP algorithms. They abstract implementation details (such as utilization of parallelism) from the algorithm designer. However, DSP tools that currently exploit data-flow techniques are unambitious in what they try to achieve.

5.5   AREAS FOR FURTHER WORK

SPADE can be extended to cover the whole life-cycle for DSP device code development. This includes the higher-level issues relating to requirements capture and specification, through prototyping and down to final implementation.

Over the next year, tools which support parallel and multiprocessing architectures will become available and these may be integrated into SPADE. The efficiency of DSP device code generation from high-level specifications will increase due to the use of functional specification languages. SPADE has currently been used to demonstrate the integration of event and sample-driven simulations. More work will be carried out to further explore these areas.

There are currently, however, few useful high-level notations for expression of parallelism within DSP algorithms, and communication between DSP tasks. Available tools may provide a means of scheduling and simulating an algorithm, but there is no automated path to efficient DSP code. Because the DSP tasks are typically hand-coded in the low-level assembly language, the message-passing routines must be constructed in the same way.

The standard of DSP development software lags very far behind the impressive capabilities of DSP hardware. This is what prevents multi-DSP from achieving its full potential.

The graphical and textual applicative notations are ideal for describing DSP algorithms in such a way that does not constrain them to sequential realization. The technological barrier that needs to be overcome is that of efficient compiler techniques to perform the translation to DSP executable code.

5.6   CONCLUSIONS

A signal processing application development environment (SPADE) is established to cater for the prototyping of digital signal processing applications. By creating a prototyping environment such as SPADE it has been possible to introduce an element of consistency to design method. This has the advantage of generating a pool of expertise and creates libraries of reusable functions. The benefits such as hierarchical specification and documentation cannot be over-emphasized. The block diagram approach to algorithm specification is a natural form of expression for the digital signal processing algorithm designer.

Simulations within SPADE are usually run on a non-real time basis. With the DSP development boards supported, it is possible to run the simulation on a DSP device. This still need not be real time and indeed can be an advantage since it allows a 'what if' approach to be tried without regard to interrupt routine execution time. Simulation data may be recorded from a real-life situation, thus giving confidence to the simulation results. Simulation results may be captured using a variety of methods and analysed. The results from a full floating-point simulation may be compared with those of a DSP device implementation. The designer may choose the precision at each stage.

SPADE is a modular environment. As such, tools may be added or removed. As much or as little of SPADE may be used as required. Entry points are provided at all levels. New tools will continue to be added to SPADE to accomplish this and some tools currently supported may be removed. SPADE is evolving and the extension will ensure that this evolution continues to track the needs of the DSP community within BT in a timely fashion.

REFERENCES

1. Ip S F A, Knowles R J and Easton P W: 'Digital signal processing system design methodology', IEE Colloquium Digest No. 1990/120, IEE Computing and Control Division, pp 2/1-4 (1990).

2. Knowles R J, Ip S F A and Easton P W: 'Structured DSP design methodology', IASTED International Conference on Signal Processing and Digital Filtering, Lugano (1990).

3. Ip S F A, Westall F A, Knowles R J, Benyon P R and Easton P W: 'A prototyping environment for digital signal processing systems', to be published in Computing and Control Engineering Journal.

4. Lee E A et al: 'Gabriel: a design environment for DSP', IEEE Micro, pp 28-45 (October 1990).

5. Lauwereins R et al: 'GRAPE: a CASE tool for digital signal parallel processing', IEEE ASSP, pp 32-42 (April 1990).

6. Knoll A and Nieberle R: 'CADiSP — a graphical compiler for the programming of DSP in a completely symbolic way', Proceedings ICASSP, V5.7, pp 1077-1080 (1990).

7. Leathley B A: 'Human factors guidance for the development of SPADE', Internal BT memorandum (October 1990).

8. Benyon P: 'An exercise in SPW code optimisation', Internal BT memorandum (January 1991).

9. Barrett P: 'An end to end network modelling environment for packetized speech transmission', Internal BT memorandum (September 1990).

10. SPECTRON MicroSystems Inc: 'Application programming guide, TMS320C30 version' (1989).

11. Ip S F A: 'The use of SPOX in an echo control rapid-prototyping environment', Internal BT memorandum (July 1990).

12. Knowles R J and Easton P W: 'Report on the integration of a DSP within the advanced signal processing project', Internal BT memorandum (January 1990).

13. Meyer B: 'Object oriented software construction', Prentice Hall (1988).

6

A LOW-COMPLEXITY HIGH-SPEED DIGITAL SUBSCRIBER LOOP TRANSCEIVER

N G Cole, G Young, N J Lynch-Aird, K T Foster and N J Billington

6.1   INTRODUCTION

BT's local network of unscreened twisted pair cables is used extensively for transmission of analogue telephony or data at relatively low speeds (e.g. 19.2 kbit/s modems or 64 kbit/s KiloStream) yet is theoretically capable of carrying data at rates of 2-3 Mbit/s. This capacity for data transmission coupled with the ubiquitous nature of existing cables is attracting increasing interest in the search for low-cost high-speed access to digital networks. Business access systems at fractional primary rates, 384 kbit/s compressed video and digital multichannel pair-gain systems are all possible applications of this technology.

The transceiver described in this chapter incorporates echo cancellation to enable full duplex operation at 480 kbit/s over a single pair. The design is of low complexity relative to other methods which have been proposed for high-speed digital subscriber loop (HDSL) applications. However, the system has good tolerance to crosstalk and achieves a performance of only 1-1.5 dB less than similar systems with 50% more computational complexity. The extensive efforts to achieve reductions in complexity have allowed an implementation based on currently available DSP microprocessor devices, yet the implementation has an overall power consumption low enough to permit line powering if required.

The overall line bit rate of 480 kbit/s supports seven 64 kbit/s channels, a framing signal and spare capacity for maintenance and supervisory purposes. Since the implementation is DSP based it is relatively straight-forward to modify the line bit rate.

6.2   SYSTEM OVERVIEW

A range of transmission techniques has been proposed for HDSL applications [1, 2] but many of the more complex methods offering higher performance are being dismissed in preference for techniques that can be deployed rapidly to achieve early exploitation. Figure 6.1 shows a block diagram of the transmission system described in this chapter. The pulse amplitude modulation (PAM) architecture is similar to that used for ISDN basic rate 144 kbit/s transceivers (e.g. Colbeck et al [3]) which are now available on single integrated circuits.

Data formatting and maintenance functions are performed by the digital interface and maintenance blocks. Data to be transmitted is scrambled and encoded to an 8-level line signal and frame synchronization words are inserted. The 8-level 160 kbaud transmit stream is then fed to line via a digital-to-analogue converter (DAC), low-pass filter, line transformer and hybrid circuit.

The line hybrid provides some suppression of the local transmitted signal in the receive path. The received signal passes through an anti-aliasing filter and is sampled at the baud rate by a 14-bit analogue-to-digital converter (ADC). These samples are passed to the DSP devices which implement the receiver function. The sampling instant of the received signal is controlled by a timing adaptation algorithm to ensure sampling occurs near to the peak of the incoming pulses. Decisions from this circuit control the receiver timebase.

In the receiver a digital filter with fixed coefficients provides a degree of fixed linear equalization. This is followed by an adaptive decision feedback equalizer (DFE) which removes post-cursor values due to inter-symbol interference (ISI). The echo canceller (EC) subtracts echo signal which leaks through the hybrid into the receive path. Both the DFE and EC adapt to suit the particular channel response over which the system is operating. The signal is quantized to 1 of 8 levels in the quantizer block and the 8-level signal is converted back to binary.

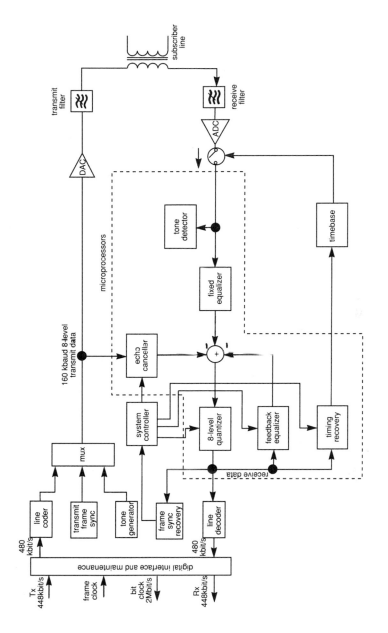

Fig. 6.1 Block diagram of transceiver.

6.2.1   Twisted-pair channel equalization

The frequency response of local network cables causes attenuation of high frequencies and a nonlinear phase response below about 10 kHz. These two characteristics result in a dispersed asymmetrical pulse response like that shown in Fig. 6.2. On long loops the pulse may be dispersed over many symbol intervals (typically more than 100) resulting in severe ISI. A linear equalizer in the form of an adaptive transversal filter may be cascaded with the channel to provide the inverse frequency response resulting in zero ISI at the sample instants. However, the linear equaliser enhances not only the attenuated higher signal frequencies but also the high-frequency crosstalk noise from other adjacent transmitters, preventing an acceptable level of performance from being achieved.

An alternative to linear equalization is the DFE [4]. This is also an adaptive transversal filter but is placed in a feedback loop around the receiver's detector. It uses previous decisions by the detector to build up a model of the tail of the pulse response so that the post-cursor ISI can be subtracted at the input to the detector. It adapts to the pulse response of a particular channel using the same algorithm (LMS) employed by the EC described in section 6.2.4. Because the transversal filter is fed by detected data and not received signal values there is no noise enhancement and the internal arithmetic and delay-line storage elements are simpler. The main problem with the DFE is the required complexity (i.e. length). To span the slowly decaying pulse tail shown in Fig. 6.2 would require in the order of

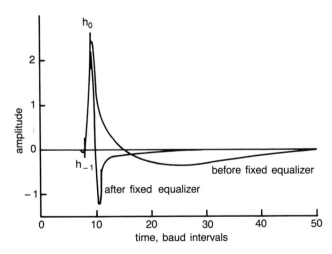

Fig. 6.2   Typical twisted-pair pulse response before and after convolving with fixed equalizer.

100 coefficients while the whole equalization operation (filter convolution and coefficient adaptation) must be performed in less than 6.25 μs.

To achieve the required ISI suppression with limited complexity, a combination of linear and decision feedback equalization is used which exploits *a priori* knowledge of the nature of the channel [5].

The tail of a transmission pulse can be accurately modelled by three real poles (i.e. the sum of three exponentially decaying components). Two of the poles vary with cable length but one is fixed with a time constant that is a function of line transformer inductance L and termination resistance R. A simple two-tap non-adaptive linear equalizer $E(z)$ may be cascaded with the channel to provide the zero that will cancel this pole:

$$E(z) = 1 - k.z^{-1}$$

where $k = e^{-RT/L}$
T = baud interval

The remaining tail energy from the two other poles can be removed by a DFE of reasonable length. Typically k has a value slightly less than unity resulting in a linear equalizer giving white noise enhancement of around 3 dB. The noise gain may be reduced by distributing the tail-reduction function to a greater number of smaller coefficients. For example, a longer $N+1$ tap finite impulse response (FIR) filter may be formed with all but the first (unity) coefficient having the value A given by:

$$A = (\Sigma\, k^{-i})^{-1} \qquad \text{for } i = 1 \text{ to } N$$

The white noise enhancement G_n of such a filter is given by:

$$G_n = 10.\log(1 + N.A^2)$$

For $N=4$, the distributed pole cancellation coefficients result in a white noise gain of less than 1 dB. The resultant smaller channel post-cursor magnitudes will also result in smaller DFE coefficients leading to reduced error propagation characteristics. Figure 6.3 shows a simplified diagram of the equalized channel with pulse responses and eye-diagrams produced by computer simulation.

Other methods have been used to reduce transversal filter computational complexity [6,7] at the expense of longer wordlength requirements.

6.2.2   Timing adaptation

The exchange unit and remote unit will have clock sources which are different in frequency. Clock drift between interworking units must therefore be countered by controlling the receiver clock phase. After convolving a loop response with the fixed FIR pole canceller the loop response will be similar to that of Fig. 6.2. A sample phase may be found that minimizes the first precursor (h_{-1}) and places the cursor (h_0) near the pulse peak on short loops and slightly before the peak on longer loops. The sample height loss on longer loops is typically less than 0.5 dB. A suitable sample phase may be located by arranging for the filtering to provide a small undershoot prior to the rising edge of the pulse. An adaptive precursor coefficient can then provide the phase error for a phase locked loop (PLL) that will attempt to drive this coefficient (h_{-1}) to zero, sampling the pulse one baud interval later.

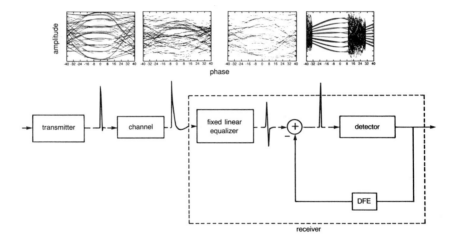

Fig. 6.3   Pulse responses and signal eye-diagrams produced by DSP simulation.

6.2.3   Line code

The transceiver uses an 8-level zero-redundancy octal line-code 3B1O which maps 3 binary digits on to 1 octal transmitted symbol. The 3B1O code is not the optimum line code for this application (see Fig. 6.4). Better performance could be achieved with codes such as 2 bits per symbol (e.g. 2B1Q). However, this difference is small (1-1.5 dB) due to the flat nature of the curves.

The use of 2 bits per symbol instead of 3 bits per symbol implies a 50% increase in symbol rate for any given bit rate. The DSP program for the transceiver has to be executed each symbol interval and thus the DSP microprocessor program rate would need to be increased by this same percentage. Use of 3 bits per symbol thus allows a large reduction in DSP program rate whilst tolerating sub-optimum performance.

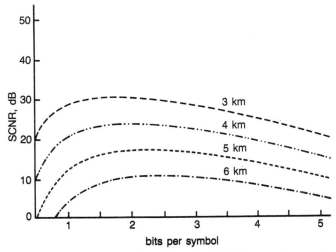

Fig. 6.4   Signal to near-end crosstalk noise ratio (SCNR) versus degree of symbol rate reduction, computed for various lengths of 0.5 mm gauge cable.

6.2.4   Echo cancellation

Simultaneous transmission of data in both directions is achieved using adaptive echo cancellation [8]. A linear echo canceller with one coefficient per symbol interval is implemented which, in conjunction with the line hybrid, achieves in excess of 65 dB of echo suppression and is adapted by means of the widely used least mean squares (LMS) algorithm:

$$c_k(n+1) = c_k(n) + \beta.e(n).d_k(n)$$

where β is the step size or gain constant, $e(n)$ is the error due to misadjustment and $d_k(n)$ the transmitted data; $c_k(n)$ is the kth filter coefficient at time nT.

Linear echo cancellation demands high levels of linearity in the analogue signal processing, the line-driver and the ADC. The fixed FIR equalization described in section 6.2.1 is also effective in reducing the large pulse tail of the echo response and thus reduces the complexity (i.e. length) requirement of this adaptive element.

6.2.5 Start-up sequence

The DFE, EC and timing extraction are fully adaptive and, once converged, will adapt to slow variations in the channel characteristics due to environmental changes such as temperature. However, at power-on, the system must perform a start-up sequence of transceiver states to ensure rapid and reliable convergence of each adaptive element. This sequence can be activated from either end and activation is signalled to the opposite end by a short tone-burst. The tone-burst is detected by a correlation algorithm implemented at power-on by the DSP devices. Upon detection of the tone the DSP devices switch to different programs to implement the digital transceiver. The start-up sequence of transceiver states is controlled by a simple embedded microcontroller which ensures that reliable activation is achieved even on the longest loops in the presence of crosstalk.

6.2.6 Peripheral transceiver functions

6.2.6.1 Maintenance and management

Once the transceiver activation sequence is completed the microcontroller becomes available for simple, real time processing for maintenance and management purposes. The DSP structure adapts to suit the characteristics of the line (e.g. length, gauge) and all the coefficients from the DSP devices can be made available to the microcontroller. For example the received signal magnitude and residual error signal can be used, whilst the transceiver is operating, to produce estimates of signal-to-noise ratio (SNR) at the quantizer or the line attenuation. The benefits of these techniques have not yet been fully quantified but there appears to be potential to supplement existing methods for digital subscriber loop maintenance and management.

6.2.6.2 Scramblers

In order to ensure adequate de-correlation between the data transmitted in each direction and thus robust adaptation of the adaptive filters, the data is scrambled before transmission with a different polynomial for each direction of transmission.

$$P_1(x) = 1 + x^{-18} + x^{-23} \text{ (from customer)}$$

$$P_2(x) = 1 + x^{-5} + x^{-23} \text{ (to customer)}$$

6.2.6.3   Framing

The frame structure on the line is programmable at the digital interface though there are constraints set by the timebase. In normal mode, data is input and output from the transceiver within a partially filled, standard 2 Mbit/s 125 μs frame structure of 32 channels. On the line in normal mode a 160-octet (480-bit) primary frame contains 8 blocks of 56 bits (8 bits for each of the 7 channels) plus a 7-octet frame synchronization pattern and spare supervisory capacity.

Since the peripheral functions consist of many varied blocks of generic logic a field-programmable gate array (FPGA) device was used for implementation. This method achieves a good level of integration whilst retaining a degree of reprogrammability and allows future migration to low-cost mask programmed devices.

6.3   SIGNAL PROCESSING DESIGN

6.3.1   Channel modelling

Reliable equalizer design requires fast and flexible access to a representative database of loop responses which may have to be equalized. To avoid vast numbers of time-consuming measurements an overall channel response can be constructed by cascading transmission matrices which accurately represent cables, line-hybrids, transformers, bridged-taps, and transmit and receive analogue low-pass filters.

Open- and short-circuit impedance measurements are taken for a cable sample from which it is straightforward to derive the characteristic impedance Z_0 and propagation constant γ. A transmission matrix for a chosen length of this cable can then be generated using standard *ABCD* parameters. Time domain responses can be generated by using the inverse fast Fourier transform on the frequency response obtained from the transmission matrix for the overall channel.

In order to specify and dimension the overall signal processing function the multistage design process of Fig. 6.5 was used.

6.3.2   Static analysis

After the requirements capture stage (Fig. 6.5), the design process consists of selecting the system architecture and dimensioning the DSP requirements. Initial performance estimates are obtained through static analysis. This involves processing simulated channel responses and noise spectra in a variety of ways to produce specific performance metrics such as the residual ISI and echo with a given DFE, EC and line code or the D/A and A/D converter resolutions required. The dynamic aspects of the system such as timing and coefficient adaptation are not included at this stage, greatly reducing the processing requirements and allowing many design options to be explored rapidly.

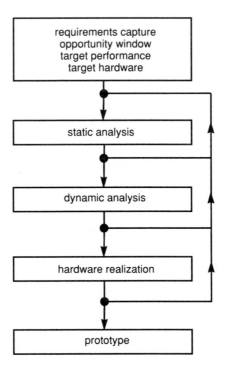

Fig. 6.5   Design process.

Because the static analysis is used to perform advance studies the software suite comprises a large number of small programs performing individual analyses. This allows greater flexibility in application and more rapid development of new routines than could be achieved with a single large program. Complex studies are rapidly assembled using shell scripts.

6.3.3   Dynamic simulation

Once a basic system architecture is defined, the dynamic behaviour must be assessed. A simulation of the entire local network link including transceivers, channel and impairments is required to complete the design and fully characterize the system performance. The transceiver includes three adaptive elements, namely the DFE, the EC and the timing acquisition loop which interact in a complex fashion. Parameters that are designed from the dynamic simulation include the gain constants of the adaptive filter algorithms, the timing loop gain and bandwidth, and the start-up sequence. Performance measures that may be gleaned from such a simulation include signal-to-noise ratio, bit error rate (BER) and start-up time.

6.3.3.1   A custom DSP design environment

Channel modelling, system design and simulation are all performed on a Hewlett Packard A845 minicomputer. This environment incorporates a standard file format and a common range of utilities and libraries. These analytical and simulation tools have evolved from earlier versions employed in the design of ISDN basic access transceivers. The latest versions are extremely flexible and run time efficient.

The dynamic simulation is performed in a Unix® environment and is controlled by three external files. The 'environment' file sets the simulation environment in terms of transceiver variables such as filter lengths, adaptive filter gain constants, timing loop parameters, baud rate and start-up sequence files. It also sets the channel variables such as cable length, crosstalk level, frequency offset and initial phase offset. Two other files define the state machines which control the start-up sequence of the transceivers at each end of the link. These are the 'function' file that control the operation of the receiver DSP elements and the transmitter together with a 'sequence' file that determines the sequence of state transitions needed to train up the transceivers. The state transitions depend on the outputs of various state transition detectors such as tone, synchronization word, signal threshold and constant data value detectors together with a state timer. These are all simulated within the main dynamic simulation program and are examined once per frame to determine the path taken by the start-up sequence state machine.

Unix® shell scripts are used for top-level control of the simulation, by putting parameters from the environment file in control loops around the simulation program enabling the simulation to be run over a range of parameters such as cable length, noise level or frequency offset. The transceiver has a multidimensional performance surface that is a function of many variables. Controlling the simulation via shell scripts enables the contours of this performance surface to be mapped out and subsequently the design may be optimized.

The simulation produces a variety of output files which, when plotted, provide a powerful aid to design and diagnostics. Examples include adaptive filter mean square error convergence, adaptive filter coefficient trajectories, sample phase, status of transition detectors during start-up and eye diagrams.

6.4   DSP IMPLEMENTATION

The transceiver could be integrated on a single integrated circuit but this may not be justified until market volumes grow. Initial applications may therefore need to rely on board-level transceiver solutions at higher unit costs, but requiring lower development investment. The transceiver described in this chapter has been implemented on a single extended Euro-card (220 mm × 100 mm) as shown in Fig. 6.6.

Two Motorola MC56001 40 MHz DSP microprocessor devices implement all of the digital signal processing functions and, in parallel, complete one full execution of a program each transmitted symbol interval (6.25 μs). The overall digital signal processing functions are partitioned from the top down to achieve efficient function implementation and passing of variables. Ideally the interprocessor communications for this type of parallel processor application require common RAM space which can be easily and very rapidly accessed by any processor, allowing simple passing of variables to and from the processors and any peripheral circuitry. Such a need could be fulfilled by a multiport RAM with very fast access times and 8-16 common address locations. However, the absence of a suitable, commercially available technology meant that a solution had to be developed using an FPGA.

Given the very short DSP program execution time, DSP algorithm control using inherently inefficient test and branch instructions was avoided by exploiting host command vectors. One of several different programs is selected by an addressing vector which is formed by the embedded microcontroller every symbol period.

The serial interface ports of the DSP devices are utilized to provide communications with an external desktop computer allowing full access to internal variables. This provides a powerful debugging tool and allows external post-processing for performance analysis and displaying of transmission parameters. Furthermore, correct operation of the DSP programs can be verified against finite wordlength dynamic computer simulations.

Fig. 6.6   Photograph of DSP based transceiver.

6.5   PERFORMANCE

6.5.1   Crosstalk tolerance

Near-end crosstalk (NEXT) from 49 adjacent identical systems in a 50-pair cable can be synthesized by applying filtered white noise transversally across the line at the customer's end using the method of combination noise injection [9]. The noise filter is designed to spectrally shape Gaussian white noise to accurately synthesize the desired power spectral density at frequencies from 100 Hz to 640 kHz (four times the baud rate) as shown in Fig. 6.7. The amount by which this noise level can be increased (i.e. margin) whilst still achieving a BER of 1 in 10^7 can be determined for various loop topologies. With a 3 dB margin against self-NEXT, deployment is expected to be possible over 85% of loops. This is equivalent to 4.5 km of 0.5 mm gauge cable.

Computer simulations have shown that NEXT from other services (e.g. speech, 144 kbit/s ISDN basic rate systems) is less severe that self-NEXT.

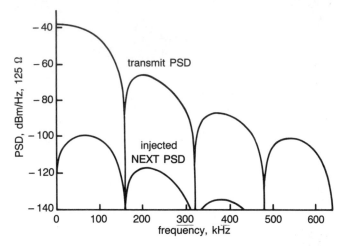

Fig. 6.7   Transmitter power spectral density (PSD) and injected near-end crosstalk PSD produced by static analysis.

6.5.2   Attenuation limit

The crosstalk limited performance assumes that all adjacent pairs in the cable carry similar systems, which is pessimistic. It might be expected that very often adjacent cable pairs will carry other services (e.g. speech or 144 kbit/s ISDN). In any sparse deployment the system may experience little or no NEXT. In these cases the system could achieve a much greater range which will be limited by attenuation and often the A/D converter performance.

6.5.3   Performance versus bit rate

It was stated earlier that the system can be operated at lower bit rates since the DSP implementation simply scales. Figure 6.8 summarizes the performance if the transceiver were to be operated at lower rates.

6.6   APPLICATIONS AND FUTURE DEVELOPMENTS

This DSP transceiver technology can be applied to a number of applications three of which are illustrated in Fig 6.9. Multichannel digital pair gain systems would enable several customers to be served with analogue telephony over

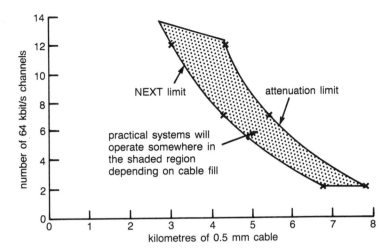

Fig. 6.8   Number of channels versus reach for HDSL systems.

a single pair. Data services such as those at 384 kbit/s (e.g. videoconferencing) or partially filled 2 Mbit/s (e.g. MegaStream leased line or ISDN30) could be delivered quickly over the existing copper cables of the local network. In the USA there is great interest in the delivery of 1.5 Mbit/s primary rate data service using two pairs, each carrying data at 784 kbit/s (dual-duplex) [10].

6.7   CONCLUSIONS

A DSP based high-speed digital subscriber loop transceiver has been described. A DSP design methodology has been outlined which has enabled large reductions in the complexity of the transceiver without compromising performance. The transceiver has been implemented using DSP technology on a circuit board of 100 mm × 220 mm and can be used in a range of applications.

The widespread geographical depoloyment of the existing copper network is set to attract ongoing interest in data transmission for access to digital networks, with interest in the future at even higher data rates. Advances in DSP technology combined with rigorous DSP design methodologies should lead to rapid and wider exploitation of the vast copper network asset.

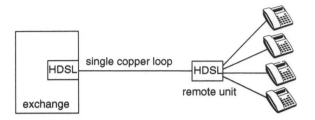

(a)   Multichannel digital pair gain for telephony.

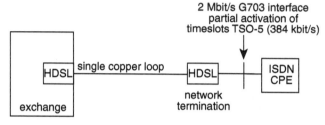

(b)   Fractional 2 Mbit/s data delivery.

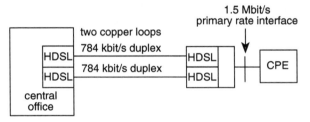

(c)   Transport of 1.5 Mbit/s primary rate (USA) using two copper pairs.

Fig. 6.9   Example applications of high-speed digital subscriber loop transceivers.

REFERENCES

1.   Zervos N A: 'High-speed carrierless passband transmission over the local network cable', Proceedings of the Global Communication Conference, pp 1188-1195, San Diego, USA (December 1990).

2.   Tu J C et al: 'Crosstalk limited performance of a computational efficient multichannel transceiver for high rate digital subscriber lines', Proceedings of the Global Communications Conference, pp 1940-1944 (1989).

3.   Colbeck R P et al: 'A single chip 2B1Q U-interface transceiver', IEEE Journal of Solid State Circuits, 24 , No 6, pp 1614-1624 (December 1989).

4.   Belfiore C A and Park J H: 'Decision feedback equalization', Proceedings of the IEEE, 67 , No 8, pp 1143-1156 (August 1979).

5. Young G and Cole N G: 'Design issues for early high rate digital subscriber loop', Proceedings of the Global Communication Conference, pp 1177-1182, San Diego, USA (December 1990).

6. Young G: 'Reduced complexity decision feedback equalisation for digital subscriber loop', IEEE J Select Areas in Commun, 9, No 6, pp 810-816 (August 1991).

7. Lynch-Aird N J: 'Review and analytical comparison of recursive and non-recursive equalization techniques for PAM transmission systems', IEEE J Select Areas in Commun, 9, No 6, pp 830-838 (August 1991).

8. Lee E A and Messerschmitt D G: 'Digital communication', Kluwer Academic Publishers (1988).

9. Cook J W and Foster K T: 'Crosstalk injection when testing transmission systems', IEE Electronics Letters, 26, No 15, pp 1181-1183 (July 1990).

10. Starr T J J, Waring D L and Werner J J: 'High bit rate digital subscriber line (HDSL): an expedient broadband access', Proceedings of the IXth International Symposium on Subscriber Loops and Services, pp 418-424, Amsterdam, The Netherlands (April 1991).

7

AN APPLICATION OF DSP
TO VOICEBAND MODEMS

J H Page, D A Smee, D Pauley, P J Hughes and R G C Williams

7.1   INTRODUCTION

This chapter has been stimulated by the successful development of the 'BT Laboratories datapump' using digital signal processing (DSP) technology. The datapump is now an important component in the BT 'Network Solutions'^{TM} range of modems. At this point it would be useful to define the term 'datapump' as used in this chapter. The datapump performs the processes required to transform data into an analogue line signal and the analogue line signal back into data. Other than the functions of the datapump, a modem includes user-interface, modem configuration, network management facilities and the data processing facilities of error correction and data compression. This chapter concentrates on the datapump aspects of the modem. The rest of this section introduces some of the basic concepts which are used by modems and the international standards to which they relate.

7.1.1   Modem objectives

The basic objective of a voiceband modem is to send digital data over an analogue voiceband communications channel at the lowest possible error rate. This is shown diagrammatically in Fig. 7.1. The data is normally accepted via a V.24 [1] interface in serial form, either synchronously or asynchronously depending on the modem type. The data sources and sinks are usually computers and data terminals, but complex networks can be built up using multiplexers and other items of communications equipment.

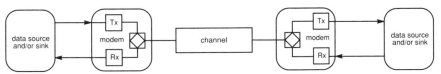

Fig. 7.1 Data communications systems over an analogue channel using modems.

The transmitter (Tx), or 'MO' (short for modulator), part of the modem converts the data signal into a band-limited analogue signal suitable for transmission over the public switched telephone network (PSTN) or over analogue leased lines. The receiver (Rx), or 'DEM' (short for demodulator), part of the modem takes the received signal which has been degraded by the network and, by means of large amounts (for the higher speed modems) of signal processing, converts the analogue signal back into a data signal. It should be pointed out that the processing requirements of a high-speed modem (i.e. V.32 [1]) are demanding and require a processing system capable of delivering between 10 and 20 MIPS depending on the facilities being offered. To put this in perspective it has considerably more processing than is being performed by a common 286 PC which is rated at ~1 MIPS.

7.1.2 Modem systems and standards

There is a wide range of modem standards (see Table 7.1) which cover data rates from 300 bit/s to 14400 bit/s. Some are for operation over leased lines and others for use on the PSTN, with some offering full-duplex operation on a single pair of wires.

Various modulation systems are used in these standard voiceband modems. For the lower speed modems, where the signalling or symbol rate (measured in baud) is equal to the bit rate, bandwidth efficiency is not of high importance and simple FM systems are used. Here binary '1' is signalled with one frequency and binary '0' is signalled with another. This type of modulation is commonly known as frequency shift keying (FSK), and is used for data rates of up to 1200 bit/s (the V.23 standard) as well as lower data rate systems such as V.21 which operates at 300 bit/s. These types of modem can operate in a full-duplex manner, i.e. they send data in both directions at the same time by using different frequencies for the send and return directions of transmission. However, the V.23 standard only supports 75 bit/s in the return direction.

Table 7.1   V.series modem standards.

CCITT Standard	Speed bit/s	Leased circuit	PSTN	Duplex	Included in datapump
V.21	300/300	Yes	Yes	Yes	Yes
V.22	1200/1200	Yes	Yes	Yes	Yes
V.22bis	2400/2400 (1200)	Yes	Yes	Yes	Yes
V.23	1200/75	Yes	Yes	Yes	Yes
V.26	2400	Yes	No	No	Yes
V.26bis	2400	Yes	Yes	No	Yes
V.26ter	2400/2400	Yes	Yes	Yes	No
V.27	4800	Yes	No	No	No
V.27bis	4800	Yes	No	No	No
V.27ter	4800	Yes	Yes	No	No
V.29	9600 (7200,4800)	Yes	No	No	Yes
V.32	9600/9600 (4800)	Yes	Yes	Yes	Yes
V.32bis	14400/14400 (12000,9600,7200,4800)	Yes	Yes	Yes	Yes
V.33	14400 (12000)	Yes	No	No	Yes

Note: Speeds in brackets are fallback speeds. Two speeds separated by a '/' indicates the send and return full-duplex speeds.

To provide greater data throughput it is necessary to have a more bandwidth efficient modulation system. This is achieved by using phase modulation. In the V.22 standard each symbol is represented by one of four carrier phases. These four phases of modulation of the carrier signal can be represented by the points *ABCD* in a two-dimensional signal constellation diagram shown in Fig. 7.2.

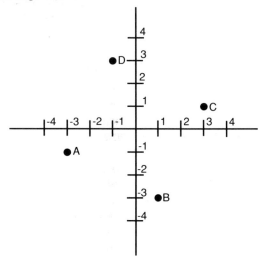

Fig. 7.2   A 4-point phase/amplitude or signal constellation diagram.

Each phase symbol represents two bits of customer's data and consequently a baud rate of only 600 symbols per second is needed to achieve a data rate of 1200 bit/s. This modulation system is extended to full quadrature amplitude modulation (QAM) in V.22bis where the carrier signal is modulated in both amplitude and phase to give 16 different symbols. This is shown in the signal constellation diagram in Fig. 7.3.

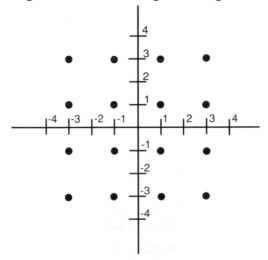

Fig. 7.3   A 16-point QAM signal constellation diagram.

With 16 different symbols each one can represent four bits of data, there being 16 different arrangements of four binary bits, and thus the V.22bis standard gives a data rate of 2400 bit/s.

V.22 operates at 1200 bit/s full-duplex, the better bandwidth efficiency of the phase modulation system allowing send and return channels to be squeezed into the standard telephone bandwidth. V.22bis operates at 2400 bit/s full-duplex using the same bandwidth, its extra data throughput rate being achieved at the expense of less noise immunity due to the closer spacing of the larger number of signal points. V.22bis modems use adaptive equalizers to reduce the intersymbol interference to recover some of this lost performance.

At still higher data rates, there are the V.32 and V.32bis standards which offer data rates between 4800 and 14 400 bit/s full-duplex. These systems also use QAM for the modulation scheme but they operate at a symbol rate of 2400 symbols per second. This higher symbol rate requires the full bandwidth of the channel and this means that the effect of the channel needs to be equalized. The telephone network and the associated channel impairments are described in section 7.2, whilst the techniques for equalization are described in section 7.3. To achieve the V.32bis data rates with the symbol rate of 2400 symbol/s, between two and six bits of data per symbol are transmitted using signal constellations with 4, 16, 32, 64 and 128 points respectively. Because most of the available telephone bandwidth is used, full-duplex operation is obtained by using echo cancellation techniques to enable the use of the same bandwidth for both directions of transmission. These techniques are described in section 7.4. The larger signal constellations required for the higher data rates mean that the spacing of the symbols within the constellation is significantly reduced for a given signal power (64 points are required for 6 bits per symbol), thus increasing the signal-to-noise ratio required for a given error rate. To counteract this, coded modulation is introduced at the higher data rates. This is described in more detail in section 7.5. Finally, section 7.6 discusses the features, facilities and performance of the BTL datapump and section 7.7 describes its implementation.

7.2   THE TELEPHONE NETWORK AND SIGNAL DEGRADATIONS

7.2.1   The switched telephone network

The basic task of the telephone network is to provide full-duplex communication between two telephones. Over short distances this may be achieved using a single pair of wires, but for longer distances the circuit losses may be such that amplification or regeneration may be required. This is not possible using a 2-wire circuit and so the telephone network has developed as a combination of 4- and 2-wire circuits as shown in Fig. 7.4.

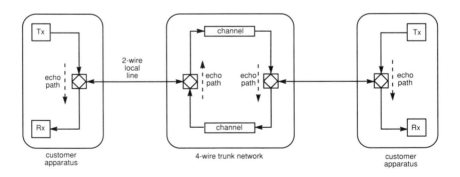

Fig. 7.4   Schematic of a 4-wire telephony link.

The main trunk network is kept as a 4-wire system, allowing efficient transmission techniques to be used, but connection between the local exchange and each customer is a 2-wire system (commonly referred to as a local line), providing a significant reduction in cabling costs. Owing to attenuation local lines are rarely more than a few miles long.

The conversion between the 2- and 4-wire circuits is performed by a device known as a 2-to-4-wire hybrid at the periphery of the 4-wire trunk network. This separates the send and return signals such that all the energy from the customer apparatus is coupled into the forward transmission path, and all the received energy is coupled into the local line. In practice complete separation is not achievable since this requires perfect impedance matching. There will also be a 2-to-4-wire hybrid in the customer apparatus to isolate the receiver from the locally transmitted signal.

7.2.2   Forward transmission channel characteristics

The forward channels in the network may consist of a large number of tandem transmission systems, all of which will degrade the signal. These degrading effects on the signal fall into four categories — linear distortion, nonlinear distortion, frequency translation and transmission echoes.

7.2.2.1   Linear distortion

This covers both phase and amplitude distortion in the transmission channel. It comes from a number of sources:

- the local line exhibits progressively greater attenuation at the higher frequencies, resulting in 'amplitude slope' distortion;

- channel filtering in frequency division multiplex (FDM) and pulse code modulation (PCM) transmission systems cause group delay and amplitude distortion, notably at the band edges;

- loaded junction cable gives group delay distortion at higher frequencies.

There is also a propagation delay through the channel, which ranges from a few milliseconds for a local call, up to 0.6 s for an international call requiring two tandem satellite links.

In order to preserve stability in the 4-wire network, the loop gain must be less than unity and so in addition to the local line loss there will be some loss in the 4-wire system. The maximum designed end-to-end loss in the UK network is 40 dB.

Linear distortion causes intersymbol interference, but the effects may usually be overcome using appropriate equalization techniques. These are described in section 7.3.

7.2.2.2   Nonlinear distortion

This arises from analogue nonlinear distortion in transmission systems, amplifier overload, etc, quantization distortion in digital systems, and controlled digital distortion in speech coders such as adaptive differential PCM (ADPCM).

Nonlinear distortion is difficult to overcome, but the degree of distortion is usually small. Some forms of speech coding can seriously impair the performance of echo cancellers and adaptive equalizers.

7.2.2.3   Frequency translation

Frequency translation is used in many transmission systems and can cause frequency offset due to mismatched up-and-down conversion. The maximum specified offset is ± 6 Hz. Unstable oscillators can give rise to both slow frequency variations and phase jitter. Frequency translation effects are usually compensated for by using phase locked loops.

7.2.2.4   Transmission echoes

Transmission echoes arise because the 2-to-4-wire hybrid in Fig. 7.4 relies on a good impedance match into the local line and this is never perfect, so signals leak across and give rise to the 4-wire loop problems of instability and echoes. There will also be a substantial echo across the 2-to-4-wire hybrid in the customer apparatus, traditionally called side tone. There may also be smaller echoes reflected from any impedance mismatches in the local line. The receiver at both ends therefore sees echoes of its own transmitter signal, which is called talker echo, and echoes of the received signal from the far end caused by double echoes in the 4-wire network. These are called listener echoes.

Since the channel delays the signal, there will generally be two distinct talker echoes. The first comes from the hybrid circuits in the customer apparatus and its entry to the 4-wire circuit. This is called near echo. The other comes from the hybrid at the distant end and is called far echo. For a double satellite link this can be delayed by up to 1.2 s — the echo time through the satellite links and associated terrestrial links.

An additional feature of the far echo arises if the transmission equipment modulation and demodulation frequencies, in the send and return transmission paths, are not exactly matched, thus causing the distant echo to be offset in frequency — an effect commonly called phase roll. This is not a feature in the UK national network since the up-and-down converters are locked, but on international routes there is often a frequency offset, typically less than 1 Hz, but sufficient to cause data transmission problems unless adequate steps are taken to track it. The techniques for dealing with echo are explained in section 7.4.

7.2.3   Noise

Additive noise in the system will be both continuous and impulsive, and can emanate from a number of sources — contact and joint resistance in the local line, circuit switching, crosstalk, unremoved echoes and digital transmission errors.

Noise will always be present. However, its effects may be reduced by filtering to prevent out-of-band noise reaching the receiver, and by using sophisticated channel coding as discussed in section 7.5.

7.3   INTERSYMBOL INTERFERENCE AND ADAPTIVE EQUALIZATION

This section begins by considering the transmission of high-speed synchronous digital data through band-limited channels which have linear distortion, as previously described, and introduces the concept of intersymbol interference. Digital signal processing techniques used within voiceband modems to mitigate the effect of intersymbol interference are then presented.

7.3.1   Intersymbol interference

The passage of data signals through the telephone network results in signals arriving at the receiver that have been subjected to a wide range of impairments. The range and nature of these impairments was described in the preceding section. In the telephone network the most serious impairments to high-speed data transmission are normally phase and amplitude distortion, commonly referred to as group delay and amplitude slope distortion. Such impairments can cause time dispersion of the received signal which, if not accurately compensated for, will seriously degrade the performance of the receiver.

The time dispersion effect of a channel is described by its impulse response $h(t)$. A typical channel impulse response is illustrated in Fig. 7.5.

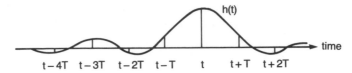

Fig. 7.5   A typical channel impulse response illustrating the channel's time dispersive nature.

It can be seen that the response extends over many symbol intervals. When a sequence of data symbols is transmitted, with signalling interval, T, the symbols arrive at the receiver overlap, with the result that the received signal $r(t)$ is the summation of the impulse responses for each of the transmitted symbols:

$$r(t) = \sum_m A_m h(t - mT) + n(t) \qquad \ldots (7.1)$$

where A_m is the sampled series of transmitted symbols and $n(t)$ is additive white Gaussian noise. At a given time, $t = kT + t_0$, where t_0 represents the channel delay and sampler phase:

$$r(kT + t_0) = A_k h(t_0) + \sum_{m \neq k} A_m h(kT + t_0 - mT) + n(kT + t_0) \qquad \ldots (7.2)$$

The first term above is the desired signal and the final term the noise. The remaining summation describes the contributions from all other transmitted symbols, due to the non-zero tails of the impulse response spaced at intervals of T away from t_0. These summation terms are known as intersymbol interference (ISI).

7.3.2   Reducing the effect of intersymbol interference

From equation (7.2) it can be seen that the ISI is zero if the channel impulse response has zero value at all T spaced intervals except for $t = t_0$. If the overall system filtering (including the channel) can be constructed to yield such an impulse response then the effect of ISI can be minimized. However, in most practical transmission systems, and the telephone network in particular, the channel response is unknown prior to transmission. Indeed, slight variations with time are also often encountered. Any form of fixed filtering to compensate for the ISI can therefore only be a compromise based on average channel characteristics. Early synchronous voiceband modems such as those based on the V.22 and V.26 Recommendations (1200 full-duplex and 2400 half duplex bit/s respectively) used these techniques but their transmission rates were severely limited by the degree of ISI. To achieve higher data rates demands that the ISI be countered more effectively. Modern high-speed modems, which are constructed with today's powerful processing devices, utilize digital signal processing filter structures that automatically adjust to compensate for the effect of the channel. These structures are known as adaptive equalizers.

7.3.3   Equalizers

Because of its simplicity the most common equalizer structure is the transversal or finite impulse response (FIR) filter. This structure, illustrated in Fig. 7.6, consists of a delay line made from connected signal storage elements with taps at intervals, T. The received signal $r(t)$ enters the delay line (see Fig. 7.6) and is shifted one position every symbol period (T).

The output $y(kT + t_0)$ is calculated every symbol interval and is formed by the addition of the weighted samples from each element such that:

$$y(kT + t_0) = \sum_{n=0}^{N-1} C_n r(kT + t_0 - nt) \qquad \text{... (7.3)}$$

where C_n are the tap weights or coefficients and N is the number of taps. This can also be expressed more concisely in vector form as:

$$y(k) = c^T(k)r(k) \qquad \text{... (7.4)}$$

where $c(k)$ is the column vector of N tap weights, $c^T(k)$ denotes the transpose of $c(k)$, and $r(k)$ is the column vector of received samples in the equalizer delay line at time kT.

The tap coefficients give the equalizer its flexibility and they are adjusted to minimize a suitable criterion. The most commonly used adaptation criterion is the minimization of the mean square error (MSE) which is defined as:

$$\text{MSE} = E[e^2(k)] = E[(y(k) - m(k))^2] \qquad \text{... (7.5)}$$

where $E[.]$ denotes the expectation operation, $y(k)$ represents the equalizer output and $m(k)$ the desired output.

With this criterion the equalizer seeks to minimize not only the ISI terms but the noise as well. An equalizer applying this criterion is known as a least mean square (LMS) equalizer. The derivation of the automatic adjustment or coefficient update of the LMS equalizer is well documented (see Chapter 3) [2]. The update algorithm is defined as:

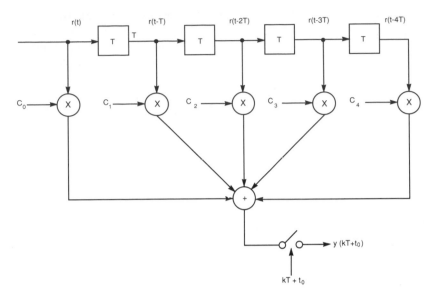

Fig. 7.6   5-tap FIR structure.

$$c(k+1) = c(k) - \mu e(k)r(k) \qquad \qquad \ldots (7.6)$$

where μ is a positive update constant. This is shown diagrammatically in Fig. 7.7.

7.3.4   QAM equalizers

Modern high-speed modems almost exclusively use QAM as their modulation method. This is because QAM, which is the superposition of two double sideband suppressed carrier amplitude modulated signals, provides a transmission system that is as bandwidth efficient as single sideband amplitude modulation [3], whilst easing the implementation of carrier and timing recovery schemes [4,5].

In a QAM receiver the signal is typically first demodulated and filtered to give complex (real and imaginary) baseband samples. The equalizer that follows also assumes a complex format consisting of separate delay lines for the real and imaginary components together with complex tap coefficients. The equalizer real and imaginary outputs are formed by cross-coupling of the signals in the delay lines with the complex coefficients. This is expressed concisely as in equation (7.4) but all variables are now of a complex nature.

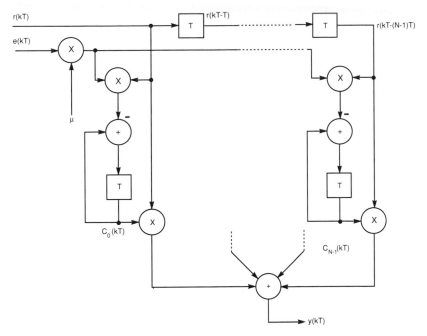

Fig. 7.7 N tap equalizer illustrating tap update procedure.

The complex arrangement also leads to a modification of the LMS update algorithm for the coefficients so that:

$$c(k+1) = c(k) - \mu e(k)r(k)^* \qquad \qquad \dots (7.7)$$

where all variables are complex and * denotes the complex conjugate.

It is worth noting here the degree of computational complexity involved with a complex equalizer implementation. The output formation requires four multiplications per complex tap and four multiplications per tap update. With a voiceband modem equalizer generally requiring in excess of 30 complex taps it can be seen that the processing load is considerable.

The theoretical convergence speed of the equalizer taps to their optimum setting is dependent upon a number of factors. These factors are well documented [6,7] and can be summarized as:

- the spectral 'flatness' of the folded receiver power spectrum (the eigenvalue spread of the autocorrelation matrix of the receiver input);

- the size of the tap update constant, μ;

- the number of tap coefficients.

In practice convergence is also critically dependent upon the accurate formation of the error signal, $e(k) = y(k) - m(k)$ and thus relies on correctly deciding which of the possible transmitter symbols has been received. At the onset of transmission, this is likely to prove difficult with the level of distortion commonly found on telephone channels. In QAM systems this problem is resolved by starting transmission with a known pseudo-noise sequence called the training sequence. At the receiver a local synchronized copy of this sequence is generated and used as the receiver decision for the error formation. This process is known as local reference training. At the end of the training sequence the equalizer is sufficiently converged for the receiver to make the decisions itself with a high probability of success. Thus the equalizer can continue adjusting during data transmission and is able to track small changes in the channel. This is known as decision-directed operation.

7.3.5   Fractional-spaced equalizers

The equalizers described so far have all used the same basic structure consisting of delay elements with coefficients at a spacing equal to the symbol period, T. These types of equalizer are known as T-spaced equalizers. A fundamental disadvantage of this structure is that the folded (aliased) spectrum caused by symbol rate sampling is created before the equalizer. The equalizer is therefore unable to suppress noise outside the roll-off frequency, π/T radians per second, and thus cannot achieve matched filtering. Additionally, because the equalizer cannot operate independently on the regions either side of the roll-off frequency, it is possible that with certain symbol clock phases the folded frequency components may add destructively to create a null. If this happens all the equalizer can do to compensate is to insert a large gain. This causes severe noise enhancement to occur at those frequencies and poor performance results.

Despite these disadvantages (timing recovery schemes are available that provide good symbol timing phase [5]), T-spaced equalizers were dominant in voiceband modem products until the mid-1980s. Since then the fractional-spaced equalizer has become widely used. In a fractional-spaced equalizer, instead of having T-spaced coefficients, the coefficients are spaced at an interval less than T. The output however is still calculated at the symbol rate and the taps adjusted according to the LMS algorithm. Equalizers with taps spaced $T/2$, $T/3$ and $3T/4$ have been reported [8,9,10]. The advantage of the fractional spacing is that equalization is performed before the symbol frequency aliasing. This means that folding about the roll-off frequency does not occur so that independent control of the regions either side of that frequency can be exercised. Creation of nulls in the roll-off region by timing phase effects (which lead to noise enhancement) is therefore prevented and the sensitivity of the degree of equalization to the clock phase is removed. Additionally, the ability to function like a matched filter in suppressing out-of-band noise means that the fractional-spaced equalizer (with an infinite number of taps) can be shown to be equivalent to the optimum linear receiver [11].

An interesting feature of the fractional-spaced equalizer is that, unlike the T-spaced equalizer which has a global optimum tap setting, it has an infinite number of tap weight solutions which give rise to similar values of mean square error. The taps therefore exhibit a 'wandering' phenomenon, which, if not countered, can lead to overflow in limited precision digital circuits. The most common way of achieving tap stability is to include an additional term in the tap update that limits the growth of large taps [12]. This is known as tap leakage and leads to a combined LMS and tap leakage update algorithm such that:

$$c(k + 1) = (1 - \partial)c(k) - \mu e(k)r(k) \ * \qquad \ldots (7.8)$$

where ∂ is a very small positive constant of the order of 10^{-6}.

7.4 ECHOES AND ECHO CANCELLATION TECHNIQUES

7.4.1 The effect of echoes on transmission

From the point of view of the receiver, all echo components are unwanted signals, i.e. noise, and therefore contribute to the signal-to-noise ratio at the receiver input. The extent of the problem will vary depending on the type of signal being transmitted, the echo power relative to the received signal level, and delay in the channel.

For satisfactory error performance, high-speed modem receivers operating with large signal constellations need a signal-to-noise ratio of around 25 dB. If full-duplex operation is required, then talker echo will appear as a constant noise source at the receiver input. With a large near echo and a high channel loss, this echo may be up to 40 dB greater than the received signal. Hence the echo removal techniques must have a dynamic range of at least 65 dB.

7.4.2   Principles of echo cancellation

For high-speed full-duplex data transmission it is necessary to use the entire bandwidth continuously, rather than the frequency or time division techniques employed at lower data rates. To achieve this the talker echo is removed at the receiver input using an echo canceller. Figure 7.8 shows a typical echo cancellation system, the principle of which is very simple — the practical implementation less so.

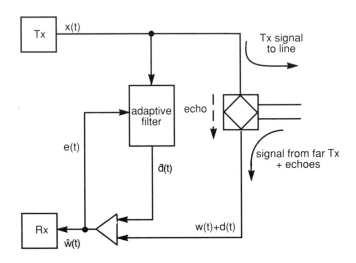

Fig. 7.8   Schematic diagram of echo canceller.

It is desired that the talker echoes, $d(t)$, from the local transmitter be removed so that only the wanted signal, $w(t)$, from the distant transmitter is present at the input to the receiver. This is achieved by passing the local transmitter output, $x(t)$, through an adaptive filter whose transfer function is ideally adapted to be identical to that of the transmission line echo path. The adaptive filter output, $\hat{d}(t)$, is then subtracted from the received line signal to leave $w(t)$ uncontaminated by echoes.

In practice complete cancellation is rarely achieved. The echo canceller performance is usually measured as echo return loss enhancement (ERLE), which is defined as:

$$\text{ERLE} = 10.\log \frac{\text{echo power after cancellation}}{\text{echo power before cancellation}}$$

Since the receiver sees the residual echo as noise, the required ERLE is dependent on the receiver sensitivity for the required data rate and the relative levels of $d(t)$ and $w(t)$. Typically a 'poor' transmission line might give: $w(t)$ = -43 dBm, $d(t) = -10$ dBm; with a receiver sensitivity of 25 dB an ERLE of $-43 + 10 - 25 = -58$ dB is required.

7.4.3 Data echo cancellers

Taking the input to the adaptive filter from the line signal as shown in Fig. 7.8 has several drawbacks. First, it is a highly correlated signal due to the transmitter pulse shaping filters. This does not provide for efficient adaptation behaviour as described in Chapter 3. Secondly, to satisfy the Nyquist sampling theorem it must be sampled at a rate relative to the line bandwidth rather than the data symbol rate and requires at least 12 bits of resolution. This produces very intensive processing requirements.

A more practical class of echo canceller structures is called data-driven echo cancellers, where the input is taken directly from the data sent to the transmitter, rather than from the line signal. These can overcome all the problems mentioned above, but require a longer echo canceller since all echoes will appear convolved with the transmitter filtering. For detailed description of these devices see Messerschmitt [13], Weinstein [14] or Buckley [15].

In order to implement this in a QAM modem the transmitter structure must be converted into a more suitable form. This is shown by starting with the balanced modulator of carrier frequency, f_c, shown in Fig. 7.9.

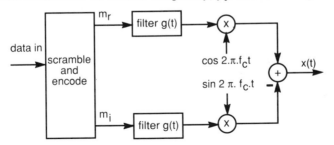

Fig. 7.9   Balanced modulator.

The data is entered into a QAM encoder which converts it into a series of multilevel complex symbols $m(r)$ (real part) and m_i (imaginary part), at $1/T$ symbol/s. These are then converted to the line sampling rate by over sampling at a rate L/T, where L is the sampling conversion factor and usually an integer. This is filtered by an N tap FIR baseband filter, $g(t)$, to provide suitable pulse shaping, modulated by a carrier and the real part is sent to line.

It is convenient to use sampled data notation and to use the exponential form for the carrier component. The output then is given by:

$$x(k) = Re\left[\sum_{n=0}^{N-1} m(k-n)g(n)e^{(j2\pi f_c kT/L)}\right]$$

... (7.9)

where $x(k)$ = output at time kT
$m(k)$ = complex data sample at time kT.

Equation (7.9) may be converted into a more appropriate form by noting that:

$$e^{(j2\pi f_c kT/L)} = e^{(j2\pi f_c nT/L)} \times e^{(j2\pi f_c(k-n)T/L)}$$

... (7.10)

Then equation (7.9) becomes:

$$x(k) = Re\left[\sum_{n=0}^{N-1} m(k-n)e^{(j2\pi f_c(k-n)T/L)} g(n)e^{(j2\pi f_c n \ T/L)}\right]$$

... (7.11)

which represents a series of modulated data symbols multiplied by a fixed complex filter. This enables the modulation to be performed at the symbol rate rather than the line sample rate, and because L is usually an integer the filter only requires N/L, rather than N multiplications per line sample interval. Furthermore, if f_c is arranged to have a suitable relationship to T, then the modulation reduces to simple rotations of the data symbols which may easily be incorporated into the encoder.

In the CCITT V.32 echo cancelling modem specification [1] f_c = 1800 Hz and $1/T$ = 2400 symbol/s; this gives a line spectrum of 600-3000 Hz. Therefore, in a practical implementation, L is set to either 3 or 4, allowing sampling rates of either 7200 Hz or 9600 Hz.

The echo canceller may be adapted using one of the schemes mentioned in Messerschmitt [13]. In most cases the least mean squares (LMS) algorithm, as described in Chapter 3, is employed since adequate adaptation may be achieved in a few seconds, and during full-duplex operation an adequate ERLE may be maintained by simply reducing the coefficient update gain.

7.4.4   Far echoes and frequency offset

The echo from the distant end may be delayed by up to 1.2 s (for a double satellite link), and may have a frequency offset. An echo canceller having a span of this magnitude would be expensive in both processing time and memory required. A solution is to configure the system as a near echo canceller (NEC) to handle the local echoes, and use a variable delay to position far echo canceller (FEC) such that it spans the echoes from the distant end. The frequency offset can be compensated by combining the data driven echo canceller with a digital phase locked loop (DPLL) as shown in Fig. 7.10. The output of the FEC is rotated using phase information derived from the far echo and the rotated FEC output. The NEC is updated with the error directly, but the error for the FEC must be de-rotated in order to maintain correct LMS operation. This system has the virtue of being able to handle large frequency offsets, but is computationally expensive since the FEC must be a fully complex system. The processing may be reduced by up to 50% if the rotation is transferred to the FEC input, but the order of the PLL will be increased by an unknown amount due to the delay through the FEC. This will impair the phase tracking abilities of the system. Phase-locked loops and frequency offset compensating echo cancellers are described in more detail in Lindsey and Chie [16] and Wang and Werner [17].

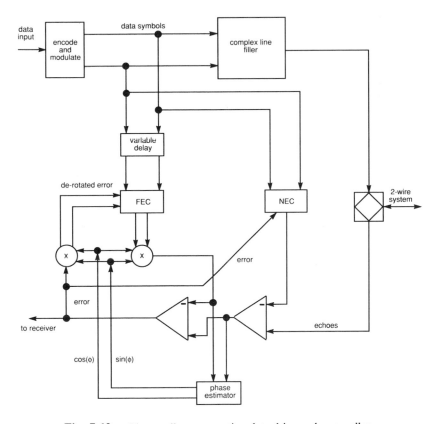

Fig. 7.10   Phase roll compensating data-driven echo canceller.

7.5   NOISE AND CODING TECHNIQUES

7.5.1   Noise

Adaptive equalization and echo cancellation are aimed at specific types of channel impairment. The equalization removes the effects of linear distortion in the channel; the echo cancellation removes the effects of mismatched impedance in the channel. After the removal of these impairments the received signal is a copy of the transmitted signal except for some corruption brought about by channel disturbances. Crosstalk, poor contacts, exchange noise and many other corruptions of a perfect channel contrive to pervert the signal on its journey from the transmitter to the receiver. All of these remain in the signal after the echo cancellation and the equalization have been done. This corruption of the signal is referred to as noise. A signal that has been corrupted in this way is said to be noisy and can be represented by a point, on the signal constellation diagram, just as the transmitted symbols are (Figs. 7.2 and 7.3). The receiver's task is to determine which symbol of the signal constellation is represented by the received point.

At low data rates the noise that is added to a signal during its transmission is insignificant (especially compared to echo and ISI). Also the distance between symbols is so large that the transmitted symbol is highly likely to be represented by the symbol nearest to the received point. As the data rate is increased, the number of required symbols increases. Because of transmitter power limitations, they necessarily become closer. The probability that the transmitted symbol is represented by the symbol nearest to the received point becomes unacceptably small. In this case a method of counteracting the noise is needed.

7.5.2   Coding

For each received point the receiver must determine which symbol was transmitted. The choice becomes unclear when the received point lies equally close to more than one constellation symbol. In such a case the receiver must be aided by further information. There must be some means of restricting its choice. One approach to this problem is to allow only certain sequences of symbols to be transmitted. Now the receiver is helped in its decision making because it is looking for the closest symbol that is part of an allowed sequence. The sequence of symbols that is transmitted is produced from a code.

A drawback to this approach is that the transmitter is now restricted in what it can send. This results in the need for a greater number of symbols to transmit the same amount of data. Although this seems to compound the receiver's problem, codes can be found that are powerful enough to give a performance advantage in spite of this.

The effectiveness of using a code to combat noise depends on the way in which the receiver can use the extra information. Ungerboeck [18] formulated a method of combining coding schemes with modulation techniques. This method is known as coding with set partitioning. Given a signal constellation this method partitions the symbols into sets whose members are far from each other. This principle is demonstrated in Fig. 7.11 for a 32-point signal constellation.

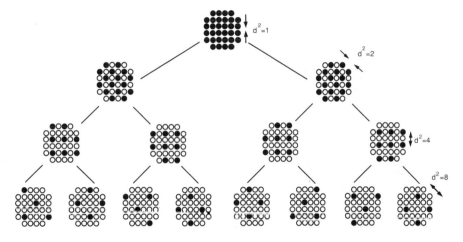

Fig. 7.11   Partitioning of 32-point constellation into eight subsets.

The constellation is partitioned into eight subsets. If the transmitted symbol is known to be a member of a particular subset then it is likely that the transmitted symbol is the nearest symbol in that subset to the received point. Therefore the coding only needs to specify sequences of symbol subsets. This reduces the complexity of the code considerably and is a very important concept.

The receiver is now being helped in two ways to decide which symbols were transmitted. The code helps the receiver choose the correct symbol subset. The distance between symbols in each subset enables the receiver to choose the correct symbol from the chosen subset.

7.5.3   Decoding

The main complexity in this sort of procedure is in decoding the code, i.e. taking the received points and finding the legal sequence of transmitted symbols that is closest to them. In this case the decoding can be done by using an algorithm devised by Viterbi [19]. The distance from each received point to the nearest symbol in each subset is found. These distances are then fed into the Viterbi decoder. The decoder uses the distances to update the information it holds on the most likely sequences to date. Where sequences merge, the decoder only retains the sequence with the lowest cumulative distance. In this way the decoder keeps the minimum number of most likely sequences. The one with the lowest cumulative distance of all is found. This has a maximum likelihood of being the transmitted sequence. The decoder traces this sequence back to produce a delayed output equivalent to a subset in the transmitted sequence. The symbol in this subset closest to the appropriate received point is the symbol that is decided upon.

The Viterbi algorithm itself is fairly arithmetically intensive. When it is used in the case described above the same operations are performed in each symbol interval. This is ideal for a DSP device. When the received signal has been demodulated and freed from echo and ISI, a pair of co-ordinates representing the noisy received point are passed to the decoder. The decoder runs these co-ordinates through the algorithm and produces an output symbol which it passes on to the descrambler. The decoder then waits for the next pair of co-ordinates to be given to it before repeating the process.

7.5.4   The V.32 code

The code chosen for the V.32 modem standard [1] (and subsequently the V.32bis standard) was found by Wei [20]. It constrains the transmitter to send the next symbol from one of four subsets depending on the sequence of the subsets that have just been transmitted. It is a convolutional code and can be represented by the circuit shown in Fig. 7.12.

The two input bits determine one of four outcomes depending on the state of the delay elements. These delay elements are in a state that represents the recent history of the transmitted sequence. The code was specially developed so that it can be made 90° rotationally invariant. With the aid of differential encoding on the two input bits, only two symbol errors will occur if the constellation is suddenly rotated by a multiple of 90° [20]. This is necessary since no other modem function will detect such a rotation.

The help given to the receiver in determining which subset was transmitted is equivalent to 7 dB performance gain. Unfortunately the scheme needs all 32 symbols of Fig. 7.11 to transmit four bits of information. This degrades the receiver's performance by a factor of two (3 dB).

Given the correct subset the receiver is eight times more likely (9 dB) to choose the transmitted symbol than without this subset information (due to the squared distance between member points — not the number of subsets).

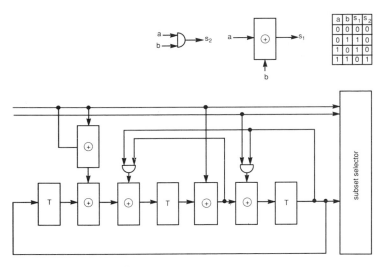

Fig. 7.12   Subset sequence generating circuit.

The performance gain in this case is therefore limited by the receiver's ability to choose the correct subset. The total performance gain over transmitting (uncoded) with 16 symbols is 4 dB.

7.6   THE BT LABORATORIES DATAPUMP — FEATURES AND PERFORMANCE

In the earlier sections of this chapter, the telephone network, its impairments and the various techniques used to overcome them have been described. The following sections give a more detailed description of the BT Laboratories datapump and its implementation, which is based on the ideas and techniques that have been described.

7.6.1   Features and facilities

The BT Laboratories datapump is used in the BT 'Network Solutions'™ range of datacommunications products which is an integrated series of datacommunications equipment and systems. The BT Laboratories datapump provides the signal processing requirements of the modems that are available in the range. It is able to support QUAD (V.21, V.23, V.22 and V.22bis), QUIN (V.21, V.23, V.22, V.22bis and V.32) and universal modem types in all their various implementations such as dial modems and networking modems.

The complete modem consists of the BT Laboratories datapump together with a modem motherboard. The datapump provides all the signal processing functions of the modem, while the modem motherboard provides all the non-signal processing functions, the main ones of which are as follows:

- datapump control;

- user configuration, including configuration menus;

- V.42 and V.42bis error control and data compression using the BT Lempel Ziv algorithm;

- V.25bis and Hayes protocols;

- automatic terminal interface configuration;

- network management facilities and diagnostics handling;

- line interface.

The BT Laboratories datapump, shown in Fig. 7.13 along with a photograph of a networking modem (Fig. 7.14), is able to support all the current CCITT standard modem types as well as previous BT proprietary modes.

Fig. 7.13   The BT Laboratories datapump.

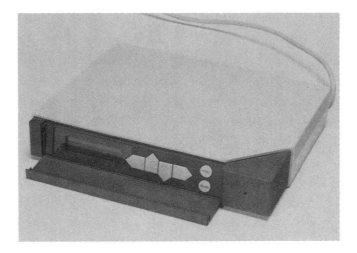

Fig. 7.14   Networking modem.

It consequently has sufficient processing power to implement the most demanding mode (V.32bis). A list of all the modes currently supported is given in the Appendix. The datapump being a software-based product can have additional modes and facilities added to its repertoire as they are required.

The datapump supports full international operation as the full-duplex echo cancelling modes have both near end and far end cancellers with far end phase roll removal. The far end canceller can be positioned at up to 1.2 s from the near end canceller. This allows operation over a link with two satellite hops which caters for any international link.

The datapump includes many additional features and facilities apart from the long list of modem modes supported. Firstly there is full automoding between all PSTN 2-wire full-duplex modes, i.e. the modem user can set up a link to any of these modem types automatically without user intervention.

In V.32bis and the proprietary BT V.32 modes there is automatic initial rate selection and in-data rate adaptation between 4800 bit/s and 14 400 bit/s. This facility means that the optimum throughput for any line will be achieved automatically even if the noise level on the line changes during a call.

In coded modes there is a 75 bit/s secondary channel which allows non-intrusive network management messages to be sent over the link without affecting the high-speed data channel.

Other more normal facilities include tone detection and generation for international PSTN progress tones, analogue and digital test loops and a loudspeaker for call progress monitoring.

7.6.2   System performance

The BT Laboratories datapump is a high performance product. Examples of the error performance are given below. The system performance is obtained by measuring the error performance of the complete modem under varying quantities of noise. The results of these tests are given in Fig. 7.15, which shows a plot of block error probability (Pe) versus signal-to-noise (S/N) ratio of the datapump in V.33 type mode operating at all the available rates. It should be noted that the 4800 bit/s rate does not use coded modulation and hence offers little error rate performance improvement over 7200 bit/s which is a coded rate.

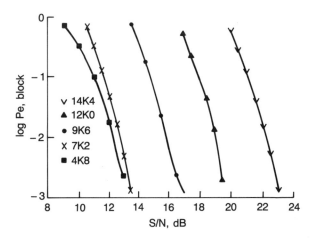

Fig. 7.15   Datapump block error performance versus signal-to-noise ratio.

The 3 dB steps in error performance between the four coded rates, as one would expect from theory with the doubling of the number of constellation points with each increase of 2400 bit/s, should also be noted.

Also shown is a performance comparison (Fig. 7.16) with the 4142TCX modem which is the predecessor to the BT Laboratories datapump. The datapump shows a performance improvement of about 1 dB due to the more sophisticated signal processing that it uses.

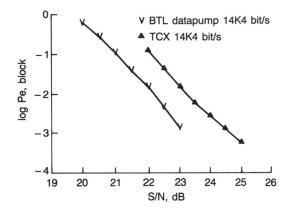

Fig. 7.16   Comparison of BT Laboratories datapump and the 4142TCX modem.

7.7   BT LABORATORIES DATAPUMP IMPLEMENTATION

7.7.1   Review of implementation techniques

Developments in voiceband modems follow a similar pattern to that of many other electronic products — a gradual migration from analogue to digital implementation techniques. Since voiceband modems operate in a restricted bandwidth environment, with the highest frequencies being less than 4 kHz, the processing sample rates are relatively low. This eases the migration to digital techniques where lower sample rates mean that more processing can be achieved within the sample period. However the processing requirements of the higher speed modems are considerable compared with most applications running in the same bandwidth environment. This has meant that the processing power of standard microprocessors has been insufficient to enable an early migration to a programmable environment. The use of programmable components in commercial modems has had to wait for the advent of the DSP microcomputer (see Chapter 1).

Early modems were made using largely analogue techniques but as the sophistication of the products increased digital solutions became both more attractive and necessary. FSK modems are easily realised using analogue techniques, but, with the introduction of adaptive equalization, digital techniques began to offer real advantages in ease of application. Early digital design made use of TTL logic chips. These worked well but were expensive, large and power hungry. The next step was the use of custom-designed large scale integrated circuits. In the early 1980s this was the only way to get a high-speed (9600 bit/s) modem into a reasonably sized box even though production volumes could only marginally justify the development effort involved. This type of implementation technique was used to produce the successful BT 4962X modem which was the world's first commercial V.32 modem. The introduction of programmable single chip digital signal processing devices has been the next step in the development of implementation techniques. Up to eight of the early generation devices have been used to implement V.32 modems, which also required significant quantities of mixed analogue and digital circuitry to provide for the analogue-to-digital and digital-to-analogue conversion processes required in this type of modem. BT's 4142TCX used a mixture of custom LSI and early DSP devices to produce, when it was launched, the world's fastest full-duplex PSTN modem.

In the current datapump implementation all signal processing functions are provided by two DSP devices together with an 8-bit microcontroller. However there is still a requirement for specialized hardware and consequently two ASICs are also used in this design. One ASIC provides the special clock generation requirements of the modem, whilst the other provides the control and data interface to the modem motherboard and the asynchronous/synchronous conversion.

7.7.2   Datapump hardware architecture

A block diagram of the hardware architecture of the BT Laboratories datapump is shown in Fig. 7.17.

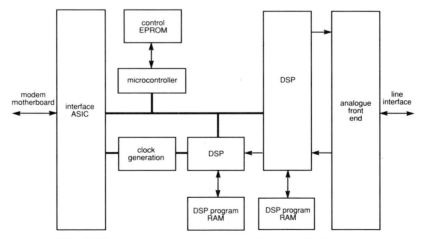

Fig. 7.17   BT Laboratories datapump hardware block diagram.

7.7.2.1   Interface ASIC

The interface ASIC was designed to provide a standard interface for a range of datapump products. This split between datapump and motherboard allows one motherboard design to be used with a range of different datapumps.

Its primary features are:

- dual-port 8-bit microprocessor interface;

- memory-mapped control registers;

- fully programmable synchronous/asynchronous and asynchronous/synchronous converters;

- serial or parallel primary channel data interface;

- secondary channel data interface;

- bit-symbol synchronization for TDM multiplexers;

- timing recovery system for low-speed modems;

- 84-pin PLCC package.

7.7.2.2   Timing ASIC

The timing ASIC was designed to provide all of the bit and symbol clock requirements for all modem types up to a maximum bit rate of 19 200 bit/s with symbol rates of 600, 1600 and 2400 baud being provided.

Its primary features are:

- provision of the transmitter and receiver bit and symbol clocks;

- provision of transmitter and receiver sample clocks;

- provision of a phase and frequency controlled oscillator, controllable by the receiver timing recovery processor;

- derivation of the transmitter clocks from an internal oscillator or an external clock or from the recovered receiver timing;

- clocks required for synchronization and MUX applications;

- 68-pin PLCC package.

7.7.2.3   Analogue interface

The analogue interface was implemented using standard commercial devices rather than a single integrated device in order to achieve the necessary performance from the A/D and D/A converters and the analogue filters at reasonable cost.

The primary features of the analogue interface are:

- D/A conversion of the transmitted signal;

- filtering of the analogue line signal;

- switching of 2-wire/4-wire and analogue loop modes;

- optional 4-wire to 2-wire conversion via an active hybrid for 2-wire modes;

- filtering of the analogue received line signal;

- A/D conversion of the received signal.

7.7.2.4   Control microprocessor

The control microprocessor is an 8-bit general-purpose device. All program storage is in external EPROM to ease facility upgrades. The primary functions of the control microprocessor are:

- physical interfacing between the interface ASIC and the DSP devices;

- running the control program.

Further details of the controller functions are given in section 7.7.3.

7.7.2.5   Digital signal processing devices

The DSP software is implemented in two Motorola MC56001 devices. These devices have been designed to maximize throughput in data intensive DSP applications and provide many features that enable the efficient implementation of the algorithms used in voiceband modems. These features include:

- dual independent memory spaces (both RAM and ROM) that support complex (real and imaginary) data manipulation;

- flexible and parallel structure that allows an arithmetic logic unit (ALU) operation, data transfer with address update and instruction fetch from program memory, all within one instruction;

- comprehensive instruction set that includes low overhead looping facilities;

- 24-bit wide data paths and 56-bit intermediate results;

- on-chip peripherals and general purpose input/output.

The software program for the two DSP devices is held in the controller 8-bit wide EPROM and downloaded on reset into 24-bit wide RAMs sitting on the external memory bus of each DSP device. This arrangement provides great flexibility and low maintenance cost as it avoids the use of mask DSP devices and allows product modification to suit individual customer requirements by the changing of a single EPROM. The software is conveniently partitioned between the two DSP devices by splitting the transmitter/echo canceller and receiver functionality.

7.7.3   Control software architecture

The major functions provided by the control processor are:

- downloading of the DSP programs and data from the master EPROM;

- interpretation of the interface ASIC control registers;

- transfer of data;

- scrambling and encoding of the transmitted data;

- generation of training segments required for all modes;

- control of the DSP devices to provide all the required modes;

- control of all training sequences, rate sequence generation and detection;

- rate selection according to line conditions;

- automatic in-data rate revision according to line conditions;

- control of a DFT to provide automatic mode discrimination;

- updating of interface ASIC with mode and rate parameters.

7.7.3.1   Operating system

The processing is driven by the transmitter and receiver symbol clock interrupts. These clocks are of nominally the same frequency but their relative phases change slowly with time at the rate of the frequency difference. The slight difference in the frequency of these clocks (<1 Hz) means that a sophisticated operating system is necessary to ensure that high priority tasks are always performed in time regardless of the order or relative time at which the interrupt occurred.

7.7.3.2   Scheduling

When a configuration change is flagged by the interface ASIC, the controller runs a scheduling process to determine which software module is required to be run. The scheduling process continues to manage the running of the software modules required until the datapump is returned to its idle mode.

7.7.3.3   Modules

The control program code is structured into individual modules for each mode of operation. This allows each mode to be considered independently and written by different programmers. Each module has a simple common interface and once running has almost unrestricted control of the resources of the datapump. This gives each programmer considerable freedom to deal with the peculiarities of each mode.

7.7.3.4   DSP control

Host vectored interrupts are employed by the control processor to communicate with the DSP devices. By this means the controller can modify DSP device control words or vector for information to be delivered into the host interface registers, to be read later by the controller.

7.7.4   DSP software architecture

Next are described the functions implemented in each of the DSP devices for the V.32 mode of operation as this mode is the most computationally intensive.

7.7.4.1   Transmitter/echo canceller DSP software

The transmitter/echo canceller DSP (TECdsp) software is controlled by a combination of hardware sample rate interrupts and software vectored interrupts from the datapump controller. The functional structure of the TECdsp software is shown in Fig. 7.18 and can be divided into two parts — transmitter functions and the received signal operations that operate at the transmitted signal sampling rate, such as the echo canceller. The transmitter operations are described first.

- Modulation — the TECdsp receives symbol co-ordinates from the controller and modulates them into the channel passband using a pair of phase quadrature 1800 Hz carriers. The complex modulated symbols are stored in delay lines for subsequent use in the echo canceller.

- Passband filtering — the modulated symbols enter a complex passband FIR filter that implements square-root spectral shaping to reduce ISI. Only the real part of the output is calculated as this is all that is required to be transmitted.

- Compromise transmit equalization — variable compromise spectral shaping of the transmitted signal is provided as a user option. It can be used to help counter the amplitude slope distortion effects of a long connection to the local exchange. The level of equalization can be preset to one of eight settings or alternatively linked to the transmitted signal level.

- Output scaling — the level of the transmitted signal to line can be set and is adjustable from 0 dBm to −15 dBm in 1 dBm steps.

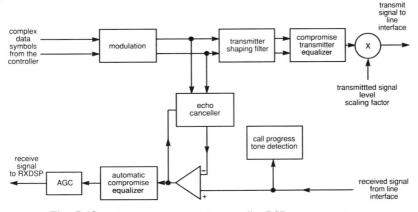

Fig. 7.18   Transmitter and echo canceller DSP arrangement.

The received signal from the hybrid circuit is also fed into the TECdsp where the following operations are performed.

- Echo cancellation — the echo canceller implementation is described in section 7.4.3. It allows for operation with both near and far echoes over links using multiple satellite hops and also deals with frequency offset echoes.

- Compromise receive equalization — after cancellation the received signal is passed through a filter which is adjusted during the V.32 start up handshake to reduce the amplitude slope distortion of the signal entering the adaptive equalizer in the receiver. This has the effect of reducing the equalizer adaptation time on difficult lines and allowing immediate high-speed transmission on a higher proportion of connections.

- Automatic gain control (AGC) — the signal passing to the receiver has to be maintained at a constant level in order to make best use of the precision available in the DSP device. The AGC is an adjustable gain function that performs this operation over the required line signal dynamic range of 0 dBm to −48 dBm.

In addition to the main signal processing functions described above there are further supervisory and diagnostic operations performed in the TECdsp.

- Call progress tone detection — filters are provided that allow detection of network supervisory tones used during call set-up such as dial and ring tones. Provision is made for operation on international circuits by allowing user selection of a range of tone filters.

- Automoding — the CCITT has defined in V.32bis, Annex A, a procedure by which the correct modem standard, or mode, is automatically selected on a dialled connection. This procedure is fully implemented in the BT Laboratories datapump and allows the BT 'Network Solutions'™ modem to call or be called by another modem conforming to any of the defined CCITT PSTN modem standards (see the Appendix to this chapter). The TECdsp runs a discrete Fourier transform (DFT) process to provide a sophisticated and highly selective tone detector in order to implement the CCITT procedure.

- Channel quality monitoring — routines are provided that monitor the channel parameters. Examples of these parameters include received signal level, near and far echo levels and a count of amplitude hits. The controller can read these values and all the DSP device memory locations via a vectored interrupt. Provision for writing to all DSP device memory locations is also made.

- Self test — diagnostic routines are performed on power up that test all internal and external memory together with the DSP device functionality.

7.7.4.2 Receiver DSP software

The receiver DSP (RXdsp) software is controlled by a combination of hardware sample and symbol rate interrupts together with software vectored interrupts from the datapump controller. The functional structure of the RXdsp software is shown in Fig. 7.19.

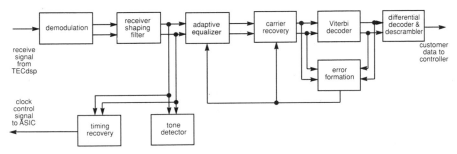

Fig. 7.19 Receiver DSP arrangement.

The input to the RXdsp is provided by the analogue interpolation path that re-samples the signal from the TECdsp at the receiver sampling rate. The main functions provided in the RXdsp are as follows.

- Demodulation — the real passband received signal is shifted to the baseband and made complex by demodulation with a pair of nominal 1800 Hz phase quadrature carriers.

- Baseband filtering — the real and imaginary parts of the complex demodulated signal are each filtered with a square-root shaping FIR filter. This helps limit ISI and remove spectral repetitions caused by sampling.

- Timing recovery — the sampling and data rate clocks of the receiver have to be locked to the clocks used at the distant transmitter in order to maintain synchronization. This is achieved by narrowband filtering and nonlinear operations on the baseband filtered data signal which generate an error signal for use in a DPLL. The output of the DPLL is a control signal to the timing ASIC that alters the frequency and phase of the receiver clocks.

- Adaptive equalization — equalization of the complex baseband signals is performed by a fractional-spaced equalizer. This deals mostly with the effects of group delay along with any residual amplitude slope distortion left by the adaptive compromise equalizer in the TECdsp.

- Carrier recovery — the phase quadrature carriers used for demodulation have nominal (or fixed) frequency. Due to component tolerances and network effects, this carrier frequency may differ to that of the received signal by an amount up to ± 7 Hz. Any difference that does exist is compensated for by a decision-directed carrier recovery DPLL that rotates the complex equalizer outputs. The DPLL also features an adaptive bandwidth algorithm that allows tracking of carrier phase jitter, if present.

- Viterbi decoding — the rotated equalizer outputs are passed to the Viterbi decoder section which provides decisions on a maximum likelihood basis. The decoder section also performs de-mapping of the symbols into data. Decisions from the decoder section are used to construct the error signals for both the adaptive equalizer and carrier recovery DPLL.

  The Viterbi decoder is also responsible for delivering secondary channel data and provides a secondary channel capacity of 75 bit/s.

- Descrambler — the final operations on the data output by the decoder are differential decoding and descrambling. These are performed in accordance with the V.32 Recommendation. The resulting customer data is then passed to the controller in byte format.

Like the TECdsp, the RXdsp includes supervisory and diagnostic functions in addition to the main signal processing operations described above.

- Tone detection — the V.32 handshake sequence depends on the accurate detection of various tones and phase changes within these tones. Filters are included in the RXdsp to detect and inform the controller of the occurrence of such signals.

- Channel quality monitoring — routines are provided that monitor additional channel parameters to those reported by the TECdsp. Examples of these parameters include carrier and timing offsets, phase jitter level and the receiver mean square error. The controller can read these values and all DSP device memory locations via a vectored interrupt. Provision for writing to all DSP device memory locations is also made.

- Self test — diagnostic routines are performed on power-up that test all internal and external memory together with the DSP device ALU. The operation of the timing ASIC is also checked.

7.8 CONCLUSIONS

The telephone network and its weaknesses for data transmission have been described, along with signal processing techniques used to overcome these weaknesses in voiceband modem applications. The implementation of the BT Laboratories datapump, presented here, demonstrates the progress that has been made in the implementation of complex signal processing systems with the introduction of modern digital signal processing devices. These devices have enabled the implementation of a V.32 datapump which is about three times smaller, yet has a larger list of facilities and modes and has a significant improvement in performance over previous generation products. The high-speed voiceband modem is probably the most sophisticated example of a signal processing product that is being sold into a large market, and in consequence, owing to the competitive forces, it provides one of the most demanding and intensive uses of digital signal processing technology.

APPENDIX

Datapump modem modes

The BT Laboratories datapump offers the following CCITT and proprietary operating modes:

V.32bis*	2W full-duplex	4.8k, 7.2k c, 9.6k c, 12.0k c, 14.4k c
V.32*	2W full-duplex	4.8k, 9.6k, 9.6k c
BTV.32*	2W full-duplex	4.8k, 7.2k c, 9.6k c, 12.0k c, 14.4k c
V.33	4W full-duplex	4.8k, 7.2k c, 9.6k c, 12.0k c, 14.4k c
V.29	4W full-duplex	4.8k,7.2k,9.6k
V.26bis*	2W half-duplex	2.4k
V.26	4W full-duplex	2.4k
V.22bis*	2W full-duplex	1.2k, 2.4k
V.22*	2W full-duplex	1.2k
V.23*	2W asymmetric duplex	1200/75
	4W full-duplex	1200/1200
	2W half duplex	1200/1200
V.21*	2W full-duplex	300
Bell 212A*	2W full-duplex	1.2k
Bell 103J*	2W full-duplex	300

c = trellis coded modulation and Viterbi decoding
* = automoding to these modes

Full-duplex echo cancelling modes have both near end and far end cancellers with far end phase roll removal. The far end canceller can be positioned at up to 1.2 s from the near end canceller. Additional facilities include:

— full automoding for all PSTN 2 W full-duplex modes (see * in Table above);
— automatic initial rate selection and in-data rate adaptation in BT V.32 and V.32bis modes;
— 75 bit/s secondary channel when in coded modes;
— tone detection and generation for international PSTN progress tones;
— analogue and digital test loops;
— loudspeaker for call progress monitoring.

Interfaces

Line interface

The line signals are presented in unbalanced form. Line transformers, protection, ringing detection and line hold circuitry, etc, are required to complete the connection to line. Switching is provided for 2 W and 4 W circuit connections.

Control interface

Control is via an 8-bit control bus to a dual port RAM device. This interface gives full control of all modes and allows complete access to all datapump variables. Full diagnostics are provided.

The following V.24 control bits are provided as memory locations: 104, 105, 106, 107, 109, 112, SC105, SC106 (SC = secondary channel).

Data interface

V.24 circuits are provided at TTL levels as follows:

— 103, 104, 105, 106, 107, 109, 113, 114, 115;
— synchronous and asynchronous data is accepted;
— data can also be accessed in 8-bit parallel form via the control interface.

Physical characteristics

Dimensions — 122 mm × 160 mm max
Power consumption — typically less than 3 W

REFERENCES

1. CCITT Blue Book Volume VIII — Fascicle VIII.1 Data Communication over the Telephone Network, Series V Recommendations.

2. Widrow B and Hoff M E: 'Adaptive switching circuits', IRE WESCON Conv Rec, pt 4, pp 96-104 (August 1960).

3. Qureshi S U H: 'Adaptive equalization', IEEE Proc, 73 , No 9, pp 1349-1387 (September 1985).

4. Falconer D D: 'Jointly adaptive equalization and carrier recovery in two dimensional digital communication systems', Bell Syst Tech J, 55 , No 3, pp 317-334 (March 1978).

5. Lyon D L: 'Envelope derived timing recovery in QAM and SQAM systems', IEEE Trans Com, COM—23 , pp 1327-1331 (November 1975).

6. Ungerboeck G: 'Theory on the speed of convergence in adaptive equalizers for digital communication', IBM J Res Devel, 16 , pp 546-555 (November 1972).

7. Gersho A: 'Adaptive equalization of highly dispersive channels', Bell Syst Tech J, 48 , pp 55-70 (January 1969).

8. Qureshi S U H and Forney G D: 'Performance and properties of a T/2 equalizer', National Telecom Conf Rec, Pt I, pp 11:1/1-9 (December 1977).

9. Ling F and Qureshi S U H: 'Convergence and steady state behaviour of a phase splitting fractionally spaced equalizer', Proc IEEE ICC'88, pp 12.4.1-12.4.5 (June 1988).

10. Ungerboeck G: 'Fractional tap spacing equalizer and consequences for clock recovery in data modems', IEEE Trans Com, COM—24 , pp 856-864 (August 1976).

11. Gitlin R D and Weinstein S B: 'Fractionally spaced equalization: an improved digital transversal equaliser', Bell Syst Tech J, 60 , pp 275-296 (February 1981).

12. Gitlin R D, Meadors H C and Weinstein S B: 'The tap leakage algorithm: an algorithm for stable operation of a digitally implemented, fractionally spaced adaptive equalizer', Bell Syst Tech J, 61 , No 8, pp 1817-1839 (October 1982).

13. Messerschmitt D G: 'Echo cancellation in speech and data transmission', IEEE Selected areas in Comms, SAC—2 , No 2, pp 283-297 (March 1984).

14. Weinstein S: 'A passband data-driven echo canceller for full duplex transmission on two wire circuits', IEEE Trans on Comms, COM—25 , No 7, pp 654-666 (July 1977).

15. Buckley P M: 'Line echo cancellers for two-wire full-duplex data transmission', Internal BT memorandum (1977).

16. Lindsey W C and Chie C M: 'A survey of digital phase locked loops', Proceedings of the IEEE, 69 , No 4, pp 410-431 (April 1981).

17. Wang J and Werner J: 'Performance analysis of an echo-cancellation arrangement that compensates for frequency offset in the far echo', IEEE Trans on Comm, COM-36 , No 3, pp 364-372 (March 1988)

18. Ungerboeck G: 'Channel coding with multilevel/phase signals', IEEE Trans, IT—28 , pp 55-67 (January 1982).

19. Forney G D: 'The Viterbi algorithm', Proc IEEE, 61 , pp 268-278 (March 1973).

20. Wei L F: 'Rotationally invariant convolutional channel coding with expanded signal space — Part II: nonlinear codes', IEEE, JSAC—2 , pp 672-686 (September 1984).

8

BLOCK CODING FOR VOICEBAND MODEMS

R G C Williams

8.1   INTRODUCTION

In 1984 trellis coded modulation (TCM) was included in the international recommendation for full duplex modems achieving 9600 bit/s over the general switched telephone network (CCITT — V.32). This was the culmination of several years' work on the emerging field of coded modulation. The V.32 standard [1] combined a convolutional code with quadrature amplitude modulation (QAM). The scheme, based on the principle of set partitioning introduced by Ungerboeck [2], was chosen as the best from several other candidate schemes. An explosion of research interest in coded modulation followed with the majority of work being concentrated around the idea of set partitioning. A major step forward in combining block codes with modulation schemes was taken by Cusack [3] who combined Reed-Muller codes with QAM to achieve asymptotic performance gains comparable to those of Ungerboeck. Later Cusack's method was generalised by Sayegh [4] who showed the flexibility that block codes could provide.

This chapter outlines the development of block coded modulation (BCM) and describes its theoretical advantages and drawbacks. After presenting solutions to the drawbacks, the chapter looks at BCM schemes in a practical light. The problems of using a BCM scheme for a voiceband modem are solved and an implementable BCM scheme is given. The final sections of the paper deal with the integration of the chosen BCM scheme with an existing voiceband modem before conclusions are drawn and discussed.

8.2   BLOCK CODED MODULATION

8.2.1   Set partitioning

In QAM the amplitudes of two orthogonal signals (carriers) are modulated. Plotting the possible combinations of carrier amplitudes on a plane results in a signal constellation. A 16-point signal constellation is shown in Fig. 8.1. Each constellation point can represent four bits. These bits are used to label the point. To transmit a given four bits of information the transmitter will modulate the amplitude of each carrier by the amount required to produce the point labelled by the four bits.

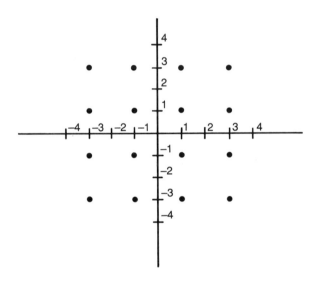

Fig. 8.1   A 16-point signal constellation.

Information is extracted from the received signal by deciding which signal point was transmitted and then reversing the bit mapping process. To do this the amplitude of each carrier is found and the nearest allowed combination is assumed to be the transmitted point. As the transmission is subject to noise the received signal will not normally lie on a valid signal point. Therefore signal points are prone to be mistaken by the receiver for their neighbours. Error control coding is used to increase immunity to noise. If the points of the constellation are labelled arbitrarily then a coding scheme cannot be applied efficiently. Set partitioning is a method of assigning bit labels to the points of a signal constellation so that efficient coding is possible.

In set partitioning each bit in a signal point's label carries information about that point's relative position. An example of set partitioning is given in Fig. 8.2. The least significant bit of each label differs for points next to each other. This means that any two points, the labels of which share the same least significant bit, cannot be neighbours. The labelling partitions the constellation into sets. Two sets of neighbours are given by the least significant bit of the labels. Similarly, each of these sets have neighbours differentiated by the next significant bit of the labels.

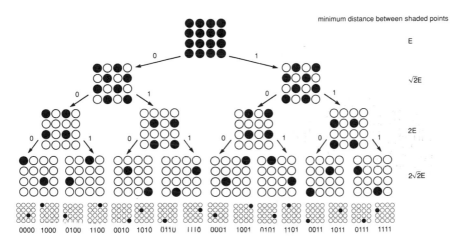

Fig. 8.2   Partitioning tree for 16-QAM.

Such a labelling can easily be used by a coding scheme. While the receiver can differentiate signal points that are far from each other (those with labels that only differ in their more significant parts) it has difficulty with points that are close to each other. Since specific bits in the point labelling determine the relative separation of the points, these bits can be coded to help the receiver decide on the transmitted point. If the least significant bits of the labels are constrained to fit into a certain pattern then once the receiver has found the pattern it stands a better chance of finding the correct points.

This is the basis of any coded modulation scheme that uses set partitioning. The V.32 standard uses a 32-point constellation partitioned into eight subsets (each containing four points). A sequence of subsets is defined by a convolutional code. This helps the receiver decide upon the correct subset for each point. The distance between points in each subset is then sufficient for the receiver to determine the correct point unaided. The principle is extended in the later V.32bis standard to a 128-point constellation where each of the eight subsets has 16 points. This scheme achieves an asymptotic performance advantage of nearly 4 dB.

8.2.2   Combining block coding with QAM

This chapter describes how block codes may be used in coded modulation schemes based on set partitioning. A (linear, binary) block code consists of 2 k codewords, each with n bits, arranged such that no two codewords differ in less than d places. Such a code is referred to as an (n,k,d) block code. It is desirable to have k and d large while keeping n small. Basic bounds for these parameters are given in MacWilliams and Sloane [5]. All the codes used in this chapter are from the class of Reed-Muller codes which are dealt with extensively in MacWilliams and Sloane [5].

Let two distinct points from a 2^m point QAM signal constellation have labels $a_{m-1},..,a_1,a_0$ and $b_{m-1},..,b_1,b_0$. Under set partitioning the distance between the two points is at least $(\sqrt{2})^i E$ where i is the lowest number for which $a_i \neq b_i$ and E is the Euclidean distance between neighbours in the constellation. Coding can be used to help the receiver distinguish between points where i is low. A powerful block code (one with a large minimum distance, d) is needed to define a pattern of bits for the least significant position in each constellation point label. As more significant bits are considered the euclidean distance between points that differ first in this position increases and therefore the receiver needs a less powerful block code (one with a smaller minimum distance) to help it distinguish between them.

To match block codes to the point labels in this way a codeword array is formed. The rows of the array consist of bits forming codewords from the block codes. Each column of the array represents the label of a point to be transmitted. The top row contains the least significant bit of each point label and is therefore a codeword from the most powerful block code. The power of the block codes decreases as the rows contain the bits that are more significant in the bit labels. An example of such an array is given below.

Point label
↓
```
1  1  1  1  1  1  1  1
1  0  0  1  0  1  1  0    ← Codeword from
1  0  0  0  1  0  0  0      a block code
0  1  0  0  0  0  0  0
```

This is one codeword array from the set of $2^{20} = 1\ 048\ 576$ codeword arrays that are depicted by:

(8,1,8)
(8,4,4)
(8,7,2)
(8,8,1)

Generally the minimum distance of the code in the ith row will be d_i. Any two codewords from this code differ in at least d_i places. Since any two points with labels first differing in the ith position are a squared euclidean distance of at least $2^i E^2$ apart, the total squared distance due to the ith row is at least $d_i 2^i E^2$. Therefore, any two codeword arrays that first differ in their i^{th} row must be a squared euclidean distance of at least $d_i 2^i E^2$ apart. Taking all the rows into account it is seen that the minimum squared distance between two valid codeword arrays is $D^2 \geq \min_i (d_i 2^i E^2)$. The distance advantage of a codeword array can now be defined as $10\log_{10}(D^2/E^2)$ dB. This figure does not give the performance (coding) gain of the codeword array but it is a useful guide to the power of a particular scheme.

In the example above $D^2 = 8E^2$. This gives a distance advantage of 9 dB. Unfortunately each codeword array can only transmit 20 bits of information using this scheme. The uncoded constellation is able to transmit 32 bits of information in the same time. This loss of information rate must be compensated for, causing the performance gain to be less than the distance advantage. If the compensation is in the form of a symbol rate increase (transmitting the data at 1.6 times the speed and therefore incurring more noise) there will still be an asymptotic performance gain of 7 dB. Since it is not always possible to increase the symbol rate it may be necessary to increase the data rate by expanding the constellation used. If this method of rate compensation is used the asymptotic performance gain will be 4 dB.

8.2.3   BCM compared to TCM

BCM has several potential advantages over TCM. The most obvious difference in the two techniques is that BCM uses blocks of symbols, whereas TCM uses a sequence of symbols. This means that BCM cannot suffer from large amounts of error propagation and it has a fixed (often short) decoding delay. Another advantage of BCM is that the range of block codes available for use is much larger than the range of trellis codes. This adds flexibility to the application of BCM and also provides BCM with higher asymptotic coding gains for a given amount of effort.

The example codeword array in section 8.2.2 shows the potential of BCM. It is short and so has a small decoding delay. It is powerful with the minimum asymptotic gain being equivalent to the V.32 code. It consists of easily decodable block codes making it simpler to implement than the V.32 code.

Unfortunately this codeword array also shows the disadvantages of applying block codes in this way. The first disadvantage is the decoding of the codeword array. Although each row contains a standard block code and all have simple decoding algorithms, it is difficult to decode all the rows together in an optimum manner. The accepted method for decoding such an array is suboptimum. It uses a staged decoder similar to the one given in Imai and Hirakawa [6]. When the codeword array is received it is decoded one row at a time. After a row has been decoded that row is assumed to be correct. Each row is therefore decoded assuming that the rows above it have been correctly decoded. This method is illustrated in Fig. 8.3. It is suboptimal because there is no way to change the decision made about an earlier row. The distance between the chosen points and a lower row could possibly be decreased if one of the higher rows had been decoded to a different codeword. There is no provision for this case but thankfully it is only a problem on the worst channels.

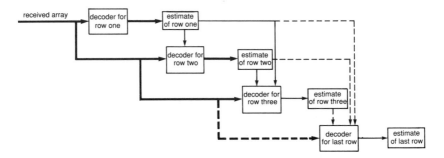

Fig. 8.3   Structure of a staged decoder.

Another problem of BCM is also illustrated by the simple example array in section 8.2.2. This is the problem of nearest neighbours. When a receiver is determining the transmitted signal from the received signal it is most likely to mistake points that are neighbours. This effect translates into the case of a receiver trying to decide which codeword array was transmitted from several received signals. The most likely errors that occur are the receiver decoding to a codeword array that is next to the actual transmitted one. When block codes are applied as in the example the number of neighbours for a codeword array can be very large. Generalizing the work by Lunn [7] to QAM, the number of nearest neighbours for a typical codeword array in the example is 3960 (495 per symbol). This is a large number of nearest neighbours and will degrade the performance of an optimum receiver by about 1.4 dB at working error rates [8]. Lunn points out that if the decoder used is a staged decoder it only considers the nearest neighbours for each row individually. This considerably increases the overall number of nearest neighbours that the receiver has to cope with. In the example a staged decoder will have 69 240 nearest neighbours (8655 per symbol) putting the estimated

degradation up to 2.2 dB. A modified decoder is given in Lunn [7] with
a reduced number of nearest neighbours. In this decoder the codeword array
is decoded once for each codeword allowed in the top row. This effectively
uses a staged decoder only on the lower rows of the codeword array.
Unfortunately this is only practical if the code used for the top row of the
array has very few (e.g. two) codewords.

The problems of decoding and many nearest neighbours are specific to
this method of BCM. When using TCM there is an optimum decoding
algorithm available in the form of the Viterbi algorithm [9]. Although the
number of nearest neighbours in a TCM scheme is a problem, it can be
controlled to a certain extent by the trellis code used. Solutions for the BCM
case are presented in the next section where an attractive BCM scheme for
voiceband modems is given.

8.3   IMPROVED BCM

The two disadvantages mentioned in the last section are a consequence of
the way block codes have been combined with QAM. An extension of work
to solve these problems [10,11] is presented here as an effective, low-
complexity application of block codes to QAM.

QAM is formed by modulating the amplitude of two orthogonal carrier
signals. If each carrier is thought of individually, QAM reduces to a method
of sending two 1-D amplitude modulated symbols at once. Figure 8.4
illustrates this point of view.

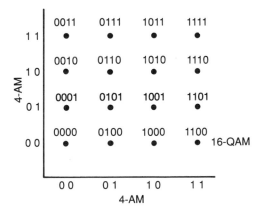

Fig. 8.4   16-point QAM viewed as the combination of two 1-D symbols.

The 16-point constellation that was considered earlier can be thought of as two 4-point constellations transmitted at once. A codeword array that matches this has half the number of rows but twice as many columns. Another difference is in the distance profile of the rows. Any two labels that differ first in the ith bit represent points that are at least a squared euclidean distance of $4^i E^2$ apart. This enables the power of the block codes to drop faster than before as the rows go down. The outcome of this is that an equivalent codeword array to the last example is:

(16,5,8)
(16,15,2)

Both examples transmit 20 bits in each block, both are 8 QAM symbols long and both have a 9 dB distance advantage. Each codeword array has 495 nearest neighbours per symbol. However, although a staged decoder in the original example resulted in 8655 nearest neighbours per symbol, in this example it only results in 975 nearest neighbours per symbol. The degradation due to these nearest neighbours is only 1.6 dB compared to 2.2 dB for the first example.

This new method is worth considering on the strength of this alone. It also has other advantages over the method of Cusack [3]. Since each QAM symbol now carries two columns of the codeword array, the number of symbols transmitted is only half of the length of the block codes used. This means that more powerful block codes can be used for the same amount of decoding delay. These can be employed either to increase the final asymptotic performance gain or to increase the information rate required while leaving the asymptotic performance gain the same. The number of block codes required is halved as well. This means that fewer decoders are needed for each codeword array. The complexity of each decoder is also reduced. With this method of combination the distances that the decoder uses to calculate the closest codeword are 1-D. In the original example they were 2-D and more difficult to use.

Both the examples given so far are purely illustrative in the sense that they would not be practical for use in a voiceband modem. The coding scheme in a voiceband modem has further design constraints placed on it. In section 8.2.2 it was mentioned that there is a drop in information rate associated with each coding scheme. This is unavoidable and must be counteracted. In the V.32 standard the coding scheme needed one redundant (not information) bit per symbol. This was compensated for (with a penalty of 3 dB) by transmitting the coded signals on a constellation with twice as many points as an uncoded signal would need. To get a BCM scheme to use only one redundant bit per QAM symbol the information rate of the block codes used must be increased. For both examples ⅚ of the bits are information bits.

This must be increased so that ¾ of the bits are information bits. To send extra information, uncoded rows may be added to the codeword array. These do not affect the asymptotic performance gain of the scheme which will still use only one redundant bit per symbol.

The distance advantage of the code used in the V.32 standard is 7 dB (reduced by 3 dB penalty for constellation expansion to give a 4 dB asymptotic performance gain). When considering BCM for use in voiceband modems it would be convenient to find a scheme with at least this asymptotic performance gain. The two examples given both promise such an asymptotic performance gain from a distance advantage of 9 dB. Therefore a distance of $D^2 = 8E^2$ seems attractive.

Finally, it can be shown that the problems of decoding complexity, nearest neighbours and decoding delay are all increased with the increasing length of the block codes used. The shortest Reed-Muller based BCM scheme with a distance of $8E^2$ and an information rate of (nearly) ¾ is:

(32,16,8)
(32,31,2)

This scheme seems an ideal candidate to compare with the V.32 code in a practical environment. The BCM scheme is 16 QAM symbols long. This matches the decoding delay of 16 QAM symbols needed to achieve the asymptotic performance gain of the V.32 code. The distance advantage of the BCM scheme is 9 dB. Counteracting the drop in information rate will use 3 dB, with an estimated 2.1 dB being lost due to the 4991 nearest neighbours per symbol. These figures bring the expected gain to around the 4 dB of the V.32 code. Before the BCM scheme can be tried as a substitute for the V.32 code some further practical problems must be addressed.

8.4   PRACTICAL PROBLEMS

8.4.1   Synchronization

In any scheme where the transmitted symbols must be decoded together their position is important. This class of schemes includes BCM and multi-dimensional TCM [12]. It does not include the TCM scheme in the V.32 standard. In V.32, if a symbol is inadvertently dropped or added to the transmission stream the effect will be felt for a limited time (the decoding delay). In schemes where the relative position of the symbols is important such an occurrence would be catastrophic if undetected. In BCM, for instance, the decoder would be trying to decode blocks that start and finish in the wrong place. It would not be able to produce accurate data from them. Therefore a method is needed to ensure that this situation is easily noticed and corrected.

To ensure that synchronization is being maintained some easily recognizable signal must be sent periodically. In order not to interrupt the data stream it is proposed that a symbol is regularly sent from outside the normal signal constellation. This can be achieved by having a second set of points around the first set that are only used for synchronization. When a synchronizing pulse is to be sent the data that would have been transmitted on a normal point is instead transmitted on one of the extra points. This method results in a slight increase in the average power of the signal constellation as well as its peak-to-mean power ratio. However, the extra points can be recognized with high reliability.

If an extra point is noticed in the wrong place (synchronization may have been lost) the number of symbols until the next one is monitored. If this number is the number expected between synchronization pulses then the symbol stream is resynchronized. If the number is not the number expected then nothing is done since one of the extra points could have been caused by channel impairments. This method of synchronization is very robust.

A small modification to the process will enable extra data to be transmitted. Instead of sending one of the extra points on every synchronization symbol this may be determined by the extra data. In the case of the data being 1, an extra symbol set point is sent. If the data is 0, a standard point is sent. Since the receiver knows when the synchronizing symbol is due it can extract this data depending upon which point it receives. The normal data transmitted in that symbol may be represented by either the extra point or the standard point and so is not affected. The extra points are now only used for synchronization probabilistically. This may result in the need to send a synchronizing symbol more often. In the case of the BCM scheme proposed for comparison with the V.32 code a symbol can be used for synchronizing in every block. This provides fast and highly reliable synchronization as well as the extra information bit per block needed to get ¾ of the bits to be information bits.

8.4.2   Phase shift immunity

Standard QAM constellations are symmetrical under rotations of 90°, 180° and 270°. This creates a problem for the receiver. If the signal constellation suffers such a rotation at any time, no receiver function will detect it. In this case the transmitter and the receiver will be using different quadrants to each other and there will be no accurate data passed between them. To get around this problem the data is differentially encoded before it is sent

[13]. The signal that is sent gives the receiver information about the difference between the quadrant last used and the quadrant being used now; the receiver does not have to find the absolute phase of the signal. The quadrant shift between the last and present symbols is used to choose the quadrant from which the present symbol is transmitted. Thus the operation rotates the present symbol before it is transmitted. The quadrant information is turned back into absolute data by differential decoding in the receiver. This ensures (along with careful labelling of the constellation points) that any phase shifts that are a multiple of 90° will not affect the data produced (except from the symbol in which the phase shift occurs and the one following it). Incidentally, the fact that the QAM constellations have this rotational symmetry is actually useful to other functions in the receiver (such as carrier tracking) as well as being power efficient.

The differential encoding is quite straightforward for the TCM used in the V.32 standard. The quadrant shift is calculated between each symbol that is transmitted. This information is then sent as part of the coded sequence. The sequence remains valid and so can be decoded before the differential decoding takes place. This is clearly desirable since the quadrant information needs as much immunity to noise as all the other data.

Similarly in BCM the decoding should take place before the differential decoding. This makes it difficult to differentially encode symbol by symbol. A method was therefore devised to differentially encode BCM schemes block by block. The condition that the block must still be a valid codeword array, after it has been rotated by the differential encoding, places further restrictions both on the codes used in the BCM scheme and on the constellation labelling.

Generalizing the labelling of the constellation in Fig. 8.4 the effect of rotating a codeword array:

x	y

can be seen as:

90° phase shift

y	x'

180° phase shift

x'	y'

270° phase shift

y'	x

where c' is part of a codeword array whose entries are the complements of the corresponding entries c. The codes used must be such that rotated versions of the original codeword array produce other codeword arrays. This restriction is obeyed by each code used in the BCM scheme that is considered for implementation. If a codeword array is used to code each dimension separately [10] the restriction on the codes is reduced. Each code used only needs to contain the complement of each of its codewords. This is true of any linear block code that contains the all 1 codeword. If a system designer wishes to use a code that only satisfies this weaker condition then twice the decoding delay is needed. Reduced nearest neighbours and potentially increased power are still available.

Quadrant bits are given by the most significant bit of each dimension's label. These bits can be used for differential encoding. They are taken from the first symbol of each block and these are then considered to be the quadrant bits for the entire block. The differential encoder finds the necessary rotation from the quadrant bits of this block and the previous block. The rotation is then applied to all the remaining bits of the block. The differentially encoded bits are placed unchanged in their positions in the first symbol and the whole block is transmitted. At the receiver the block is decoded. This is possible in spite of any phase shifts that may have occurred on the channel. The differentially encoded bits are then used to reverse the rotation of the decoded block before being differentially decoded and the result replaced in appropriate position in the first symbol. The process is represented schematically by Fig. 8.5.

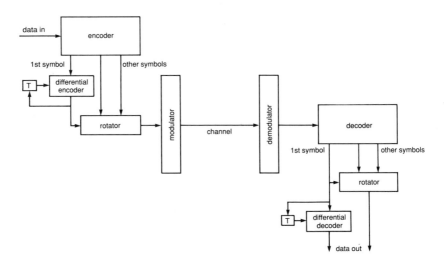

Fig. 8.5   Block diagram of the differential encoding operation.

This process provides the required immunity of 90° phase shifts without adding any extra delay. The small performance penalty suffered due to relying on the quadrant bits of each block is comparable to that suffered by the V.32 method of achieving phase shift immunity. There is a slight difference between the differential encoders used. The differential encoder/decoder pair used for the BCM schemes must be of the Gray type (as used for the uncoded mode in the V.32 Recommendation) rather than the modulo-4 type used for most TCM schemes.

8.5 IMPLEMENTATION OF THE BCM SCHEME

8.5.1 The scheme

Having addressed the remaining practical problems of BCM schemes, the scheme chosen in section 8.3 is ready to be implemented. It is a block of 16 symbols taken from a 256-point square QAM signal constellation. The last symbol in each block is taken from a 512-point cross QAM signal constellation. The block that is transmitted is built from the following codeword array by sending the first half using one dimension and the second half using the other.

(32,16,8)	1
(32,31,2)	
(32,32,1)	
(32,32,1)	

The first row is a codeword in the (32,16,8) Reed-Muller code; the second row is a codeword in the (32,31,2) Reed-Muller code (an even parity check code); the last two rows are uncoded. An extra bit is transmitted on the last symbol of each block. If this bit is a 0 then the last symbol is transmitted from the same 256-point constellation as all the others. If the extra bit is a 1 then the last symbol is transmitted from the outer 256 points of a 512-point signal constellation (with the same point separation as the 256-point constellation used for the other symbols). This is used both to increase the information rate of the scheme and to provide synchronization.

The scheme transmits seven information bits per symbol. This represents a data rate of 16.8 kbit/s at a symbol rate of 2400 sym/s or a data rate of 19.2 kbit/s at a symbol rate of 2743 sym/s. These data rates could be achieved using a 128-point constellation. However this scheme is capable of transmitting the information four times more reliably for a given average signal power. This represents an asymptotic coding gain of 6 dB. This is reduced by 0.25 dB due to the synchronizing symbol being transmitted on a larger constellation than the rest. Therefore the best coding gain that can be hoped for from this scheme is 5.75 dB.

A voiceband modem has been produced (see Chapter 7) that works to the V.32 standard (2400 sym/s). All the DSP code used to implement the TCM scheme in this modem was stripped out and new code written for the implementation of this BCM scheme. The rest of the modem's functions were left the same.

8.5.2   The encoder

The V.32 encoding was implemented in the general purpose 8-bit microprocessor used mainly for the modem's control. As the TCM encoding is straightforward and done on a symbol-by-symbol basis this was a small piece of code. The BCM encoding is slightly complicated by the fact that it is done on a block by block basis. Each symbol of the block is encoded using a different part of the codeword array's generating matrix. This means that a lot more microprocessor code needs to be written and a counter is needed to step through the symbols in each block. However each symbol's processing is straightforward. Therefore the time taken for the encoding is similar in both cases. The main difference is in the amount of microprocessor code used. The extra amount of microprocessor code needed for the BCM encoder did not cause any problems.

8.5.3   The decoder

As in the encoder, the main difference between the TCM scheme of V.32 and the BCM scheme here is that the TCM scheme can be processed on a symbol-by-symbol basis. The receive DSP device in the modem takes the incoming signal and produces a pair of co-ordinates for the received point. In the V.32 version these are passed to the TCM decoder where they are quantized to the nearest valid signal point. This information is passed to the receiver to enable it to update its adaptive equalizer taps with the relevant error information. Meanwhile, the decoder will use the received point to calculate distance metrics with which it can update the decoding paths in the Viterbi algorithm. Having done this it traces the most likely path back to a delayed decision. The data from this decision is given back to the receiver to be descrambled and passed on to the user.

This process happens regularly with each received symbol. The code needed to implement this can be made quite efficient. Even so the required processing takes a significant proportion of each symbol interval.

The BCM decoder must fit into the broad scheme of the receiver invisibly. In other words, in each symbol interval it must take the received co-ordinate pair and produce a quantized version of this co-ordinate pair as well as a symbol of data to be passed to the user. This must be done for each symbol while the decoder is actually decoding the blocks on a block-by-block basis. To do this two areas were set up in the DSP device's memory, one for the last decoded block and the other for the last received block. The processing of the decoder is shared between the 16 symbol intervals that each block is allowed for decoding. In every symbol interval a received signal is taken in and quantized on to the nearest valid signal point. This operation is the same in each of the symbol intervals except that the last symbol comes from the larger constellation. The result of this is fed to the receiver for the update of the adaptive equalizer taps. The received signal is then stored in the last decoded block's area and the displaced information is the delayed data that is passed on to the user through the receiver. This information is calculated from the last decoded block in the first symbol interval of each decoding cycle.

The rest of the decoder's time is spent decoding the last received block. The hardest part of this is the decoding of the (32,16,8) Reed-Muller code. The trellis based algorithm suggested by Forney [14] is used. This algorithm takes the (32,16,8) Reed-Muller code and breaks it down into a four-section trellis. Each section is made up from the (8,7,2), the (8,4,4) and the (8,1,8) Reed-Muller codes. The trellis is pictorially represented by Forney as in Fig. 8.6.

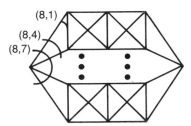

Fig. 8.6   Defining trellis for the (32,16,8) Reed-Muller code.

The entire trellis is built up from eight sub-trellises (this is denoted by the (8,7) arc in Fig. 8.6 and arises from the eight cosets of the (8,4) code in the (8,7) code). Each sub-trellis is in four sections. The first section is a coset representative of any of the eight cosets of the (8,1) code in the (8,4) code. The next two sections are another pair of any of these coset representatives. These choices confine the choice of the fourth section to be a specific coset representative. Due to these choices each sub-trellis uses nine information bits (three from each choice of one out of the eight cosets).

A total of four information bits determine which codeword from the (8,1) code is used in the chosen coset in each section of the sub-trellis (1 bit for each section). The final three information bits carried by the (32,16) code are provided by the choice of sub-trellis in the entire trellis.

To decode this trellis the decoder finds the most likely path through each version of the sub-trellis. Therefore the decoder spends 8 of its 16 symbol intervals working out the most likely path through each of the sub-trellises. First it needs to calculate the distance measures for each of the received symbols. This is done in the 2nd symbol interval of the decoding cycle. In the 3rd to the 10th symbol intervals the decoder calculates the relevant distances to an (8,7)/(8,4) coset for each of the sections and works out the most likely path through the sub-trellis. This path is stored by defining the coset along with the nodes where the sections join. The decoder uses this information in symbol interval 11 to extract the most likely path through the entire trellis. The 12th symbol interval checks the decoded top row against the received data and makes any corrections that are necessary. The 13th symbol interval is used to do this for the special case of the last symbol where the larger constellation is in use.

The top row has now been decoded and the received data has been changed accordingly. The staged decoder therefore progresses on to the second row. It is very simple to check and correct the even weight code and all the processing for this is done in the 14th symbol interval. The decoding is now complete, so the 15th symbol interval is free except for the input/output requirements. In the 16th symbol interval the quantization of the received signal is slightly different as mentioned before. This symbol interval is also used to place the block of received data into the last received block area and the decoded block into the last decoded block area. The decoder is now ready to receive the next block and to decode the block it has just received. The operations in each symbol are shown in Fig. 8.7.

As the description suggests, the time that the decoder processing takes in each of the 16 symbol intervals in a decoding cycle varies greatly from symbol to symbol. This is illustrated in Fig. 8.8. This is a photograph of an oscilloscope trace depicting the processing time that the receiver typically takes in each symbol interval. The top trace is the symbol interval counter which is reset when the first symbol of a block is received and is incremented as each subsequent symbol is received. The lower trace depicts the receiver's processing time. The trace is raised when the receiver starts processing and lowered when the receiver finishes processing. The symbol interval counter is set some time after the receiver starts its processing because this is done by the codeword array's decoder which runs after the initial receiver functions

Symbol interval	:	Action
1	:	Input noisy co-ordinates
	:	Output immediate decision
	:	Differentially decode last decoded block
	:	Extract and store information from last decoded block
	:	Output delayed data
2	:	Input noisy co-ordinates
	:	Output immediate decision
	:	Calculate distances for decoding received points
	:	Output delayed data
3—10	:	Input noisy co-ordinates
	:	Output immediate decision
	:	Offset distances for relevant coset
	:	Find most likely path through sub-trellis
	:	Store path if most likely to date
	:	Output delayed data
11	:	Input noisy co-ordinates
	:	Output immediate decision
	:	Form the codeword from the likeliest path information
	:	Output delayed data
12	:	Input noisy co-ordinates
	:	Output immediate decision
	:	Change received points in line with codeword
	:	Output delayed data
13	:	Input noisy co-ordinates
	:	Output immediate decision
	:	Change last point in line with codeword
	:	Output delayed data
14	:	Input noisy co-ordinates
	:	Output immediate decision
	:	Check and correct the even weight code
	:	Output delayed data
15	:	Input noisy co-ordinates
	:	Output immediate decision
	:	Output delayed data
16	:	Input noisy co-ordinates
	:	Output immediate decision
	:	Output delayed data
	:	Swap information around memory space

Fig. 8.7   Decoder operations in each symbol interval of the decoding cycle.

(demodulation, filtering, equalization, carrier recovery, etc) have been used. The variations in processing time are solely due to the decoder since the other receiver functions take roughly the same time in each symbol interval.

It is clear from Fig. 8.8 that the majority of the processing done by the decoder is during the 3rd to the 10th symbol intervals. This is expected owing to the chosen method of implementation. If a system were to be built specifically for a BCM scheme (rather than fitting the BCM scheme into an existing system) the decoder's processing would probably be shared equally between the symbol intervals. In this case the processing in each symbol interval would take a similar amount of time to the V.32 processing. The DSP code used for this implementation was not optimized as it would be for a dedicated BCM system. It was about twice as large as the code used to implement the V.32 scheme.

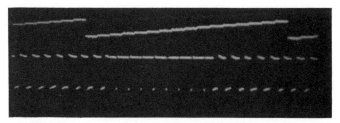

Fig. 8.8   Oscilloscope trace showing the variation of the receiver's processing time throughout the decoding cycle.

The implementation was tested in the laboratory over a channel with Gaussian noise added. The performance was compared to the same system running a version of the V.32 code delivering data at 16.8 kbit/s. The graph in Fig. 8.9 shows the results. The BCM scheme performs slightly better than the TCM scheme.

8.6   CONCLUSION

A method of combining block coding with QAM has been investigated. A BCM scheme had been found that promised a reasonably effective coding gain and this was implemented in a voiceband modem instead of the V.32 standard TCM scheme. The DSP code used for the BCM scheme took a similar average time to run as the DSP code needed to implement the V.32 standard. The amount of DSP code needed was larger in the BCM scheme's case. When the performance of the two schemes was compared the BCM scheme was found to be marginally better.

This is an encouraging result. BCM schemes have many potential areas of application. Although in voiceband modems TCM seems to be the favoured method of coding, there are several other areas where block coding techniques are more suitable. Environments where data can only be sent in

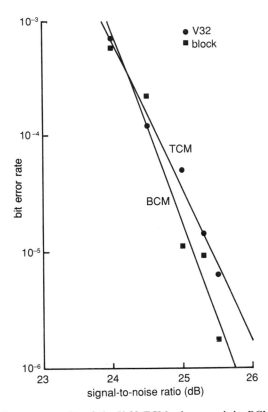

Fig. 8.9   Performance results of the V.32 TCM scheme and the BCM scheme.

small packets or where the channel is particularly prone to bursts of noise are both ideal candidates for the use of block codes. The results presented here show that it is possible to use BCM schemes to the same effect as TCM schemes with comparable complexity. Although it is easy to find BCM schemes with attractive asymptotic gains it is clear from the earlier theory in this paper that the number of nearest neighbours should be checked for any BCM scheme. It has also been noted that this figure is dependent on the implementation of the scheme.

Since the usefulness of coding techniques was first realized [15] there has been a large amount of research into the algorithmic decoding of block codes. Presently, interest is also being shown in decoding block codes using neural networks. As the processing power available to system designers increases and the methods of decoding block codes improve, BCM may become clearly more attractive than TCM. It has been shown here that the difficulties of implementing BCM in a voiceband modem are conquerable and it is expected that problems arising in other areas of application should also be readily solved. Keeping in touch with block coding may prove advantageous in the future.

REFERENCES

1. CCITT Recommendation V.32: 'A family of 2-wire, duplex modems operating at data signalling rates of up to 9600 bit/s for use on the general switched telephone network and on leased telephone-type circuits', (1984).

2. Ungerboeck G: 'Channel coding with multilevel/phase signals', IEEE Trans, IT—28 , pp 55-67 (January 1982).

3. Cusack E L: 'Error control codes for QAM signalling', Elec Letts, 20 , pp 62-63 (19th January 1984).

4. Sayegh S I: 'A class of optimum block codes in signal space', IEEE Trans, COM—34 , pp 1043-1045 (October 1986).

5. MacWilliams F J and Sloane N J A: 'The theory of error-correcting codes', North Holland Publishing Co (1978).

6. Imai H, and Hirakawa S: 'A new multilevel coding method using error-correcting codes', IEEE Trans, IT—23 , pp 371-377 (May 1977).

7. Lunn T J: 'Error rate bounds for block coded modulation using M-ary PSK', Internal BT memorandum (October 1990).

8. Forney G D Jr: 'Coset codes I: introduction and geometrical classification', IEEE Tans, IT—34 , pp 1123-1151 (September 1988).

9. Forney G D Jr: The Viterbi algorithm', Proc IEEE, 61 , pp 268-278 (March 1973).

10. Williams R G C: 'Low complexity block coded modulation', PhD Thesis, University of Manchester (August 1988).

11. Farrell P G, Williams R G C, Borelli W C and Lee L H C: 'Codulation techniques with block and convolutional codes', Invited paper presented at the Int Symp on Information and Coding Theory, Campinas, Brazil, (ISICT '87) (July 1987).

12. Wei L F: 'Trellis-coded modulation with multidimensional constellations', IEEE Trans, IT—33 , pp 483-501 (July 1987).

13. Wei L F: 'Rotationally invariant convolutional channel coding with expanded signal space — Part II: Nonlinear codes', IEEE, SAC-2 , pp 672-686 (September 1984).

14. Forney G D Jr: 'Coset codes II: binary lattices and related codes', IEEE Trans, IT—34 , pp 1152-1187 (September 1988).

15. Shannon C E: 'A mathematical theory of communication', Bell Sys Tech J, <u>27</u> , pp 379-424 and 623-657 (July and October 1948).

9

CORRELATIVE-LEVEL Mod(N) ENCODING AND NEAR-MAXIMUM-LIKELIHOOD DECODING

S F A Ip and A P Clark

9.1 INTRODUCTION

As the technology of programmable digital signal processing (DSP) devices advances (see Chapter 1), it eases the way to realize new algorithms and techniques rapidly. In the field of data transmission, as the demand to transmit data as quickly and with as little error as possible increases, recent systems for public telephone networks have all been designed and implemented with forward error correction schemes. In the process of selecting the most appropriate error correction scheme, trellis-coded and block-coded modulation (see Chapter 8) have been both studied and compared [1-20]. However, the trellis-coded modulation technique has generally been adopted in most international standards as a bandwidth-efficient way of processing digital data [1-19]. Trellis-coded modulation is achieved by appropriately combining a process of trellis (convolutional) encoding with the modulation of the signal in a digital data transmission system. A useful improvement in tolerance to additive white Gaussian noise of up to some 6 dB at high signal-to-noise ratio can be achieved over more conventional uncoded systems [1]. The redundant data introduced by the trellis code is carried by the additional possible values of the transmitted encoded symbols, so that the data-symbol rate is not changed. Since each transmitted signal element carries one data symbol, neither the signal-element rate nor the signal bandwidth are changed as a result of the encoding process. In a trellis-coded modulation system, a soft-decision maximum-likelihood decoding process is normally applied to the received symbols in order to exploit the full advantage in performance that can be achieved by the system. The decoding process is usually implemented by means of the Viterbi algorithm [21-23].

Since Ungerboeck's paper was first published in 1976 [1,8], much effort has been spent in seeking even better codes [2-7,9-19]. Further improvements in asymptotic coding gains have been found by not restricting the shape of the expanded signal set to being equally spaced [14-16], or by using a multidimensional signal set [17-19]. The Mod(N) [15] codes described in this chapter need an appropriate combination of correlative-level coding [24-27] and some modulo-arithmetic operations [28,29]. The results of the encoding process are basically similar to those obtained by using more conventional trellis-coded modulation techniques [1-16], but the encoded signal set is no longer predetermined and is set-partitioned [1,8] automatically. The possible values of a Mod(N) encoded symbol may now be unequally likely and unequally spaced. The number of possible values of the transmitted encoded symbols are no longer restricted to double that of the corresponding uncoded data symbols [8]. It has previously been reported that the Mod(N) encoding technique gives an advantage in tolerance to additive white Gaussian noise, over the corresponding uncoded system, of about 1 to 2 dB at a bit error rate of 10^{-4} [15]. The effect of precoding, similar to that used in the conventional correlative-level coding scheme, has been shown to reduce further the average number of bit errors per error burst [15]. This chapter presents the reader with a new class of trellis-coded modulation schemes and studies the optimum performance of this class of codes with up to 64 encoder states. This chapter also illustrates that some of Ungerboeck's codes [1,8] can be generated by combining correlative-level coding with modulo arithmetics; consequently, it gives a new description of trellis-coded modulations.

Codes with a longer memory (or larger number of encoder states) can potentially achieve a better tolerance to noise, at very high signal-to-noise ratios, than shorter codes. However, the corresponding optimum decoding process using the Viterbi algorithm becomes increasingly complex. This is because of the exponential growth in equipment complexity of the Viterbi decoder with the code memory. Thus, the advantage gained by using a long memory code diminishes. In recent years there have been developed a number of decoders that come close to achieving maximum-likelihood decoding, without, however, requiring nearly as much computation and storage as does the Viterbi decoder [29-38]. These are known as near-maximum-likelihood decoders. The decoders aim to achieve performances as close to that of the corresponding Viterbi decoder as possible, while operating with a fraction of the equipment complexity required by the latter. The near-maximum-likelihood decoders generally require less storage and a smaller number of operations per decoded symbol than the corresponding Viterbi decoder. Hence, the development of near-maximum-likelihood decoding processes has made possible the use of codes with long memories. This chapter then goes on to study how to combine a long memory encoder with a simple decoder in order to gain advantage in tolerance to noise over a more conventional system with a Viterbi decoder. Extensive computer simulation tests have been carried out to compare performances and trade-offs of various arrangements of the encoders and decoders.

Section 9.2 describes the model of the data-transmission system. Section 9.3 describes the quaternary correlative-level Mod(N) encoding scheme and studies a set of Mod(N) codes with 16 and 64 states. Section 9.4 then presents Ungerboeck's encoder (at the same information rate as the Mod(N) codes) using a description based on the correlative-level coding and modulo arithmetic. Section 9.5 describes a Viterbi-algorithm decoder and a near-maximum-likelihood decoder. Section 9.6 presents and discusses the results of the computer simulation tests.

9.2   MODEL OF THE DATA TRANSMISSION SYSTEM

The model of the encoded serial synchronous data-transmission system is shown in Fig. 9.1. The original binary symbol sequence (not shown in Fig. 9.1) is converted into the sequence of quaternary symbols $\{s_i\}$, using a Gray coding. The quaternary symbols, $\{s_i\}$ at the input to the encoder, are statistically independent and equally likely to have any one of their possible values 0,1,2 or 3. The sequence $\{s_i\}$ is then encoded and the encoded symbol sequence $\{q_i\}$ is produced at the input of the linear baseband channel.

The linear baseband channel may include linear modulation and demodulation processes together with a bandpass transmission path. The

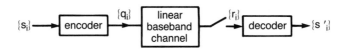

Fig. 9.1   The data transmission model for a quaternary encoded system.

channel introduces no distortion, attenuation or delay. Additive white Gaussian noise (AWGN) is introduced at the output of the transmission path such that, at time $t = iT$, the received symbol at the input of the receiver is given as:

$$r_i = q_i + w_i \qquad \qquad \text{... (9.1)}$$

where the $\{w_i\}$ are statistically independent Gaussian random variables with zero mean and variance σ^2.

The decoder operates on the received symbols $\{r_i\}$ to give the decoded data symbols $\{s'_i\}$. The decoder can either carry out a maximum-likelihood decoding process by means of the Viterbi algorithm or alternatively be a near-maximum-likelihood decoder. At the output of the receiver, the $\{s'_i\}$ are converted back to a binary symbol sequence using Gray decoding, which, in the absence of noise, is the same as the binary symbol sequence fed to the transmitter.

9.3 THE QUATERNARY CORRELATIVE-LEVEL Mod(N) ENCODER

The encoder in Fig. 9.1 is shown in detail in Fig. 9.2 and is the correlative-level Mod(N) encoder, known simply as the Mod(N) encoder. The signals shown in Fig. 9.2 are those occurring at time $t = iT$. The data symbols $\{s_i\}$ are 'precoded' to form the symbols $\{u_i\}$ and the latter are 'correlative-level encoded' to produce the symbols $\{p_i\}$. The $\{p_i\}$ are then processed by a non-linear network which includes the use of a modulo-arithmetic operation to give the encoded symbols $\{q_i\}$. This nonlinear network carries out a set of operations which can be defined by a function, Mod(.).

In this encoding method, the term 'mod-N' is short for 'modulo-N'. For instance in Fig. 9.2, (N = 4), the mod-4 operation for the symbol u_i is given by:

$$u_i = (s_i - v_i)_{\text{mod-4}} \qquad \qquad \text{... (9.2)}$$

where v_i and hence $(s_i - v_i)$ are integers. Now, u_i is the smallest non-negative integer that satisfies $(s_i - v_i) = u_i + 4j$, where j is any integer. Thus:

$$u_i = 0, 1, 2, \text{ or } 3 \qquad \qquad \text{... (9.3)}$$

The linear feedforward transversal filter has g taps with gains $\{y_h\}$, such that its output symbol, at time $t = iT$, is:

$$v_i = \sum_{h=1}^{h=g} u_{i-h} y_h \qquad \qquad \ldots (9.4)$$

where the $\{y_h\}$, for $h \geq 0$, are integers and form the $g+1$ components of the encoding vector ($y_0 = 1$) in Fig 9.2:

$$Y = [y_0 \, y_1 \, y_2 \ldots \ldots y_{g-1} y_g] \qquad \qquad \ldots (9.5)$$

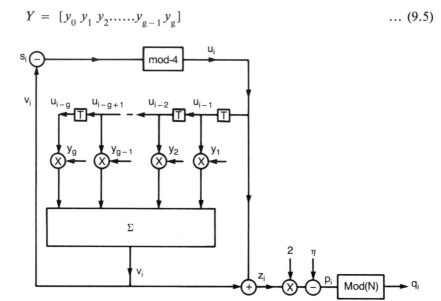

Fig. 9.2   The quaternary correlative-level Mod(N) encoder.

The quantities z_i and p_i are given as:

$$z_i = u_i + v_i \qquad \qquad \ldots (9.6)$$

and

$$p_i = 2z_i - \eta \qquad \qquad \ldots (9.7)$$

where η is the expected value of $2z_i$, so that p_i has zero mean and its possible values are integers and are spaced at intervals of 2.

The nonlinear network, defined by the function Mod(N), operates on the input symbol p_i to produce the symbol q_i, such that:

$$q_i = (p_i + N/2)_{\text{mod-}N} - N/2 \qquad \ldots (9.8)$$

or alternatively

$$q_i = p_i + jN \qquad \ldots (9.9)$$

where N need not be an integer and j is the appropriate integer that ensures:

$$-N/2 \leq q_i < N/2 \qquad \ldots (9.10)$$

Clearly, when $-N/2 \leq p_i < N/2$, $p_i = q_i$. In general, the different possible values of the encoded symbol, q_i, are not necessarily equally spaced or equally likely. Provided that $jN/2$ (where j is an odd integer) is not a possible value of p_i (a condition which must be ensured), the possible value of q_i is symmetrically spaced about zero and q_i has zero mean.

When no encoding is used, the components $\{y_h\}$ (equation (9.5)) with $h > 0$, are all set to zero, and $y_0 = 1$, so that $v_i = 0$, and the value of N in the function Mod(N) (equations (9.8) and (9.9)) is set to 8 or more. The output symbol q_i in Fig. 9.2 now becomes:

$$q_i = 2s_i - 3 \qquad \ldots (9.11)$$

where q_i is equally likely to have any of the possible values ± 1, ± 3. This uncoded system is used as a reference to compare the relative advantages in tolerance to AWGN of various correlative-level Mod(N) encoded data transmission systems.

9.3.1   Mod(N) codes with $g = 2$ or 3.

A search has been carried out to optimize the combination of Y (equation (9.5)) and N (equation (9.8)). However, due to the vast number of possible Mod(N) codes to be considered, the search is not exhaustive. Table 9.1 shows some Mod(N) codes with $g = 2$ or 3 (equation (9.5)). These codes have been found to give a good tolerance to noise at a bit error rate of 10^{-3}. The optimization was not carried out at the more typical bit error rate of, say, 10^{-4}, because the excessive computer time that would have been required. The possible values of q_i for these codes are shown in Table 9.1.

Table 9.1   Details of codes 1-6

Code	Encoding vector Y	N	Number of possible $\{q_i\}$	Possible values of the encoded symbol q_i
1	{1 2 − 1}	2.6	13	$0, \pm 0.2, \pm 0.4, \pm 0.6, \pm 0.8, \pm 1, \pm 1.2$
2	{1 2 − 1}	9.0	9	$0, \pm 1, \pm 2, \pm 3, \pm 4$
3	{1 3 − 1}	8.5	16	$\pm 0.5, \pm 1, \pm 1.5, \pm 2, \pm 2.5, \pm 3, \pm 3.5, \pm 4$
4	{1 5 − 1}	8.5	17	$0, \pm 0.5, \pm 1, \pm 1.5, \pm 2, \pm 2.5, \pm 3, \pm 3.5, \pm 4$
5	{1 1 − 5 1}	8.5	17	$0, \pm 0.5, \pm 1, \pm 1.5, \pm 2, \pm 2.5, \pm 3, \pm 3.5, \pm 4$
6	{1 13 2 − 1}	8.1	52	$\pm 0.4, \pm 0.5, \pm 0.6, \pm 0.7, \pm 0.8, \pm 0.9, \pm 1,$ $\pm 1.1, \pm 1.2, \pm 1.3, \pm 1.4, \pm 1.5, \pm 1.6,$ $\pm 2.4, \pm 2.5, \pm 2.6, \pm 2.7, \pm 2.8, \pm 2.9, \pm 3,$ $\pm 3.1, \pm 3.2, \pm 3.3, \pm 3.4, \pm 3.5, \pm 3.6$

When examining the state-transition diagram of a $g = 2$ Mod(N) code, as shown in Fig. 9.3, and assuming that precoding is employed, there are two symbols u_{i-2} and u_{i-1} held in the encoder's memory at time $t = iT$. The state of the encoder, at time $t = iT$, is S_i and is uniquely determined by the values of the symbols u_{i-2} and u_{i-1}. On the receipt of the symbol u_i, the encoder produces the encoded symbol q_i at its output. The whole encoding process can be modelled as a finite-state machine. In practice, the encoding process can be implemented by one look-up table with the state S_i, the input symbol u_i forming the pointer to it. The output of this look-up table is the symbol to be transmitted, q_i. The next state, S_{i+1}, can be easily calculated from S_i and u_i since $(u_{i-1} = (S_i)_{\text{mod-4}})$.

The vector Y (equation (9.5)) determines the state transition diagram in Fig. 9.3 and the Mod(N) function (equation (9.8)) determines the possible values of the transmitted symbols $\{q_i\}$. For a Mod(N) code with $g = 2$, the encoder has 16 distinct states corresponding to the 16 different possible combinations of the values of $\{u_{i-2}\}$ and $\{u_{i-1}\}$ in the encoder memory. There are four different state-transition branches (corresponding to the four different possible values of the input symbol u_i) diverging from each of the 16 possible encoder states. The four branches lead to four distinct states. Hence, in total, there are 64 distinct state-transitions and associated with each state transition is the corresponding value of the output symbol, $\{q_i\}$.

Figure 9.3 also shows the encoded symbol, q_i, associated with each state transition. The four branches that diverge from a given state S_i and are associated with four of the different possible values of the encoded symbol q_i. The values of these four encoded symbols are not necessarily equally spaced, contrary to Ungerboeck's codes [1,8]. For instance, as shown in Table 9.2, the code generated by $Y = [1\ 2\ -1]$ and $N = 9.0$ produces a range of possible values for q_i that are equally spaced. However the values from a particular state may not be equally spaced. From state $S_i = [00]$, the output symbols are $q_i = 3, -4, -2, 0$, when $u_i = 0, 1, 2, 3$, respectively.

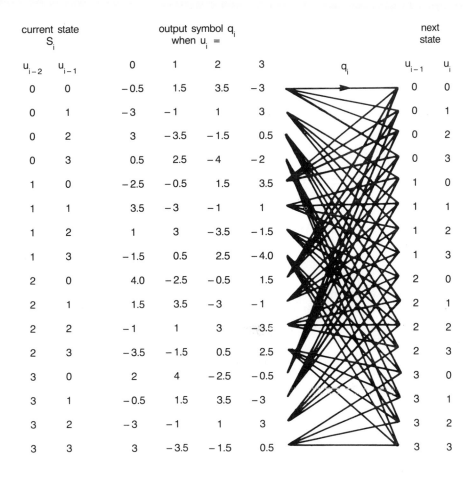

current state S_i		output symbol q_i when $u_i =$				q_i	next state	
u_{i-2}	u_{i-1}	0	1	2	3		u_{i-1}	u_i
0	0	−0.5	1.5	3.5	−3		0	0
0	1	−3	−1	1	3		0	1
0	2	3	−3.5	−1.5	0.5		0	2
0	3	0.5	2.5	−4	−2		0	3
1	0	−2.5	−0.5	1.5	3.5		1	0
1	1	3.5	−3	−1	1		1	1
1	2	1	3	−3.5	−1.5		1	2
1	3	−1.5	0.5	2.5	−4.0		1	3
2	0	4.0	−2.5	−0.5	1.5		2	0
2	1	1.5	3.5	−3	−1		2	1
2	2	−1	1	3	−3.5		2	2
2	3	−3.5	−1.5	0.5	2.5		2	3
3	0	2	4	−2.5	−0.5		3	0
3	1	−0.5	1.5	3.5	−3		3	1
3	2	−3	−1	1	3		3	2
3	3	3	−3.5	−1.5	0.5		3	3

Fig. 9.3 State-transition diagram for the correlative-level Mod(N) encoder with $Y = \{1\ 3\ -1\}$ and $N = 8.5$.

These four possible values of q_i are not equally spaced. Then, from state $S_i = [0\ 1]$, the encoded output symbols are $q_i = -2, 0, 2, 4$, when $u_i = 0, 1, 2, 3$, respectively. These four possible symbols are equally spaced.

In order to achieve a good minimum Euclidean distance between different transmitted sequences, the minimum difference between the four possible values of q_i, assigned to the four branches which diverge from, or converge into a given state, must be maximized. Thus the four values of q_i must be suitably interspersed between all the different possible values of q_i. In comparison with Mod(N) encoding, Ungerboeck's technique of signal

Table 9.2   Precoding and encoding truth tables for the correlative-level Mod(N) code with $Y = [1 \ 2 \ -1]$ and $N = 9.0$.

State S_i $\{u_{i-2}u_{i-1}\}$	Symbols: when $s_i =$ 0 1 2 3	p_i 0	p_i 1	p_i 2	p_i 3	q_i 0	q_i 1	q_i 2	q_i 3
0 0	0 1 2 3	-6	-4	-2	0	3	-4	-2	0
0 1	2 3 0 1	2	4	-2	0	2	4	-2	0
0 2	0 1 2 3	2	4	6	8	2	4	-3	-1
0 3	2 3 0 1	10	12	6	8	1	3	-3	-1
1 0	1 2 3 0	-6	-4	-2	-8	3	-4	-2	1
1 1	3 0 1 2	2	-4	-2	0	2	-4	-2	0
1 2	1 2 3 0	2	4	6	0	2	4	-3	0
1 3	3 0 1 2	10	4	6	8	1	4	-3	-1
2 0	2 3 0 1	-6	-4	-10	-8	3	-4	-1	1
2 1	0 1 2 3	-6	-4	-2	0	3	-4	-2	0
2 2	2 3 0 1	2	4	-2	0	2	4	-2	0
2 3	0 1 2 3	2	4	6	8	2	4	-3	-1
3 0	3 0 1 2	-6	-12	-10	-8	3	-3	-1	1
3 1	1 2 3 0	-6	-4	-2	-8	3	-4	-2	1
3 2	3 0 1 2	2	-4	-2	0	2	-4	-2	0
3 3	1 2 3 0	2	4	6	0	2	4	-3	0

mapping, by set-partitioning can be viewed as a special case of the Mod(N) function, applied to a set of equally spaced $\{q_i\}$ having twice as many possible values as the corresponding uncoded symbol $\{s_i\}$ [1,8].

9.4   APPLICATION OF MODULO ARITHMETICS TO THE DESIGN OF RATE 2/3 8AM/4AM CODE

The use of modulo-arithmetic gives a new description of trellis-coded modulations. This new description provides a flexible way to generate a trellis coding structure with the resultant encoded symbols automatically mapped to maximize the coding gain. Hence the new description presents an alternative means to carry out code searches. In order to gain further insight, this section examines a special case of the Mod(N) codes which are similar to Ungerboeck's rates 1/2 and 2/3 codes with amplitude modulated (AM) signals. This description of Ungerboeck's code using correlative-level coding and modulo arithmetics offers no coding advantage but to illustrate the flexibility of the Mod(N) approach. Performances of the rate 2/3 8AM/4AM codes will be used later as comparisons for the Mod(N) codes in section 9.6.

Firstly, let us consider a simple example using a binary Mod(N) encoding which is equivalent to Ungerboeck's 4AM/2AM code [1,8], assuming that $\{a_i\}$ are statistically independent binary symbols which are equally likely to have the possible values 0 or 1. The symbols $\{a_i\}$ are fed into an encoder which employs the equations (9.4)-(9.10) (with $\{a_i\}$ instead of $\{u_i\}$ when no precoding is employed). For an encoding vector $Y = [2 \ 7 \ 2]$ (equation (9.5)) and $N = 8.0$ (equation (9.8)), the resultant encoded symbols $\{q_i\}$ are as shown in Table 9.3. This code is equivalent to Ungerboeck's 4-state 4AM/2AM code with the encoder shown in Fig. 9.4 and the encoded symbols shown in the last column of Table 9.3 for appropriate mapping between $\{v_{i,0}, v_{i,1}\}$ and $\{q_i\}$. This example illustrates that Mod(N) encoding can generate the encoded symbols and performs Ungerboeck's set-partitioning automatically to maximize the coding gain for a given encoding vector.

Another approach to make use of the correlative-level coding and modulo arithmetic is to apply only equations (9.4)-(9.6) to $\{a_i\}$. The symbols $\{z_i\}$ (equation (9.6)) are then fed into a mod-4 operator (equation (9.2)) to produce the quaternary symbols $\{d_i\}$ which have the possible values 0,1,2 or 3. The symbols $\{d_i\}$ are then mapped on to the 4AM signal-set $\{q_i\}$ appropriately as shown in Table 9.3. This approach, of course, foregoes the advantage of Mod(N) which is to generate the encoded 4AM signal-set automatically. Table 9.3 summarizes the three encoding approaches.

Table 9.3 Comparisons between the Mod(N) code ($N = 8$), the binary mod-4 code and Ungerboeck's 4AM/2AM code.

Symbols			Binary Mod(N) code with $N = 8, Y = [2\ 7\ 2]$	Binary mod-4 code with $Y = [2\ 1\ 2]$	Ungerboeck's 4AM/2AM code with code generator $G(D) = [D\ \ 1 + D^2]$			
a_{i-2}	a_{i-1}	a_i	q_i	d_i $q_i = f(d_i)$	$v_{i,0}$	$v_{i,1}$	d_i	q_i
0	0	0	-3	0 $\quad -3$	0	0	0	-3
0	0	1	1	2 $\quad$ 1	0	1	2	1
0	1	0	3	1 $\quad$ 3	1	0	1	3
0	1	1	-1	3 $\quad -1$	1	1	3	-1
1	0	0	1	2 $\quad$ 1	0	1	2	1
1	0	1	-3	0 $\quad -3$	0	0	0	-3
1	1	0	-1	3 $\quad -1$	1	1	3	-1
1	1	1	3	1 $\quad$ 3	1	0	1	3

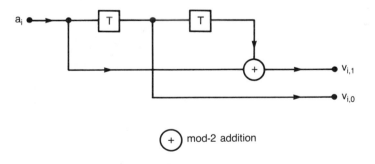

Fig. 9.4 A conventional feedforward 4-state convolutional encoder.

In order to extend the approach to rate 2/3 (same information rate as the Mod(N) studied in section 9.3), the encoded signal-set studied here is a 8AM signal as shown in Fig. 9.5. The encoder is shown in Fig. 9.6 and has at its input two information bits per symbol interval. The two bits, at time $t = iT$, are designated as, $a_{i,1}$ and $a_{i,2}$. The bits $\{a_{i,1}\}$ and $\{a_{i,2}\}$ are statistically independent and equally likely to have either values 0 or 1. The bit, $a_{i,2}$, is fed directly into an 8AM signal mapper without any encoding and is therefore called the 'uncoded bit'. Meanwhile, the other bit $\{a_{i,1}\}$ is being encoded using correlative-level coding (equations (9.4)-(9.6)) and modulo-4 arithmetic (equation 9.2)) to produce the quaternary encoded symbols $\{d_i\}$. The encoded symbol, d_i, is then fed into the 8AM signal mapper (Fig. 9.5).

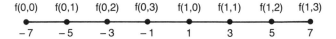

$$f(0,0) \quad f(0,1) \quad f(0,2) \quad f(0,3) \quad f(1,0) \quad f(1,1) \quad f(1,2) \quad f(1,3)$$

$$-7 \quad\quad -5 \quad\quad -3 \quad\quad -1 \quad\quad 1 \quad\quad 3 \quad\quad 5 \quad\quad 7$$

Fig. 9.5   The 8AM signal mapping function $f(a_{i,2}, d_i)$.

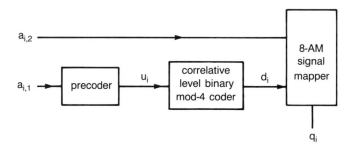

Fig. 9.6   The precoded binary correlative-level modulo-4 8AM/4AM encoder.

At the output of the signal mapper is the symbol $q_i = f(a_{i,2}, d_i)$ where $f(a_{i,2}, d_i)$ is the mapping function which uniquely determines the encoded symbol q_i, the value of the symbol q_i depending on the values of the encoded symbol d_i and the uncoded bit $a_{i,2}$. Fig. 9.5 shows a one to one relationship between the signal mapping function, $f(a_{i,2}, d_i)$ and the corresponding encoded 8AM signals, $\{q_i\}$. The effect of having the uncoded bit results in parallel transitions on the state transition diagram [1,8]. Using Ungerboeck's mapping by set-partitioning process for 8AM, the distance (difference) between the two encoded symbols $\{q_i\}$ that are associated with the same symbol d_i but different $a_{i,2}$ must have the maximum possible distance. Next, the $\{q_i\}$ that are associated with $d_i = 0,2$ or $d_i = 1,3$ are then mapped with the possible encoded signals separated by the next maximum possible distance.

Thus, the distance between two encoded symbols, $\{q_i\}$, that are associated with the same symbol d_i sets the limit for the maximum achievable asymptotic coding gain of 5.8 dB over the quaternary uncoded system. Table 9.4 provides a list of the encoding vectors and asymptotic coding gains for a set of correlative-level mod-4 codes.

At the receiver, at time $t = iT$, the decoded bit $a'_{i,1}$ is obtained using the Viterbi-algorithm (VA) (section 9.5) decoder, for a given value of the decoded bit $a'_{i,2}$, which is obtained by the threshold detector. In order to reduce the effect of error burst, a simple precoding rule to generate $\{u_i\}$ (equation (9.2)) is introduced here such that when $\{a_{i,1}\} = 0$, $d_i = 0$ or 1, and when $\{a_{i,1}\} = 1$, $d_i = 2$ or 3. As shown in Fig. 9.6, when precoding is applied, $\{u_i\}$ is encoded instead of $\{a_{i,1}\}$ using correlative-level encoding and mod-4 arithmetic. As $\{a_{i,1}\}$ can be uniquely determined by $\{d_i\}$, this means that the precoding rule is effectively the same as equation (9.2) in Mod(N) encoding. It is also found [29] that the combined precoding and correlative-level modulo-arithmetic encoding generates the same code as Ungerboeck's feedback systematic code [1,8].

Table 9.4 Asymptotic coding gains of some 8AM/4AM codes using correlative-level modulo-4 encoding over the quaternary uncoded system.

Coding memory $g(bits)$	Coding vector $Y = [y_0 \, y_1 ... y_g]$	Asymptotic coding gains (dB)
2	[2 1 2]	3.3
3	[2 2 1 2]	3.8
4	[2 2 1 0 2]	4.2
5	[2 1 2 2 0 2]	5.0
6	[2 2 0 2 0 1 2]	5.2

9.5 DECODING PROCESSES

The decoder shown in Fig. 9.1 can perform a maximum-likelihood or a near-maximum-likelihood decoding process. The decoder operates directly on the sequence of received symbols, $\{r_i\}$, to produce the decoded sequence, $\{u'_i\}$, at the output of the decoder, where u'_i is the decoded value of u_i. In the absence of noise, $u'_i = u_i$ for all i. The sequence $\{u'_i\}$ can then be converted to the corresponding sequence $\{s'_i\}$, where s'_i is the decoded value of s_i. The conversion can be achieved based on the fact that the sequence $\{s_i\}$ is uniquely determined by the sequence $\{p_i\}, \{z_i\}$ or $\{u_i\}$ (see Table 9.2 and equation (9.7)). The sequence of decoded binary symbols can then be determined from the sequence $\{s'_i\}$. Theoretically, the final choice of each decoded symbol may not be reached until the entire message $\{r_j\}$ has been received. But this usually means an unacceptably long delay in decoding. In practice, a delay of n symbol intervals is introduced into the decoding process, where n is sufficiently large to avoid any significant increase in decoding errors due to early detection. The following sections describe a Viterbi-algorithm decoder and also a near-maximum-likelihood decoder.

9.5.1   Viterbi-algorithm decoder

The Viterbi-algorithm (VA) decoder operates on the $\{u'_i\}$, in a manner now to be described. Just prior to the receipt of r_i, the VA decoder holds in its store 4^g different n-component vectors $\{X_{i-1}\}$, where

$$X_{i-1} = [x_{i-n}\; x_{i-n+1} \cdots\cdots x_{i-1}] \qquad \qquad \ldots (9.12)$$

where g is the number of symbols $\{u_i\}$ in the memory of the encoder. The component x_{i-j} takes on one of the four possible values of u_{i-j} (equation (9.3)). Clearly, a vector X_{i-1} is formed by the last n-components of the corresponding $(i-1)$-component sequence:

$$[x_1\; x_2 \cdots\cdots x_{i-1}] \qquad \qquad \ldots (9.13)$$

which is a possible value of the data sequence:

$$[u_1\; u_2 \cdots\cdots u_{i-1}] \qquad \qquad \ldots (9.14)$$

The method of selecting the stored vectors differs from one system to another. In order to select any one of the stored vectors, a measure of merit must be used to differentiate them. In this case, the squared Euclidean distance between the vector and actual received symbols is used. The larger the squared Euclidean distance, known as the 'cost', the less likely is the sequence to be correct.

Associated with each vector X_{i-1} is the cost:

$$C_{i-1} = \sum_{j=1}^{j=i-1} |r_j - f_j|^2 \qquad \qquad \ldots (9.15)$$

where f_j is a possible value of the encoded symbol, q_j. The symbol f_j is determined by encoding the sequence $\{x_j\}$, of which X_{i-1} forms the last $n-1$ components, into the corresponding sequence $\{f_j\}$, according to the encoding process employed at the transmitter.

The last g components:

$$[x_{i-g} \; x_{i-g+1} \; \cdots\cdots\cdots \; x_{i-1}] \qquad\qquad \ldots (9.16)$$

of a stored vector X_{i-1} uniquely correspond to a possible value of the state of the encoder at the transmitter at time $t = iT$. Hence, the 4^g stored vectors, $\{X_{i-1}\}$, possess the 4^g different possible states, $\{S_i\}$, of the encoder. The VA decoder works in such a way that each of the 4^g vectors, $\{X_{i-1}\}$, has the smallest cost for its corresponding one of the 4^g possible states of the encoder at time $t = iT$. Thus, for each of the 4^g possible encoder states, the decoder holds the vector with the minimum cost. It can be shown that the vector with the smallest cost is formed by the last n components of the required maximum-likelihood sequence of symbols $\{u'_i\}$.

On the receipt of the symbol, r_i, each stored vector, X_{i-1}, is expanded to form four $(n+1)$-component vectors $\{P_i\}$, where:

$$P_i = [x_{i-n} \; x_{i-n+1} \; \cdots\cdots\cdots \; x_{i-1} \; x_i] \qquad\qquad \ldots (9.17)$$

with the last component x_i taking on the four possible values of u_i, and the other components being as in the original vector X_{i-1}. The cost of each vector P_i is taken to be:

$$C_i = C_{i-1} + c_i \qquad\qquad \ldots (9.18)$$

where

$$c_i = |r_i - f_i|^2 \qquad\qquad \ldots (9.19)$$

and the $\{x_{i-j}\}$, from which f_i is determined, are given by the corresponding vector P_i. There are now 4^{g+1} expanded vectors $\{P_i\}$, together with their costs $\{C_i\}$. The decoder now selects the vector P_i with the smallest cost, for each of the 4^g different possible combinations of the last g components:

$$[x_{i-g+1} \; x_{i-g+2} \; \cdots\cdots\cdots \; x_i] \qquad\qquad \ldots (9.20)$$

The resulting 4^g vectors $\{P_i\}$ correspond to the 4^g different encoder states $\{S_{i+1}\}$. The decoder then selects the P_i with the smallest cost, and this vector is taken as the true maximum-likelihood vector at time $t = iT$. The decoder takes the first component x_{i-n} of this vector as the decoded value u'_{i-n} of u_{i-n}. The first components of all 4^g vectors $\{P_i\}$ are next omitted, without changing their costs, to give the corresponding set of 4^g n-component vectors $\{X_i\}$. Finally, the 4^g vectors $\{X_i\}$, corresponding to the 4^g possible encoder states S_{i+1}, are stored with their associated costs $\{C_i\}$. In order to avoid overloading the stores, the smallest of these 4^g costs is subtracted from each C_i. The decoder is now ready for the next received symbol.

The VA decoder is very easy to implement on a DSP device and the program space needed will be small. However, problems may be encountered as g increases. For the decoding of each data symbol, the VA decoder involves the evaluation of 4^{g+1} costs, followed by the comparison of four costs for each of the 4^g stored vectors. It also requires storage for the 4^g n-component vectors. The VA decoder achieves the maximum-likelihood detection of the received encoded symbol, under the assumed conditions, but for codes with a long memory ($g >> 2$), the decoder requires a large number of stored vectors. Since the number of stored vectors increase exponentially with g, both the amount of storage required and the number of operations per decoded data symbol become unacceptably large as g increases. This, therefore, prevents the use of the VA decoder with long memory codes.

9.5.2   Near-maximum-likelihood decoder

The decoding process studied here is adapted from corresponding detection processes developed for the detection of received signals that are distorted by intersymbol interference and AWGN in telephone channels [30-35]. These detection processes have been shown to give a very considerable reduction both in the amount of storage required and in the number of operations per received symbol, but with no serious reduction in tolerance to noise relative to a VA detector [30-35]. One of the simplest detectors is used here for the decoding of the correlative-level Mod(N) nodes. The corresponding VA decoders for codes, with memories of 2, 3, 4, ... quaternary symbols, require 16, 64, 256, ... stored vectors, respectively. The near-maximum-likelihood (NML) decoder here operates with m stored vectors where $m << 4^g$.

Just prior to the receipt of the r_i, the decoder holds in its store m different vectors, $\{X_{i-1}\}$ (equation (9.12)). Associated with each vector $\{X_{i-1}\}$ is the cost C_{i-1} (equation (9.15)). On the receipt of the symbol r_i, at the time $t = iT$, each of the m stored vectors $\{X_{i-1}\}$ is expanded to form k vectors $\{P_i\}$ (equation (9.17)), where $k \leq 4$. The first n components of each of the k vectors, $\{P_i\}$ derived from any one vector X_{i-1} are the same as in the original X_{i-1}, and the last component x_i takes on k of the four possible values of u_i for which the cost of the P_i is smallest. Since the k vectors $\{P_i\}$ derived from any particular X_{i-1} differ only in the values of x_i, they have the same value of C_{i-1} but different values of c_i (equation (9.19)), so that the k $\{P_i\}$ with the smallest $\{C_i\}$ also have the smallest c_i. The k $\{x_i\}$ for which the $\{c_i\}$ are the smallest, and hence $\{|r_i - f_i|\}$ are the smallest, are therefore stored. The cost, C_i, of each of the k $\{P_i\}$ is then evaluated.

The decoder then selects the vector P_i with the smallest cost C_i. The decoder takes the first component x_{i-n} of the vector as the decoded value of u_{i-n}, which is designated as u'_{i-n}. Next, all vectors $\{P_i\}$ for which $x_{i-n} \neq u'_{i-n}$ are discarded. This is called anti-merging technique, and it prevents any stored vector and its cost from becoming the same as another vector and its cost, which reduces the effective number of stored vectors. The first components of all vectors $\{P_i\}$ are next omitted, without changing their costs, to give the corresponding set of n-component vectors $\{X_i\}$. The decoder next selects $m - 1$ vectors $\{X_i\}$ with the smallest costs, $\{C_i\}$, from the remaining (non-selected) vectors $\{X_i\}$, to give a total of m vectors, which are now stored with their costs. In order to avoid overloading the stores, the smallest of these m costs is subtracted from each C_i. The decoder is now ready for the next received symbol.

In the NML decoder, a proper start-up procedure is required to prevent some of the stored vectors and their costs from becoming the same (or merging). The system is initialized by allocating zero cost to one stored vector and very high costs to the remaining $m - 1$ vectors. This causes all the subsequent stored vectors to originate from the original vector with zero cost. Then, provided that the anti-merging technique is used at every selection process, the decoder is constrained to operate with m distinct vectors.

For m stored vectors, the decoder evaluates km costs per decoded symbol and carries out m searches through these km costs. However, for the same number of stored vectors, the corresponding VA decoder requires m searches but with each search being carried out with the comparison of only four costs. Thus, for a given number of searches through the costs, or else a given number of cost comparisons, the NML decoder must employ fewer stored vectors than the corresponding VA decoder.

9.6 COMPUTER SIMULATION TESTS

The Mod(N) codes listed in Table 9.1 have been tested by computer simulation to measure the variation of bit error rate (BER) with signal-to-noise ratio (SNR). The BER is that in the decoded binary sequence derived from the $\{s_i'\}$. The SNR is taken as ψ dB, where:

$$\psi = 10 \log_{10}(\overline{q_i^2}/2\sigma^2) \qquad \ldots (9.21)$$

In equation (9.21), $\overline{q_i^2}$ is the mean-square-value of q_i and σ^2 is the variance of w_i (equation (9.1)). $\overline{q_i^2}/2$ is the average transmitted signal power per bit.

The performance of the Mod(N) codes 1-4, with the VA decoder using $n = 34$, is shown in Fig. 9.7. This shows that, at BER $= 10^{-4}$, code 2 with the fewest possible $\{q_i\}$ achieves an advantage of 2.2 dB in tolerance to noise over the uncoded system. The best encoded system, where the possible values of q_i are equally spaced, is code 4. At BER $= 10^{-4}$, code 4 has an advantage in tolerance to noise of 2.5 dB over the quaternary uncoded system. The best of all the $g = 2$ encoded systems studied here is code 3 which has an advantage of 2.6 dB in tolerance to noise over the uncoded system at BER $= 10^{-4}$. However, the possible values of q_i of code 3 are no longer equally spaced. It has also been observed that, for the Mod(N) codes tested, each wrongly decoded symbol s_i is accompanied by only one bit error. Thus the bit error is half of the corresponding symbol error rate. This is due to the Gray coding employed here.

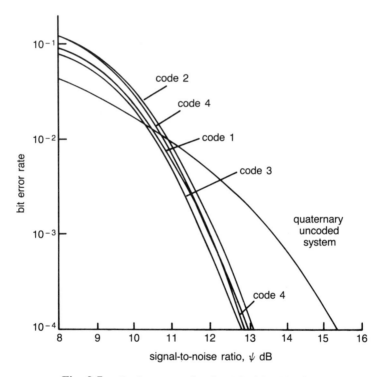

Fig. 9.7   Performance of codes 1-4 with a VA decoder.

The performance of the NML decoder, with $g = 2$ and $n = 34$ and for the Mod(N) codes 1-4, is shown in Figs. 9.8 and 9.9. It is not surprising to find that the NML decoder with $m = 16$ has the same performance as the VA decoder at BER $< 10^{-4}$. For code 3, the NML decoder, with $m = 8$ and at

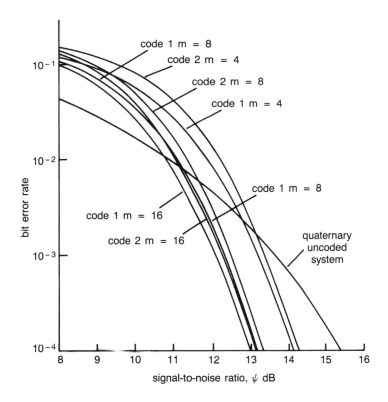

bit error rate

signal-to-noise ratio, ψ dB

Fig. 9.8   Performance of codes 1 and 2 with an NML decoder and $k = 4$.

BER $= 10^{-4}$, can achieve an advantage in tolerance to noise of up to about 2.4 dB over that of the uncoded system. For codes 1-4, the NML decoder, with $m = 8$ and at BER $= 10^{-4}$, is only about 0.2 dB inferior to the VA decoder. With $m = 4$ and at BER $= 10^{-4}$, code 3 gains an advantage of about 1.4 dB over the uncoded system. For codes 1-4, the NML decoder, with $m = 4$ and BER $= 10^{-4}$, is about 1.2 dB inferior to the VA decoder. However, with $m = 4$, the NML decoders offer a substantial reduction in complexity against the trade-off in tolerance to noise.

The degradation in tolerance to noise of the NML decoder relative to the VA decoder is mainly due to the longer error burst lengths that occur with the former, and the effect of longer error bursts on the BER increases with the BER. When $m \ll 4^g$ and $k = 4$, the degradation of the performance of the NML decoder relative to the VA decoder can be severe, even at low BER. However, at low BER, the NML decoder, with $m = 8$ and $k = 4$,

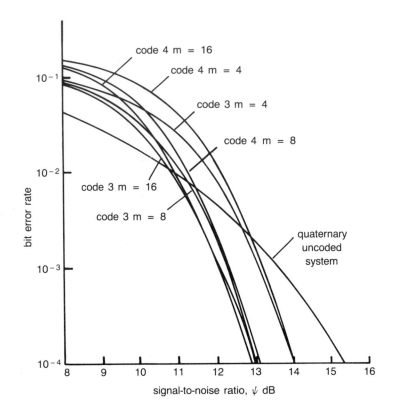

Fig. 9.9   Performance of codes 3 and 4 with an NML decoder and k = 4.

can achieve nearly the same performance as the VA decoder, with half the amount of the storage and half the number of costs to evaluate.

In the NML decoder with $k = 2$ and m stored vectors, $2m$ costs need to be evaluated and searched through, per decoded symbol. Hence, for the same value of m, the decoder needs only half of the computational complexity relative to the NML decoder with $k = 4$. In other words, for a given degree of computational complexity, the NML decoder with $k = 2$ can operate with nearly twice the number of stored vectors as the NML decoder with $k = 4$.

Figure 9.10 shows the performance of Mod(N) codes 2-3 with an NML decoder $k = 2$ and $n = 34$. The performances of the decoder with $m = 8$ and 16 are compared in Fig. 9.10. Generally speaking, for a given m, the NML decoder with $k = 2$ has the same tolerance to noise as the corresponding NML decoder with $k = 4$. This means that the two-way expansion of a stored vector in an NML decoder causes only a negligible degradation in performance relative to the corresponding four-way expansion. Hence, for codes 1-4, a

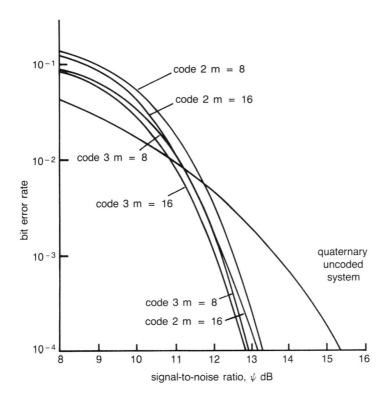

Fig. 9.10   Performance of codes 2 and 3 with an NML decoder and k = 2.

cost-effective arrangement would be to use the NML decoder with $m = 8$ and $k = 2$, the performance of which compares well with the VA decoder for half the amount of storage and a quarter the number of evaluated costs.

Figure 9.11 shows the performances of Mod(N) codes with $g = 3$, an NML decoder, $k = 4$, $m = 16$ and $n = 67$. It can be seen that the code 6 has an advantage in tolerance to noise of about 2.6 dB over the uncoded system. At BER $= 10^{-4}$, code 5 has roughly the same tolerance to noise as the best of codes 1-4. In the case of codes 5 and 6, only one bit error has been found for each symbol error, as for codes 1-4. Figure 9.11 also shows the performances of the Mod(N) codes with $g = 3$, an NML decoder, $k = 2$, $m = 32$ and $n = 67$. The number of stored vectors used here is half that required by the corresponding VA decoder which has, however, not been tested. Advantages in tolerance to noise of about 1.8 and 2.9 dB are achieved over the uncoded system at BER $= 10^{-3}$ and 10^{-4}, respectively. For equivalent complexities and at BER $= 10^{-4}$, the NML decoder with $k = 2$ and $m = 32$

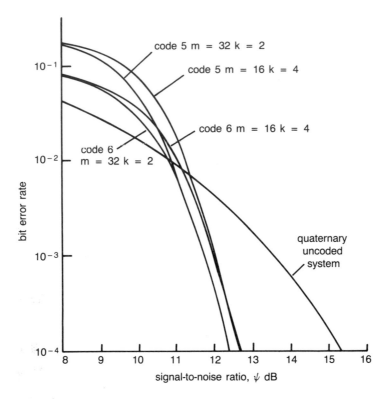

Fig. 9.11   Performance of codes 5 and 6, with an NML decoder.

has an advantage in tolerance to noise of about 0.25 dB over the NML decoder with $k = 4$ and $m = 16$. Codes 5 and 6 with an NML decoder, with $m = 32$ and $k = 2$, have roughly the same computational complexity as the VA decoder for codes 1-4. However, using this arrangement, codes 5 and 6 actually have advantages of 0.4-0.7 dB in tolerance to noise over codes 1-4 at BER $= 10^{-4}$. In other words, codes 5 and 6 with an NML decoder can achieve better tolerance to noise than codes 1-4, with VA decoding, without increasing the computational complexity. This, of course, precludes the fact that codes 1-4 can also be implemented with an NML decoder to reduce the overall system complexity.

In order to compare codes 5 and 6 with the conventional Ungerboeck's 8AM/4AM type code with precoding, Fig. 9.12 shows the relative tolerance to noise of the codes listed in Table 9.4 up to $g = 5$ using VA decoder. The 8AM/4AM code has an advantage in tolerance to noise of 2.2, 2.5, 2.7 and 3 dB for $g = 2, 3, 4, 5$, respectively, at BER $= 10^{-4}$. For Mod(N) codes 1-4,

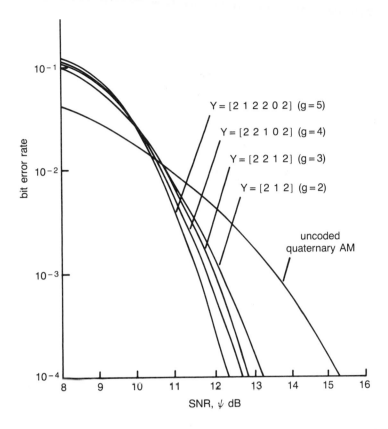

bit error rate

$Y = [2\ 1\ 2\ 2\ 0\ 2]\ (g=5)$

$Y = [2\ 2\ 1\ 0\ 2]\ (g=4)$

$Y = [2\ 2\ 1\ 2]\ (g=3)$

$Y = [2\ 1\ 2]\ (g=2)$

uncoded
quaternary AM

SNR, ψ dB

Fig. 9.12   Performances of precoded correlative-level mod-4 8AM encoders with $g = 2,3,4,5$.

for a given degree of computational complexity in terms of the numbers of stored vectors employed, the general performance in terms of relative tolerance to noise at $BER = 10^{-4}$ is inferior to the 8AM/4AM type code. However, for codes 5 and 6 which employs an NML decoder with $m = 32$ and $k = 2$, the performance is similar to the $g = 5$ 8AM/4AM code with an $m = 32$ VA decoder. Although asymptotically the 8AM/4AM codes would perform better, codes 5 and 6 have better tolerance to noise than Ungerboeck's $g = 5$ 8AM/4AM code at high BERs between 10^{-1} and 10^{-2} (or low SNR).

9.7   CONCLUSIONS

The chapter describes a class of trellis-coded modulations that is much wider than the class of conventional trellis codes and includes the latter as a special case or subset. In particular, the possible values of an encoded and transmitted symbol are automatically generated and set-partitioned, and no longer constrained to be equally spaced or integers. By employing an NML decoder in place of the conventional Viterbi decoder at the receiver, a good compromise between performance and complexity can be achieved. The use of an NML decoder enables a longer code memory to be employed, thus improving the coding gain but without unduly increasing the equipment complexity. This approach has been applied to two 64-state correlative-level Mod(N) codes and have performances comparable, under the same computational complexity, to some of the conventional Ungerboeck's 8AM/4AM codes using a Viterbi decoder. The combination of NML decoder and long trellis codes may prove to be a cost-effective way to design and implement some of the recent multidimensional trellis-coded modulation systems [17-19].

REFERENCES

1. Ungerboeck G and Csajka I: 'On improving data-link performance by increasing channel alphabet and introducing sequence coding', IEEE Int Symp on Information Theory, Ronneby, Sweden (June 1976).

2. Anderson J B and Taylor D P: 'A bandwidth-efficient class of signal space codes', IEEE Trans, IT—24 , pp 703-712 (1978).

3. Rhodes S A and Lebowitz S H: 'Performance of coded OPSK for TDMA satellite communications', Fifth Int Conf on Digital Satellite Communications, Genoa, Italy (March 1981).

4. Muilwijk D: 'Correlative phase shift keying — a class of constant envelope modulation techniques', IEEE Trans, COM—29 , No 3, pp 226-236 (1981).

5. Borelli W C, Rashvand H F and Farrell P G: 'Convolutional codes for multi-level modems', Electron Lett, 17 , No 9, pp 331-333 (1981).

6. Lebowitz S H and Rhodes S A: 'Performance of coded 8-PSK signalling for satellite communications', Int Conf on Commun Rec, Denver, CO, USA, pp 47.4/1-8 (June 1981).

7. Hui J and Fang R J F: 'Convolutional code and signal waveform design for band-limited satellite channels', Int Conf on Commun Rec, Denver, CO, USA, pp 47.5/1-10 (June 1981).

8. Ungerboeck G: 'Channel coding with multi-level/signals', IEEE Trans, IT—28 , pp 55-67 (1982).

9. Aulin T and Sundberg C C: 'On the minimum Euclidean distance for a class of signal space codes', IEEE Trans, IT—28 , pp 43-55 (1982).

10. Rhodes S A, Fang R J and Chang P Y: 'Coded octal phase shift keying on TDMA satellite communications', COMSAT Tech Rev, 13 , pp 221-258 (1983).

11. Calderbank A R, Mazo J E and Shapiro H M: 'Upper bounds on the minimum distance of trellis codes', Bell Syst Tech J, 26, No 8, pp 2617-2646 (October 1983).

12. Rhodes S A and Chang P Y: 'Coded M-ary modulation for FDMA satellite communications', IEEE Global Telecommun Conf Rec, Atlanta, Georgia, pp 726-734 (November 1984).

13. Wei L F: 'Rotationally invariant convolutional channel coding with expanded signal space: part I and part II', IEEE J on Sel Areas in Commun, SAC—2, pp 659-685 (September 1984).

14. Calderbank A R and Mazo J E: 'A new description of trellis codes', IEEE Trans, IT—30, No 6, pp 784-791 (1984).

15. Clark A P and Ip S F A: 'Modulo-m correlative level coding of baseband signals', Int J Circuit Theory and Applications', 13, pp 189-194 (1985).

16. Divsalsar D, Simon M K and Yuen J H: 'Trellis coding with asymmetric modulations', IEEE Trans, COM—35, No 2, pp 130-141 (1987).

17. Wei L F: 'Trellis-coded modulation with multidimensional constellations', IEEE Trans, IT—33, pp 483-501 (July 1987).

18. Ungerboeck G: 'Trellis-coded modulation with redundant signal sets. Part I: Introduction', IEEE Comms Mag, 25, pp 5-11 (February 1987).

19. Ungerboeck G: 'Trellis-coded modulation with redundant signal sets. Part II: state-of-the-art', IEEE Comms Mag, 25, pp 12-21 (February 1987).

20. Solomon G and van Tiboung H C A: 'A connection between block and convolutional codes', SIAM J Appl Math, 37, No 2, pp 358-369 (October 1979).

21. Viterbi A J: 'Error bounds for convolutional codes and an asymptotically optimum decoding algorithm', IEEE Trans, IT—13, No 2, pp 260-269 (1967).

22. Omura J K: 'On the Viterbi decoding algorithm', IEEE Trans, IT—15, pp 117-179 (1969).

23. Forney Jr G D: 'The Viterbi algorithm', Proc IEEE, 61, pp 268-278 (1973).

24. Clark A P: 'Principle of digital data transmission', 2nd Edition, Pentech Press (1983).

25. Lender A: 'Correlative digital communication techniques', IEEE Trans, COM—12, pp 128-135 (1964).

26. Lender A: 'Correlative level coding for binary transmission', IEEE Spectrum, 3, pp 104-115 (1966).

27. Kobayashi H: 'Correlative level coding and maximum-likelihood decoding', IEEE Trans, IT—17, pp 586-594 (1971).

28. Clark A P and Serinken M N: 'Nonlinear equaliser using modulo arithmetic', Proc IEE, 123, No 1, pp 32-38 (1976).

29. Ip S F A: 'Trellis-coded modulation for voiceband data modems', PhD Thesis, Dept of Electronic and Electrical Eng, Loughborough University of Tech (1988).

30. Clark A P, Harvey J D and Driscoll J P: 'Near-maximum-likelihood detection processes for distorted digital signals', Radio & Electron Eng, 48 , pp 301-309 (1978).

31. Acampora A S: 'Analysis of maximum likelihood sequence estimation performance for quadrature amplitude modulation', Bell Syst Tech J, 60 , No 8, pp 865-885 (October 1981).

32. Clark A P and Fairfield M J: 'Detection processes for a 9600 bit/s modem', Radio & Electron Eng, 51 , pp 455-465 (1981).

33. Clark A P, Ip S F A and Soon C W: 'Pseudobinary detection processes for a 9600 bit/s modem', IEE Proc F, Commun, Radar & Signal Process, 129 , No 5, pp 305-314 (1982).

34. Clark A P and Clayden M: 'Pseudobinary Viterbi detector', IEE Proc F, Commun, Radar & Signal Process, 131 , No 2, pp 208-218 (1984).

35. Clark A P, Abdullah S N, Jayasinghe S G and Sun K H: 'Pseudobinary and pseudoquaternary detection processes for linearly distorted multilevel QAM signals', IEEE Trans, COM—33 , pp 639-645 (1985).

36. Clark A P: 'Minimum distance decoding of binary convolutionary codes', Computer and Digital Techniques, 1 , pp 190-196 (1978).

37. Aftelak S B and Clark A P: 'Adaptive reduced-state Viterbi algorithm detector', J of IERE, 56 , No 5, pp 197-206 (1986).

38. Zhu Z C and Clark A P: 'Near-maximum-likelihood decoding of convolutional codes', IEE Proc F, Commun, Radar & Signal Process, 135 , No 1, pp 33-42 (1988).

10

A REVIEW OF DIGITAL SIGNAL PROCESSING TECHNIQUES IN SPEECH ANALYSIS

A Lowry and A P Breen

10.1 INTRODUCTION

The analysis of acoustic speech signals using digital signal processing (DSP) techniques has been a rich area of research for several decades. Progress in this field requires expertise in modelling the processes of speech production and perception in addition to signal processing. In speech processing, these separate disciplines tend to reinforce one another, as better models of production and perception lead to new signal processing algorithms to estimate model parameters. Similarly, improved analysis algorithms provide deeper insights into the nature of the signal, aiding the development of better models. It is evident, then, that any discussion of DSP techniques in speech processing should include an introduction to the mechanisms of speech production and perception.

In this chapter, a broad introduction to speech science precedes a review of signal processing algorithms for speech analysis and modelling. This review of current methods starts with minimal assumptions about the nature of speech production and perception, and progresses through algorithms which make use of more detailed knowledge of these processes. Finally, more recent developments, as yet unproven in practice, are discussed briefly.

10.2   AN INTRODUCTION TO SPEECH SCIENCE

What is speech science? This apparently straightforward question is in fact very difficult to answer as the subject embraces a large number of diverse disciplines. Simply stated, the central purpose of speech science is the scientific investigation of all aspects of spoken language.

Spoken language is communicated through air via a longitudinal pressure wave. It is the analysis of the production, composition and perception of this 'speech pressure wave' which forms the basis for most work in speech science. By its very nature, the speech pressure waveform is a complex signal. It is an extraordinarily efficient method of transferring information between humans. It has the capacity to reliably communicate complex ideas and emotions under constantly varying and hostile environmental conditions. It can also be viewed as an efficient method of 'encoding' linguistic information for verbal communication.

With only a very superficial auditory examination of the speech pressure wave, it is clear that speech is composed of a number of different sounds, each of which is, in some way, of particular linguistic importance. This fact is expressed to a greater or lesser extent in the writing systems of most languages, which consist of a set of symbols which either directly or indirectly represent the sounds in the language. However, phoneticians realised that orthography (the conventional spelling system of a language) was not an accurate enough symbolisation of the sounds in language and so constructed an alternative set of symbols, known as phonemes, that relate directly to the sounds of language. The number and type of phonemes used to describe a language differ depending on the sounds considered to be important in that language or accent. Table 10.1 lists the set of phonemes found in one accent of English.

For many years linguists viewed speech as being composed of a linear sequence of phonemes put together like a string of beads. This misleading interpretation of the speech signal led engineers to imagine that machines could recognise speech simply by decoding the individual phonemes as they appear through time, the same way as a modulated data signal is decoded.

In a similar vein, it appeared that synthetic speech could be produced by simply storing an example of each of the phonemes of a language and then sticking together the phonemes in the new word as required. Unfortunately this is not the case. While it is convenient to visualise speech as a set of discrete symbols, in reality linguistic information is not encoded in discrete packets of time as imagined by the phoneme, and traditional methods of dealing with coded signals will not work with speech.

In this respect speech is unlike any other encoded signal. Clearly the above has implications for speech researchers, but before an explanation can be given of why this is the case, it is necessary to first understand more about the nature of the signal itself.

Table 10.1   SAM-PA (speech assessment methodology phonetic alphabet) machine-readable phonetic symbols for transcribing English [1].

p	pet	t	tie	k	key
b	bag	d	dog	g	gap
m	map	n	nap	N	ring
f	five	v	via	T	thigh
D	they	s	sixs	z	zoo
S	shake	Z	lounge	l	lip
w	will	r	ring	j	you
h	hit	tS	church	dZ	jive
i	bead	I	bit	e	bed
{	bad	A	hard	Q	cod
O	court	U	good	u	food
V	bud	3	bird	@	ago
eI	day	@U	mow	aI	high
OI	toy	I@	hear	e@	rare
U@	Ruhr	aU	cow		

10.2.1   Theories of speech production and the acoustic signal

The field of phonetics is concerned with describing the speech sounds of all languages. This can be done through an analysis of the articulators used to produce the utterance (articulatory phonetics) or alternatively through an analysis of the speech pressure waveform (acoustic phonetics).

10.2.1.1   Articulatory phonetics

The place to start, when describing the production of speech through a study of the mechanics of production, is to catalogue the articulatory or vocal organs used. Figure 10.1 shows these organs diagrammatically.

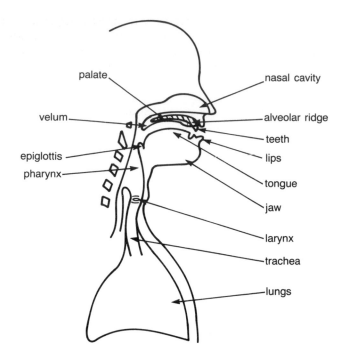

Fig. 10.1　Schematic diagram of the human vocal organs.

The way these articulators are used to produce the varied sounds found in speech will now be described. The source for most speech sounds is air expelled from the lungs through muscular action. Air leaving the lungs passes through a body of interlocking cartilage called the larynx (Fig. 10.2).

The two major components of the larynx are the thyroid cartilage and the cricoid cartilage. In men the thyroid cartilage is set at a slight angle, the front of which is commonly called the 'Adam's apple'. Within the thyroid and cricoid cartilage are two bodies of muscle and membranous tissue called the vocal folds. At the front of the larynx the vocal folds are brought together and attached to the thyroid cartilage, while at the back they are attached to a pair of small cartilages called the arytenoids. During normal breathing the vocal folds are abducted (held apart) allowing air to pass freely through the gap between the two folds (termed the glottis). During 'voiced' speech the vocal folds are repeatedly brought together (a process termed phonation).

The fundamental period of vocal fold oscillation is varied by tilting the thyroid cartilage, thus changing the tension in the folds. A fundamental period can be simplistically divided into two portions, an open glottis phase when the folds are apart allowing air to pass through the folds, and a closed glottis

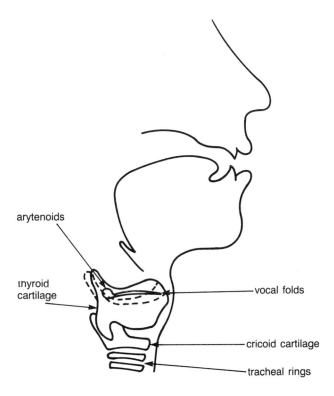

Fig. 10.2   The larynx — the thyroid cartilage has been cut away to reveal the arytenoids
and the vocal folds within.

phase when the folds are in contact, stopping the flow of air from the lungs.
These two phases are shown schematically in Fig. 10.3.

During the closed phase of the glottal cycle a pressure difference builds
up between the pressure in the lungs and trachea, and the external atmospheric
pressure. The sub-glottal pressure on the folds forces them to move apart,
allowing air to pass through the glottis. The particle velocity of the air through
the glottis is high and a Bernoulli force is induced which, in conjunction with
the muscular tension in the folds, tends to draw the folds back together,
eventually closing the glottis. This procedure is repeated over and over again.
The theory is called the myoelastic-aerodynamic theory of phonation. The
time between each closure of the vocal folds is called the fundamental period
T_0, the reciprocal of which is the fundamental frequency F_0.

In contrast to voiced sounds, voiceless sounds are produced when the vocal
folds are sufficiently abducted to allow air to pass relatively unimpeded
through the glottis.

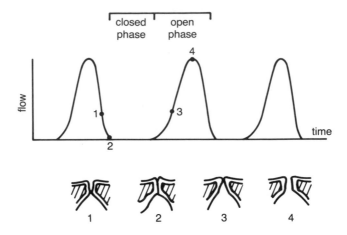

Fig. 10.3   Schematic diagram showing flow through the glottis. The open and closed phases are clearly marked. Also shown pictorially is an indication of the behaviour of the vocal folds over the larynx cycle.

The sound pressure wave modulated by the larynx, as described above, is then modified by the vocal tract in one of two ways:

- through the modification of the spectral distribution of the energy in the sound wave;

- through the generation of sound within the vocal tract.

The position of the velum, tongue, teeth, jaw and lips dictate which phoneme will be heard as the pressure wave radiates from the mouth and nose. The process of speech production just described is shown in Fig. 10.4.

In articulatory phonetics the consonant sounds of a language are described using three variables:

- voice;

- manner;

- place.

For example, consider a nasal consonant such as /n/, in 'nap'; the voice parameter indicates whether the vocal folds are used during its production while the manner of articulation describes how the sound is produced. For nasals, the oral cavity is closed completely and the velum lowered so that the sound may escape through the nasal cavity. The velum is a small flap

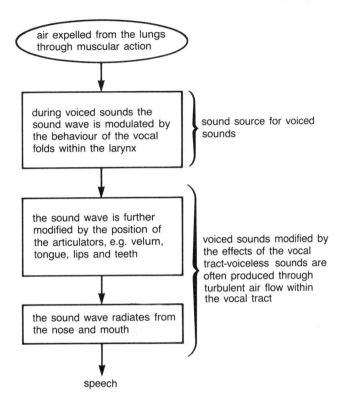

Fig. 10.4   The speech production process.

of skin which divides the oral cavity from the nasal cavity (Fig. 10.1). The type of nasal sound produced depends on the place of articulation (for nasal sounds the place of articulation describes where the oral cavity was occluded); in this example the place of articulation is at the alveolar ridge, and so an alveolar nasal is produced.

Vowel sounds are produced differently to most consonants, in that the articulators do not come into such close contact and so the air stream through the oral cavity is much less impeded. As a result vowel sounds are described in terms of the position of the highest part of the tongue in the vocal tract (i.e. front or back, and high, mid or low) and whether the lips are spread (e.g. the front, high vowel /I/ in the word 'bit') or rounded (e.g. the back, high vowel /U/ as in the word 'good').

When describing the production of any given utterance phoneticians commonly use the parameters outlined above [2,3].

10.2.1.2   Acoustic phonetics

An alternative and more tractable method of speech analysis is to examine the speech pressure waveform rather than the articulators involved in its production. There are a large number of signal processing tools used in the acoustic analysis of speech, some of the most important of which will be discussed in detail in the following sections. This section concentrates on the results produced by one such signal processing tool, the Fourier transform. Frequency domain representations have been used for a number of years in speech research, and many of the significant advances made in the study of the speech signal have been achieved through spectral analysis. The fact that spectral analysis of speech has been so important in the development of speech research is not surprising when one considers the nature of the speech signal. In section 10.2.1.1 it was demonstrated that the vast majority of speech sounds are produced through the modification of the speech pressure wave above the larynx by the articulators within the vocal tract. The vocal tract behaves as a resonating cavity with resonant characteristics that vary relatively slowly as the articulators move from one mode of articulation to another. A natural way of representing such information is in the frequency domain. For many years the short time spectral analysis of the speech signal was produced using a device called the sound spectrograph. This device was a significant advance because it was a comparatively fast method of producing short time spectra and, more importantly, because it allowed a comparatively large amount of spectral data to be presented clearly at one time. With the advent of fast computers and efficient algorithms, such as the fast Fourier transform, most spectrograms are now produced using general-purpose computing hardware. Figure 10.5 (a)-(c) shows the spectrograms produced for the vowels /i/ as in 'beet', /A/ as in 'cart' and /U/ as in 'good'. Clearly visible in these plots are the characteristic resonances of the vocal tract for each articulation. The dark black bands representing the resonances of the vocal tract are referred to as the formants of speech. Formants are labeled F1, F2, F3, etc, as they increase in frequency and are clearly visible in most portions of voiced speech.

Figure 10.5 (a)-(c) shows examples of 'wideband' spectrograms. In the context of a spectrogram produced using Fourier transforms, the term 'wideband' implies that relatively few speech samples were used to generate each transform leading to high resolution of the time-varying aspects of the speech signal at the cost of less resolution in the frequency domain. If Fig. 10.5 (a)-(c) is examined closely, a series of vertical lines can be seen running through the formants. These lines are known as 'striations'. Striations are visible during voiced speech in all 'wideband' spectrograms. They represent the individual excitations of the vocal tract as the vocal folds close. By

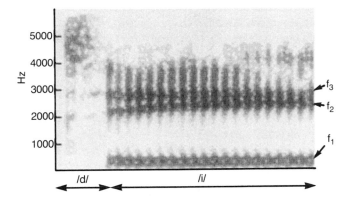

(a)   'Wideband' spectrogram of the sounds /d i/ as in 'deep'.

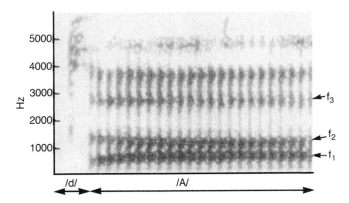

(b)   'Wideband' spectrogram of the sounds /d A/ as in 'dart'.

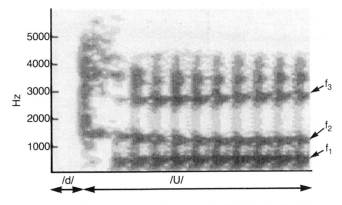

(c)   'Wideband' spectrogram of the sounds /d U/ as in 'dude'.

Fig. 10.5   Spectrograms of speech.

measuring the distance between striations it is possible to get an estimate of the fundamental period. An alternative spectral representation of the speech signal can be generated by performing Fourier transforms on much longer stretches of speech data. Spectrograms produced in this manner are termed 'narrowband' and contain more detail in the frequency domain at the cost of decreased resolution in the time domain. Figure 10.6 shows a 'narrowband' spectrogram of Fig. 10.5(b).

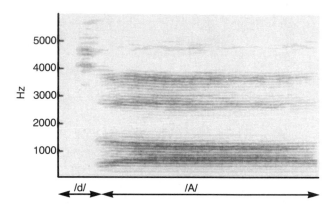

Fig. 10.6   'Narrowband' spectrogram of the sounds /d A/ as in 'dart'.

In Fig. 10.6 the formant structure of the speech signal is much less evident and the striations are no longer present as a number of larynx cycles have been smeared over time; however, a much more detailed representation of the harmonic structure of the signal is presented. For voiced speech, the harmonic structure of the speech is clearly shown as a series of harmonics running through the formants. As with the striations observed in the wide band spectrogram, it is possible to obtain an estimate of the fundamental frequency by measuring the individual harmonics.

If the process of marking the formant frequency for the first three formants of a vowel, as shown above, is repeated for a number of repetitions and different vowels, a graph representing the characteristic resonances of the vocal tract against vowel type can be produced as shown in Fig. 10.7.

It is clear from Fig. 10.7 that the different vowel sounds can be well represented by the position of the first two formants. Thus it is possible to represent the vowels alternatively as an F1 versus F2 plot as shown in Fig. 10.8.

The results presented in Figs. 10.7 and 10.8 suggest that the identity of a given vowel can be adequately expressed using the first three formants. This fact has been used in a number of perceptual studies in an attempt to simplify the auditory data presented to subjects. This characteristic of speech is examined in more detail in the next section.

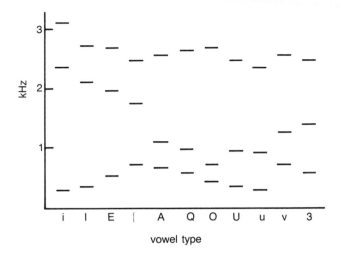

Fig. 10.7   Mean formant frequencies for first three formants (plot reproduced from Wells [4]).

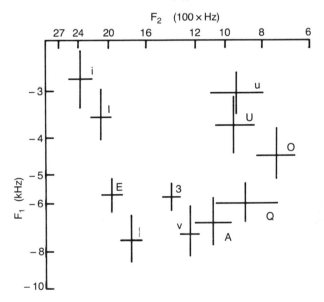

Fig. 10.8   Mean and dispersion of formant frequencies — plot of F1 versus F2. Lines extend one standard deviation to each side of the mean value (logarithmic scale calibrated in hundreds of hertz) (plot reproduced from Wells [4]).

So far the data presented has been for vowels only. In Fig. 10.9 a 'wideband' spectrogram of a much longer portion of speech is presented. It can be seen how the formants appear to move as the articulation changes, and also how the formants are less clearly defined in voiceless portions of speech; also missing in the voiceless portions of the speech signal, as might be expected, are the contributions due to the voiced source.

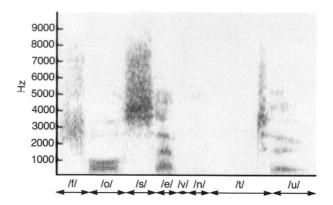

Fig. 10.9   'Wideband' spectrogram of a longer stretch of speech data.

The spectrogram continues to be an important tool in the analysis of speech, presenting speech researchers with a clear pictorial representation of the speech signal which is related to the articulation of speech sounds. While the spectrogram is a very powerful pictorial representation of the speech signal, it is not an accurate representation of the spectral pattern presented to the human auditory system once it has passed through the transduction stage of the ear. For example, consider the results of psycho-acoustical experiments designed to investigate how accurately humans can determine the relative pitch of two sinusoids [5]. In such experiments it was observed that if a human is presented with a sinusoid of 1 kHz and asked to adjust the frequency such that they hear a sound of twice the pitch, they will in fact choose a frequency of just over 3 kHz. It would appear that the perception of pitch in humans is nonlinear. In fact it is approximately logarithmic in nature. The pitch perceived by a human can be converted to the equivalent frequency in kilohertz using a conversion scale known as the 'Mel' scale. An alternative similar scaling strategy, known as the 'Bark' scale, can be derived using information gained from critical bandwidth experiments which attempt to determine the sensitivity of the peripheral auditory system to given frequencies.

Such experiments clearly warn against assuming that features which appear significant in the Fourier transform of the speech signal will remain significant once the speech signal has been processed by the ear.

10.2.2   Theories of human speech perception

The previous sections have concentrated on describing how speech is produced and examined the characteristics of such a signal. Production is however only half the story. Speech is first and foremost a means of communication between human beings. To fully understand the encoding strategy adopted by speakers, it is necessary to examine how speech is perceived by human listeners. As a first stage in understanding how humans perceive speech, it is necessary to first determine what aspects of the speech pressure waveform are considered to be perceptually significant. A number of studies have been conducted over the past 40 years which have attempted to determine exactly what parts of the speech signal are perceptually significant. Some of the most interesting of these were conducted at Haskins Laboratories during the 1950s. As observed in the previous section, the formant structure of the speech signal is an important cue to identifying vowel sounds, but the picture is not as clear for consonant sounds. As an example consider the results of experiments conducted at the Haskins Laboratories on the perception of the voiced plosive sounds /b/, /d/ and /g/ in a number of CV (consonant-vowel) combinations [6]. These experiments used a device known as the 'pattern play back' to produce synthetic speech from very stylized spectrograms. Owing to the stylistic nature of the data, the quality of the synthetic speech was very robotic; however, the intelligibility was considered sufficient to allow researchers to investigate which characteristics of the speech signal contributed to the perception of a given consonant. The results from these experiments showed that the transition of formants into the following vowels acted as a strong acoustic cue to determining which consonant was heard and, furthermore, that the type of transition observed for a given plosive changed depending on the vowel that followed. Figure 10.10 (a)-(c) shows the stylized spectrograms used to produce the voiced plosives before a number of vowels. Figure 10.5 (a)-(c) shows the spectrograms produced from real speech for the same plosive vowel combinations.

Such results clearly have ramifications for a segmental theory of speech production. The data suggests that speech is not made up from a set of concatenated phonemes. The results given above show, for example, that it is impossible to cut a /b/ sound out of the speech signal without including a portion of the following vowel. While it is possible to imagine speech to be produced by a set of phonemes it would appear that in reality the speech signal is not composed of a linear sequence of phonemes.

Individual speech sounds or phones are not generally produced in isolation, but appear as syllables which in turn may be combined to make words. It is not surprising therefore, that each sound is greatly affected by the production of its neighbours. In this sense speech sounds are said to be

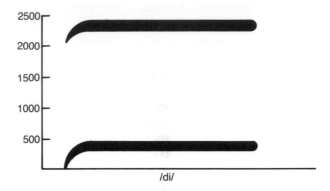

(a)   Stylized spectrogram of the sound /d i/ as in 'deep'.

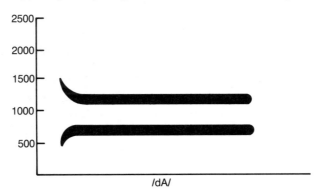

(b)   Stylized spectrogram of the sound /d A/ as in 'dart'.

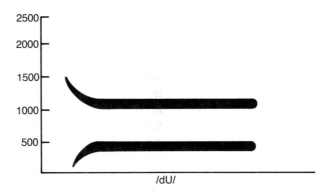

(c)   Stylized spectrogram of the sound /d U/, /U/ as in 'good'.

Fig. 10.10   Stylized spectrograms of speech data similar to those produced in Delattre et al [6].

co-articulated. The process of co-articulation affects all sounds, not simply the voiced plosives. In the previous section it was observed that vowel sounds, when produced in isolation, have a particular set of formant frequencies. However, in the production of fluent speech these 'target' frequencies are seldom achieved. In other words, the formant structure of a vowel is modified due to the surrounding speech sounds and the rate of production.

So far it has been shown that the frequency of formants is important in the identification of vowels and also that the movement of formants is significant in determining the character of a preceding voiced plosive. A number of perceptual studies have been conducted which have attempted to investigate exactly how, for example, the movement in formant position helps to signal a given phoneme such as a voiced plosive. These studies have shown that listeners appear to perceive vowels in isolation differently to vowels in CVC contexts and, to an even greater extent, differently to certain consonants, due to their manner of production [7]. It is much easier to discriminate between different vowels than it is to identify a specific vowel. This behaviour can be found in other areas of human perception. For example most people find it comparatively easy to distinguish between different shades of colour but would find it difficult to 'pin down' precisely what a given colour was. Not all speech sounds are perceived in the same way as vowels; plosives in English, for example, are discriminated no better than they are identified. Are such effects learned or are they 'preprogrammed'? Experiments using infants suggest that such categorical discrimination is present in very young infants, other work has suggested that the auditory system has specialised feature detectors for decoding speech. Stevens [8] has proposed a physiological motivation for the different way humans perceive consonants and vowels. He suggests that speech is quantal in nature, that is, there are places along the vocal tract that cause an abrupt change in the sound of the articulated speech, but that between these places of importance the accurate placement of articulators is not necessary. This quantal behaviour is only observed when the tongue is close to the roof of the mouth, e.g. during the production of plosives. When the tongue moves away from the roof of the mouth, during the production of vowel sounds, the quantal behaviour is not observed and sounds appear to be able to change smoothly from one type to another. Stevens quantal theory of speech in its prediction of the behaviour of different speech sounds is as much a theory of perception as it is an explanation for the behaviour of certain speech sounds. The method by which listeners are able to decode the information in the speech signal, ignoring semantics, pragmatics and syntax, is the subject of the last subsection.

The last part of this section on speech science examines the final stage in the perception of speech, that is, how the features observed in the speech signal are actually interpreted. In other words by what process does a human convert these acoustic characteristics in the speech signal into an abstract representation of speech, and alternatively by what process does a human talker convert the abstract description of language in their head into the acoustic signal. Such questions are important to speech synthesis and recognition as it is at this level of abstraction that any invariance in the speech signal will be present. There have been two broad approaches adopted in the theories constructed to explain speech perception — passive theories and active theories. Passive theories view listeners to be comparatively passive in their perception of speech sounds. In such models the process of speech perception is performed by a comparatively simple mapping of the features of the speech signal on to a set of acoustic phonetic features of the language. In contrast active models require listeners to take a more active role in the perception of speech sounds. In these models, it is envisaged that listeners postulate possible utterances by comparing the incoming speech signal with an internal representation of the production process they would use to produce the sounds. Examples of a passive and an active model of speech perception are given below.

In the passive model of speech perception proposed by Fant [9], the speech perception mechanism and the speech production mechanism draw on a common database of distinctive features. These distinctive features have been 'learnt' during the listener's acquisition of language. During speech perception the listener need not reference speech production to understand the speech unless the utterance is being spoken under unusual listening conditions. This process is shown schematically in Fig. 10.11.

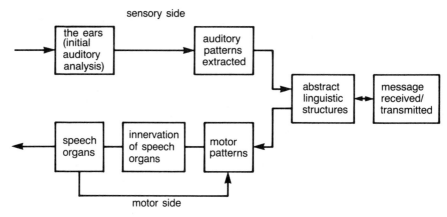

Fig. 10.11   Passive model of speech perception proposed by Fant [9].

The 'Motor Theory' of speech perception proposed by Liberman [10] is an example of an active theory of speech perception. This theory was borne out of a desire to explain why the concept of the phoneme is such an attractive abstraction. It was shown above that the acoustic signals for a plosive such as /d/ before /i/ is very different from the acoustics of the same phoneme before /u/; why then do humans appear to hear the same sound /d/ in both cases? Liberman suggests that humans 'hear' identical initial plosives because they 'decode' the speech signal using prior knowledge of the articulatory gestures used in the production of the sounds. In this theory speech is not decoded on a segment-by-segment basis, but over a larger unit such as the syllable. That is, the phonemes do not exist outside the context of the syllable in which they reside. This process is shown schematically in Fig. 10.12.

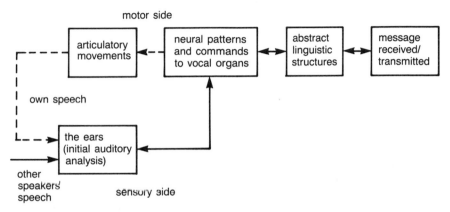

Fig. 10.12   The 'Motor Theory' of speech perception proposed by Liberman [10].

The passive theories of speech perception are attractive because they imply that there are invariances within the speech signal that can be mapped on to a set of distinctive features and that these features will be sufficient to decode the phonetic component of the speech signal. In contrast the motor theory implies that determining the distinctive features of a language may not be enough and that a further level of abstraction is performed which maps the observed features on to a set of motor patterns. The listener then internally generates a number of possible descriptions of the data, using their motor mechanism, which are matched to the observed motor patterns.

With the present state of knowledge it is not yet possible to say which of the two types of theories is closer to the truth. If active theories are closer to the true mechanism of speech perception than passive models, the attempts to find distinctive acoustic features may not yield the improvements in decoding of the speech signal hoped by many speech researchers.

The first part of this chapter has introduced a number of fundamental concepts in speech science, as it is only with a clear understanding of the nature of speech that the models and features employed in the analysis of the speech signal can be fully appreciated. The remainder of the chapter discusses signal processing algorithms which have been developed to estimate model parameters and features from the speech signal.

10.3   ACOUSTIC SPEECH ANALYSIS

Throughout this discussion, a generalized source/filter model of specch production is assumed, as shown in Fig. 10.13.

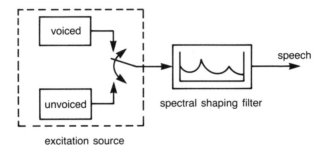

Fig. 10.13   Generalized source/filter model of speech production.

The model is generalized in that no prior constraints are made on either the spectral or time-domain characteristics of the various components. Specific versions of the model tend to be associated with different analysis techniques. For example, certain methods assume that the excitation signal is spectrally flat, and that the shaping filter combines the spectral characteristics of the glottis, vocal tract and lip radiation, whilst others combine glottal and lip radiation characteristics in the excitation, leaving the shaping filter to model only the vocal tract response.

Of the many aspects of speech analysis, this chapter covers the extraction of source and vocal tract information from the speech signal. The methods are discussed mainly in the context of acoustic models of speech production, although analysis for recognition is also mentioned.

10.3.1   Spectral analysis techniques

As discussed in section 10.2.1.2, the Fourier spectrogram is a good pictorial tool for speech research, but its usefulness in automatic speech analysis is limited. To separate the speech signal into source and vocal tract components, alternative spectral representations must be sought. This section introduces some of the more successful spectral analysis techniques applied to speech processing.

10.3.1.1 Linear prediction

Linear prediction (LP) has become one of the most powerful DSP tools in speech processing [11]. Although many LP formulations have appeared in the literature, the basic principles remain the same and are presented in this section.

The popularity of this technique can be attributed to two main properties:

- it can be related to a simple model of speech production;

- basic LP analysis/synthesis algorithms are both easily understood and efficient to implement.

The general LP model has the form:

$$s_n = \sum_{k=1}^{p} a_k s_{n-k} + G \sum_{l=0}^{q} b_l u_{n-l}, \quad b_0 = 1 \qquad \ldots (10.1)$$

where s_n is the output of the model, u_n the input, and the model parameters are $a_k: k-1, \ldots, p$, $b_l: l = 1, \ldots, q$, and the gain factor G. With a known input, the output of the model is completely predictable as a linear combination of past inputs and outputs. By transforming to the Z-domain and re-arranging, the transfer function of the model, $H(z)$, can be obtained as:

$$H(z) = \frac{S(z)}{U(z)}$$

$$= G \cdot \frac{1 + \sum_{l=1}^{q} b_l z^{-l}}{1 - \sum_{k=1}^{p} a_k z^{-k}}$$

$$= G \cdot \frac{B(z)}{A(z)} \qquad \ldots (10.2)$$

$H(z)$ has both poles (roots of $A(z)$) and zeros (roots of $B(z)$).

When modelling the behaviour of a real signal such as sampled speech, the input is generally not available, and the output of the model is at best only an approximation. In this case, the transfer function has no zeros, and is referred to as an all-pole model. This is the most familiar model in speech processing. The prediction is denoted:

$$\tilde{s}(n) = \sum_{k=1}^{p} a_k s_{n-k} \qquad \qquad \dots (10.3)$$

and the difference between the real and predicted signals is called the prediction error, or residual:

$$e_n = s_n - \sum_{k=1}^{p} a_k s_{n-k} \qquad \qquad \dots (10.4)$$

In least squares prediction, the objective is to minimize the total squared error:

$$E = \sum_{n} e_n^2 \qquad \qquad \dots (10.5)$$

over some interval (as yet undefined) by an appropriate choice of predictor coefficients a_k. By expanding equation (10.5) and setting partial derivatives to zero [11], the set of simultaneous equations:

$$\sum_{k=1}^{p} a_k \sum_{n} s_{n-k} s_{n-i} = - \sum_{n} s_n s_{n-i}, \; i = 1, \dots, p \qquad \qquad \dots (10.6)$$

is obtained, which can be solved for the predictor coefficients.

To proceed further, the analysis interval must be defined by imposing explicit limits on the summation terms in equations (10.5) and (10.6). Two alternatives are available, and the choice has consequences on both the method of computation and the properties of the resulting model. Both methods are of use in speech analysis.

In the autocorrelation method, equation (10.5) is minimized over an infinite interval, and equation (10.6) can be written as:

$$\sum_{k=1}^{p} a_k r_{i-k} = r_i, \; i = 1, \dots, p \qquad \qquad \dots (10.7)$$

where r_i is the autocorrelation function. In practice, this is achieved by windowing a finite portion of the signal. This method leads to a computationally efficient solution, and is also guaranteed to produce a stable model.

In the covariance method, the error is minimized over a finite interval with no windowing, giving equations of the form:

$$\sum_{k=1}^{p} a_k \psi_{ki} = \psi_{0i}, \; i = 1, \dots, p \qquad \qquad \dots (10.8)$$

where the ψ_{ki} terms are cross-correlations rather than autocorrelations. The name arose because the matrix of ψ_{ki} terms has the same properties as a covariance matrix, and is not connected with the standard usage of covariance to describe correlation with the signal means removed. For short analysis intervals, the covariance method is more accurate as it does not assume the signal to be zero-valued outside the analysis interval, but has the disadvantages of increased computational load and no guarantee of a stable model.

To see how least-squares LP analysis relates to the generalized production model of Fig. 10.13, the frequency-domain properties of LP must be considered. It can be shown [11] that the power spectrum of the LP model is given by:

$$\hat{P}(\omega) = \frac{G^2}{|A(e^{j\omega})|^2} \qquad \dots (10.9)$$

where $A(e^{j\omega})$ is $A(z)$ evaluated on the unit circle. Similarly, the speech power spectrum can be shown to be:

$$P(\omega) = \frac{|E(e^{j\omega})|^2}{|A(e^{j\omega})|^2} \qquad \dots (10.10)$$

where $E(e^{j\omega})$ is the LP error spectrum. Comparing equations (10.9) and (10.10), it is evident that the error spectrum is modelled by a flat spectrum of constant value G^2. The inverse filter $A(z)$ therefore attempts to flatten the signal spectrum, and produces a model $G/A(z)$ which minimizes a well-defined spectral error measure [11]. In this sense, LP analysis provides an optimal estimate of the signal spectrum. In the LP version of the production model, the excitation is assumed to have a flat spectrum (either white noise or a pulse train), and the all-pole shaping filter models the combination of glottis, vocal tract and lip radiation.

10.3.1.2 Cepstral analysis

Cepstral analysis is a particular example of the class of homomorphic signal processing techniques, and is primarily a deconvolution method. In speech processing, the objective is to deconvolve the input, or excitation, from the spectral shaping filter. LP analysis does this by assuming that the excitation has a flat spectrum and the shaping filter is all-pole. Cepstral analysis assumes that the excitation spectrum varies rapidly with frequency, whereas the response of the shaping filter varies slowly, allowing them to be separated by a special form of frequency-domain filtering.

In the speech model of Fig. 10.13, the speech signal is formed by a time-domain convolution of the excitation with the impulse response of the shaping filter, or equivalently in the frequency-domain as a product of the corresponding spectra. Deconvolution is the process of recovering the individual components from the resultant signal, and can be achieved by homomorphic filtering. The main property of homomorphic systems is that the superposition principle holds for convolution rather than addition, allowing the components to be separated in the same way that linear filtering can separate additive components (subject to certain conditions being satisfied).

Homomorphic filtering can be represented as the 3-stage process shown in Fig. 10.14.

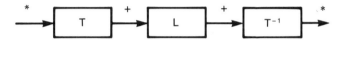

* convolution

Fig. 10.14   Block diagram of homomorphic filtering.

Firstly, the transform T, which usually contains a nonlinear element, takes the signal into a domain where convolution becomes addition. Secondly, the desired component is separated out by a linear filter L. Finally the inverse transform T^{-1} returns the isolated component to the original domain. Since convolution becomes multiplication in the frequency-domain, the cascade of DFT-LOG-IDFT is a commonly used form of the transform T, where LOG is the natural logarithm. The property:

$$LOG[AB] = LOG[A] + LOG[B] \qquad \qquad \text{... (10.11)}$$

converts the frequency-domain product to a sum, which is preserved by the IDFT. The success of homomorphic deconvolution relies on the components occupying distinct 'bands' in the transformed domain, so that they can be isolated by linear filtering. The inverse transform is then DFT-EXP-IDFT. Although theory dictates that complex logs should be used with the DFT, this creates difficulties with signals which are not minimum-phase. In practice, particularly in speech processing, only the magnitude component of the DFT is used.

From the above discussion it is apparent that the potential for confusion in terminology is considerable, e.g. although the transformed signal has units of time, it does not exist in quite the same time-domain as the original signal due to the nonlinear operation in the frequency-domain. For this reason, the term 'cepstrum' was coined for the IDFT of the log magnitude spectrum, which was said to exist in the 'quefrency-domain'. Not surprisingly, the linear filtering operation to separate the components has become known as 'liftering'.

The two components of the speech spectrum (slowly varying envelope and rapidly varying excitation) occupy low and high quefrency bands respectively in the cepstrum, and can be separated by a suitable window (cepstral lifter). For voiced speech, the excitation component is periodic in the log spectrum, and produces a spike in the cepstrum which can be used for pitch estimation and voicing classfication. A windowed speech segment, log spectrum and cepstrum are shown in Fig. 10.15. A low-pass lifter on the cepstrum removes the excitation component, effectively smoothing the log spectrum to leave the envelope component. The inverse transform T^{-1} gives an impulse response which can be convolved with the excitation to produce synthetic speech. A full discussion of such a system, the homomorphic vocoder, is given in Rabiner and Schafer [12].

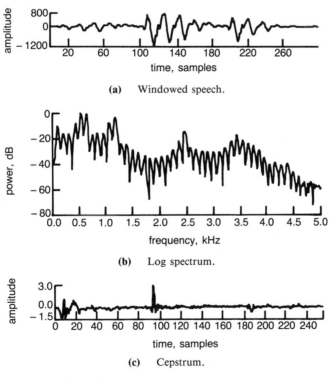

(a)   Windowed speech.

(b)   Log spectrum.

(c)   Cepstrum.

Fig. 10.15   Cepstrum of voiced speech.

10.3.1.3   Mel frequency cepstral coefficients

The spectral analysis algorithms discussed above use a linear frequency scale, although it is known that human pitch perception is a nonlinear process. The Mel scale referred to in section 10.2.1.2 is a nonlinear mapping of the frequency axis into bands which have been determined by experiment to have equal perceptual importance. The filter spacing is linear at low frequencies and logarithmic at higher frequencies. Mel filterbanks with filters spaced according to this scale should, in principle, be more appropriate for speech recognition than linear filterbanks.

A further stage of processing is applied to the logged filterbank outputs to produce MFCCs (Mel frequency cepstral coefficients), first defined by Davis and Mermelstein [13]. A DCT (discrete cosine transform) is used to decorrelate the filterbank outputs, and enable them to be reduced in number for a more compact representation of the spectrum. The DCT is used as it has been shown to approximate the optimal KLT (Karhunen-Loeve transform) for speech spectra [14,15] and has an efficient implementation. Referring to the previous section on cepstral analysis, the filterbank outputs go through a similar process of logging and transforming, although the transform is a DCT rather than an inverse DFT. This probably accounts for the 'cepstral' term in the name, although the sequence of coefficients obtained is not a true cepstrum as defined in section 10.3.1.2.

MFCCs are probably the most successful features in use at present for speech recognition, especially in noisy environments where the decorrelating transform maximizes the signal-to-noise ratio of the coefficients.

10.3.2   Formant analysis

Although the spectral analysis methods discussed above provide a means of estimating vocal tract and source information from the speech signal, they do so at a gross spectral level, making minimal use of the available knowledge on mechanisms of speech production and perception. For example, the dynamic behaviour of the formants is known to influence perception significantly (see section 10.2), and several successful speech production models are based upon explicit modelling of formants [16]. Spectral analysis alone, however, does not isolate the formants from the speech signal, or track their variation with time. More specific knowledge about the structure of speech must be incorporated in the analysis to extract this important information. Accurate formant data enables progress to be made both in understanding of speech perception and development of better production models. This in turn can lead to improvements in automatic speech recognition, text-to-speech synthesis and data rate compression.

The most straightforward formant analysis methods involve post-processing of smooth spectral estimates, and generally use some form of LP or cepstral representation. The post-processing usually has two distinct stages:

- extraction of raw formant data from the spectral estimates;

- application of temporal continuity and frequency constraints to produce formant data consistent with theories of speech production.

Only the first step is considered in this chapter. The second step generally involves the application of a set of logical rules to deal with missing and extra formants, and discontinuities in formant tracks.

Extraction of raw data from spectral estimates relies on the assumption that spectral features can be consistently related to the formants themselves. A simple method, for example, is to equate peak locations in the spectra to formant centre frequencies [17]. This method can be applied to any smoothed spectral estimate. With LP analysis, the complex roots of the LP polynomial can be equated to the resonances of the vocal tract, giving both centre frequencies and bandwidths [18]. A variety of related methods have been developed over the years, including spectral differentiation [18], and use of group-delay rather than magnitude spectra [19].

Peak-picking is basically an open-loop approach, with no feedback from the model in which the formant data is being used. Closed-loop feature extraction, on the other hand, makes use of a speech production model in a feedback loop. An initial set of parameters is obtained by open loop analysis, and used to generate synthetic speech. The natural and synthetic spectra are then compared and the difference used to iteratively optimise the parameters until a good match is obtained [21,22]. This 'analysis-by-synthesis' approach is well-suited to applications requiring faithful storage and reproduction of speech at low data rates [23], where the physical interpretation of the parameters is not of prime importance. For investigations into speech production and perception, however, the relevance of closed-loop analysis is heavily dependent on the accuracy of the production model used.

This discussion has assumed that the spectral analysis is performed at fixed intervals, with analysis frames of fixed duration. Unfortunately, this type of analysis, referred to here as 'fixed-frame', is fundamentally unsuited to the problem of extracting accurate vocal tract information from the speech signal for the following reasons.

- In voiced speech, the periodicity results in a harmonic spectrum if several periods are included in the analysis frame. For high-pitched speakers, the harmonics are widely spaced, and tend to bias the positions of smoothed spectral peaks due to the sparse information in the frequency domain. This is a problem with the analysis methods themselves. LP analysis in particular is very sensitive to the position and number of periods in the analysis frame [24].

- Formant frequencies and bandwidths can vary dynamically within each larynx cycle. Fixed-frame analysis cannot track these local variations, resulting in smeared estimates of formant behaviour.

- Since the speech signal includes contributions from the glottal source (which is not spectrally flat) and lip radiation, the analysis models these in addition to the vocal tract resonances.

A common method of reducing the combined spectral effects of glottal excitation and lip radiation is to pre-emphasize the speech with a filter of the form:

$$P(z) = 1 - az^{-1} \qquad \qquad \text{... (10.12)}$$

where a is generally in the range 0.9 to 1.0. This can only account for gross effects such as spectral tilt. To improve the accuracy of formant estimates and track their short-time variation due to source/tract interaction, analysis algorithms must make use of explicit glottal activity information.

10.3.2.1 Larynx synchronous analysis

To avoid confusion, the general class of algorithms which use glottal information will be referred to here as larynx-synchronous methods, as terms such as 'pitch-synchronous' have been applied to specific cases.

The first step towards more accurate formant analysis is to eliminate points of glottal excitation from the analysis frame. For this purpose, these can be assumed to correspond primarily to glottal closure, and, to a much lesser extent, glottal opening. All-pole LP analysis can accurately predict the response of an all-pole system to an impulsive excitation, but cannot predict the time at which an impulse will occur. This results in large error values around the points of glottal excitation. Unfortunately, as LP analysis blindly minimizes the total squared error over the frame, large error values inevitably influence the process, resulting in a model which does not accurately represent the vocal tract response and hence the formant structure. Ideally, the analysis should ignore the error signal around points of major excitation, so that the prediction is applied only to portions of the signal corresponding to the zero-input response of the vocal tract. Miyoshi et al [25] proposed a two-stage LP analysis method to achieve this, with particular application to high pitched speech. The error signal from a conventional LP analysis is examined. Those segments of the speech signal where the error is consistently small are then used for a second LP analysis by a covariance approach. An improvement in the accuracy of formant estimates was reported for both synthetic and natural speech. This method is still fixed-frame in nature, and can include an arbitrary number of pitch periods in one analysis frame. The next logical step is to analyse a single larynx cycle following glottal closure, as discussed below.

By confining the analysis to a single pitch period, transient behaviour of the formants can be tracked more accurately. Again, the major excitation due to glottal closure should be excluded. Since the number of samples in a single period can be relatively small for high-pitched speakers, the covariance method of LP analysis must be used to ensure accurate results. This mode of analysis is generally referred to as pitch-synchronous. Even at this level, however, the spectral estimate is obtained from a signal which spans a full glottal cycle, and it is known that the sub-glottal cavities can affect the observed resonances of the vocal tract when the glottis is open. Theoretically, the best estimate of the vocal tract response is obtained by LP analysis of the speech signal while the glottis is closed (closed-phase analysis). In this way, the signal contains no points of excitation, and no contribution due to sub-glottal coupling. Although fine in theory, practical problems exist with closed-phase analysis, mainly with high-pitched speakers:

- the small analysis frames can create stability problems for covariance LP analysis;

- when the pitch period is short, the response of the vocal tract to a glottal pulse may not decay significantly before the next pulse occurs, so that the resultant speech signal is the superposition of two or more responses.

Despite these problems, investigations have shown [26] that closed-phase LP analysis yields more consistent estimates of formant centre frequencies and bandwidths than either fixed-frame or pitch-synchronous methods. In this discussion, it has been assumed that glottal information has been freely available to the analysis algorithm. However, as the reader will probably appreciate by now, no information comes freely in speech analysis. Source analysis, the problem of deriving suitable glottal information, is the subject of the next section.

10.3.3   Source analysis

As with vocal tract analysis, source analysis can be performed at various levels. At the most coarse level, analogous to the fixed-frame spectral analysis methods of section 10.2.2, a fundamental frequency value can be obtained at fixed intervals. This gives no information on the fine detail of each glottal cycle such as points of closure and opening or the shape of the pulse itself, but merely represents an average pulse frequency over the analysis interval. A sequence of such values is often referred to as a pitch contour, and is used in conjunction with LP or formant-based synthesizers to produce synthetic speech at low data rates. A vast number of algorithms exist for estimation of fundamental frequency from the speech signal (see Hess [27] for a comprehensive survey), and the field is too large to be adequately covered here. In general terms, however, most of the algorithms fall into one of three categories:

- time domain methods, which rely on the short-term periodicity of voiced speech;

- frequency domain methods, which rely on well-defined harmonic structure in spectra of speech;

- hybrid methods, which normally involve a nonlinear operation in the frequency domain, followed by a return to the time domain where the fundamental period is estimated.

As discussed in the previous section, more detailed information is often required for speech analysis applications. There are two approaches to the problem:

- more intensive analysis of the speech signal;

- direct measurement of the source parameters.

The first method is the only viable approach for most telephony applications, as only the speech signal is available. The second method is more reliable, and is feasible when the speech is being collected and processed off-line. A device which gives reliable estimates of glottal closure and, to a lesser extent, glottal opening, is the laryngograph [28].

10.3.3.1   Two-channel analysis using a laryngograph

The laryngograph is a device which gives an indication of the motion of the vocal folds. Electrodes attached to the throat measure the conductance across the larynx. This conductance varies as the vocal folds oscillate, increasing when they are closed and decreasing when they are open. Figure 10.16 shows a segment of voiced male speech with the corresponding laryngograph signal and LP residual.

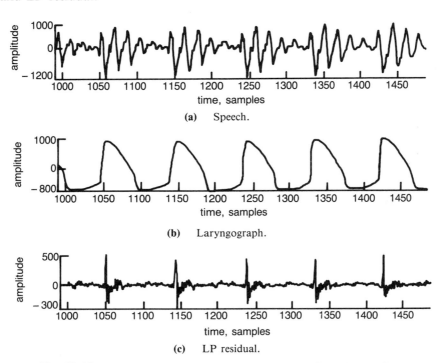

(a)   Speech.

(b)   Laryngograph.

(c)   LP residual.

Fig. 10.16   Laryngograph and LP residual waveforms for male speech.

Glottal closure is marked by a rapid increase in conductance, and opening by a more gradual decrease. Although closure is not an instantaneous process, the point of closure is usually defined as being the point of maximum slope on the rising edge of the signal. These points are easily identified as local maxima in the first differential of the laryngograph signal. Having determined the points of closure, glottal opening instants can be similarly estimated as local minima, although this procedure is less reliable. It should be noted that the shape of the glottal pulse cannot be estimated from the laryngograph signal, as a glottal pulse represents volume-velocity flow and is not easily related to conductance across the larynx. The major contribution of the laryngograph is accurate glottal closure information. When aligning the closure points with the speech signal, care must be taken to account for the time taken for the acoustic pressure wave to travel from the larynx, through the vocal tract to the microphone. This delay is usually 1-2 ms, and can be estimated by visual inspection of the waveforms, or automatically by correlation with an inverse filtered speech signal.

10.3.3.2   Estimation of glottal closure from the speech signal

In many cases it is not practical, or even possible, to use the laryngograph. To analyse larynx-synchronously in such cases, it is necessary to estimate points of glottal closure from the speech signal alone. Several algorithms have been proposed which rely on the observation that glottal closure produces some form of discontinuity in the speech signal [29-31]. Since the excitation due to glottal closure is inherently unpredictable, an obvious candidate for detection of the discontinuity is the LP residual. In strongly voiced speech, glottal closure is usually marked by large spikes in the LP error, as shown in Fig. 10.16. In order to eliminate problems caused by the phase response of the predictor (closure could correspond to positive or negative spikes, or the zero-crossing between them), Ananthapadmanabha et al [30] based their algorithm on the Hilbert envelope of the residual, which represents all these cases as positive peaks. The success of these algorithms depends on the accuracy with which the LP filter models the vocal tract, and on the vocal folds undergoing a sharp closure in the first place. When the vocal tract response is not minimum-phase, LP filtering cannot fully remove the vocal tract phase characteristic from the speech signal. This can result in a residual which, although spectrally flat, does not have the pulse-like nature required for estimation of glottal closure.

Problems are often encountered with female speakers as shown in Fig. 10.17, where the residual does not always exhibit sharp spikes during closure. From a signal processing viewpoint, this is related to the low-pass nature of the signal, as illustrated in the log spectrum in Fig. 10.17(c), where the signal-to-noise ratio is very low for higher harmonics. When the spectrum is flattened by LP filtering, the lower harmonics are swamped by noise. To produce a useful LP residual, strong harmonics are required over the full frequency range as in Fig. 10.15(b). Incomplete or slow closure of the vocal folds is often responsible for the lack of higher harmonics in the excitation. In these cases, the residual is not suitable for glottal closure analysis.

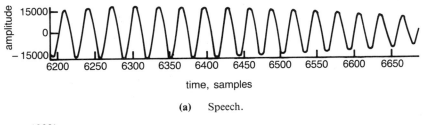

(a) Speech.

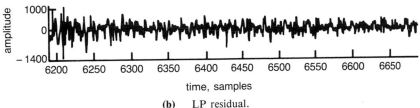

(b) LP residual.

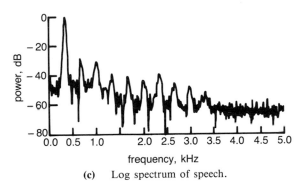

(c) Log spectrum of speech.

Fig. 10.17 LP analysis of female speech.

Estimation of glottal closure from the speech signal is not sufficiently robust to a wide variety of speakers and noise conditions as yet, and devices such as the largyngograph are still necessary for reliable results. More research is required in this area before fully automatic larynx-synchronous analysis becomes a reality, particularly with telephony speech.

10.3.3.3 Inverse filtering

The final step in this line of analysis is to estimate the shape of each glottal pulse, representing the airflow during each glottal cycle. This is usually achieved by inverse filtering to remove the spectral shaping introduced by each individual formant. This implies, of course, that accurate estimates of formant frequencies and bandwidths are available. A major problem with glottal inverse filtering is that the phase characteristic of the signal is crucial in determining the waveform shape. This means that for truly accurate inverse filtering, the phase response of the recording equipment (microphone, anti-aliasing filters, etc) should be accounted for in addition to the vocal tract response. In practice, however, inverse filtering is normally restricted to the vocal tract response. Information on glottal flow is important in characterizing different modes of voicing, e.g. modal, vocal fry, diplophonia, and so has applications in perceptual studies and improvement of speech synthesis models.

10.3.4   Alternative analysis techniques

Space does not permit a full discussion of alternative analysis methods, so attention is focused here on two areas which the authors believe will yield significant results in the future — higher order spectra and nonlinear dynamical systems. Of the techniques not covered, time-frequency transforms [33] and wavelet transforms [34] are worthy of further investigation by those interested in analysis of the speech signal.

Most of the techniques applied in speech processing are based on second order signal statistics, and assume linear models of speech production. Extending to higher order statistics and nonlinear models is attracting considerable interest at present, as these have the potential to extract more information about the speech production process.

10.3.4.1   Higher order cumulants and spectra

The autocorrelation function is the second-order cumulant of a signal [35], and its Fourier transform is the power spectrum. The properties of these functions are well understood, and they have found widespread use in all areas of signal processing. The information provided by second order statistics, however, is fairly restricted. For example, phase information is lost, and only linear dependencies in the signal are represented. Higher order cumulants and spectra preserve phase information and can reveal nonlinear dependencies which occur in many signals of interest, including speech (see Mendel [35] for a comprehensive tutorial on higher order statistics).

The third order cumulant of a signal is defined as:

$$C_3[x(n)] = E\{x(n)x(n+m_1)x(n+m_2)\} \qquad \text{all } m_1, m_2 \qquad \text{... (10.13)}$$

where $E\{\ \}$ denotes expectation. Again, this is recognizable as a correlation of three terms. For orders of four and above, however, extra lower order terms contribute to the cumulant function, and the correspondence to familiar correlations no longer holds. In practice, the expectation would be estimated by averaging over segments of real data. This gives rise to one disadvantage of higher order methods, as they require longer data segments for reliable estimates. An important area for research in speech processing is therefore the estimation of higher order cumulants from short data segments, during which stationarity can be assumed. Due to this estimation problem, effort will probably be concentrated on third order cumulants for speech.

Cumulants have several properties which could be exploited in speech processing:

- Gaussian processes (white or coloured) have zero valued cumulants for orders greater than two;

- if $x(n)$ and $y(n)$ are independent, then:

$$C_k[x(n) + y(n)] = C_k[x(n)] + C_k[y(n)] \qquad \ldots (10.14)$$

   for all k;

- as a consequence of the first two properties, cumulants of order three and above are robust to additive Gaussian noise, even if the noise is coloured.

Robustness to coloured Gaussian noise has been demonstrated in speech recognition experiments [36], and cumulants have also been applied successfully to pitch detection with speech corrupted by more realistic types of noise (aircraft, car) [37].

As pointed out by Mendel [35], it is worth revisiting many of the algorithms based on second-order techniques to see if higher order statistics will improve their performance, particularly if the signals under investigation may have nonlinear dependencies, are non-Gaussian and are contaminated by noise.

10.3.4.2 Nonlinear dynamics

The concepts of deterministic chaos and fractals have been popularized in recent years by their application in modelling natural phenomena. The beauty and complexity of the Mandelbrot set and Julia sets, and the ability to generate highly complex and seemingly random behaviour from very simple recursive mappings have fascinated the general public and academics alike. In signal processing, analysis techniques from these fields have the potential to improve identification and modelling of nonlinear systems.

Traditionally, systems have been classified as either deterministic or stochastic. The state of a deterministic system can be predicted exactly at any time, given precise knowledge of the initial state. A stochastic system is governed by probability, and produces a random output. Chaotic systems, on the other hand, can produce apparently random outputs from very simple nonlinear deterministic equations. An important feature of chaotic systems is sensitivity to initial conditions, where arbitrarily close initial states can produce totally different output sequences. Chaotic systems can therefore be highly unpredictable, regardless of their deterministic nature, if the state of the system cannot be measured or estimated with sufficient accuracy.

The dynamic behaviour of a system can be represented by plotting its state as a function of time in a suitable space, giving a graph known as a phase portrait. An unforced pendulum, for example, could be represented in a two-dimensional space of angular displacement and velocity. Regardless of the initial state of the pendulum, it eventually comes to rest vertically giving a fixed point in the phase portrait. This state is an attractor for the unforced pendulum. The concept of attractors is important in dynamical systems analysis. Two types of attractor exist for stable non-chaotic systems:

- the fixed point, as described for the pendulum;

- the limit cycle, where the system periodically repeats a sequence of states.

In chaotic systems, a third type of attractor can be found — the strange attractor. The phase portrait may have a distinct pattern occupying a compact region of the space, yet the system will never visit the same state more than once. The structured but infinitely variable nature of these attractors led to the 'strange' label.

For practical applications, methods are required which can detect and characterize chaotic behaviour by analysis of the system output alone. One method under investigation is determination of Lyapunov exponents, which 'are the average exponential rates of divergence or convergence of nearby orbits in phase space' [39]. Basically, they are indicators of sensitivity to initial conditions. Positive exponents mean that orbits with arbitrarily close starting points will diverge, giving very different output sequences — chaos — whilst negative exponents imply convergence to limit cycles or fixed points.

Another approach is to estimate the dimension of the attractor, which is generally fractal as it does not fill the phase space. A quantity known as the correlation dimension [40] has been used in experiments with speech signals [41], and evidence of chaotic behaviour has been found in pathological speech using similar techniques [42]. The application of such techniques in speech analysis is becoming more widespread, and has the potential to reveal more of the nonlinear nature of the speech production process.

10.4   CONCLUSIONS

Although the scope of this chapter is necessarily restricted, the review of basic spectral analysis techniques and their application in speech processing has illustrated the need for development of more suitable signal processing algorithms for analysis, modelling and synthesis of speech signals. The process is one of continuous feedback — improved algorithms lead to a better understanding of speech, which in turn drives algorithm development in more appropriate directions. In particular, the recent growth of interest in nonlinear dynamics and higher order statistics for signal analysis may lead to better characterization of the speech signal, as many aspects of speech production and perception appear to be nonlinear in nature.

REFERENCES

1.   Fourcin A J, Harland G, Barry W and Hazan V: 'Speech input and output assessment', Ellis Horwood, Chichester (1989).

2.   Gimson A C: 'An introduction to the pronunciation of English', Edward Arnold, fourth edition (1989).

3.   Ladefoged P: 'A course in phonetics', Harcourt Brace Jovanovich (1982).

4.   Wells J C: 'A study of the formants of the pure vowels of British English', MA Thesis, University of London (1962).

5.   Fant G: 'Speech sounds and features', Cambridge, Mass, MIT Press (1973).

6.   Delattre P C, Liberman A M and Cooper F S: 'Acoustic loci and transitional cues for consonants', JASA, 27 , No 4, pp 769-773 (July 1955).

7.   Borden G and Harris K S: 'Speech science primer', Baltimore, Williams and Wilkins (1980).

8.   Stevens K N: 'Quantal nature of speech', in David E E Jr, Denes P B (Eds), 'Human communication: A unified view', New York, McGraw-Hill (1972).

9.   Fant G: 'Auditory patterns of speech', in Wathen-Dunn (Ed), 'Models for the perception of speech and the visual form', Cambridge Mass, MIT Press (1967).

10.   Lieberman P: 'Intonation, perception, and language', Cambridge, Mass, MIT Press (1967).

11. Makhoul J: 'Linear prediction: a tutorial review', Proc IEEE, 63 , No 4, pp 561-580 (April 1975).

12. Rabiner L R and Schafer R W: 'Digital processing of speech signals', Prentice-Hall, Englewood Cliffs, NJ (1978).

13. Davis S and Mermelstein P: 'Comparison of parametric representations for monosyllabic word recognition in continuously spoken sentences', IEEE Trans ASSP, ASSP – 28 , No 4, pp 357-366 (August 1980).

14. Pols L: 'Real-time recognition of spoken words', IEEE Trans Computers, C – 20 , No 9, pp 972-978 (September 1971).

15. Ahmed N, Natarajan T and Rao K: 'Discrete cosine transform', IEEE Trans Computers, C23 , No 1, pp 90-93 (January 1974).

16. Breen A P B: 'Speech synthesis models: a review', Electronics and Communication, 4 , No 1, pp 19-31 (February 1992).

17. McCandless S S: 'An algorithm for automatic formant extraction using linear prediction spectra', IEEE Trans ASSP, ASSP – 22 , pp 135-141 (1974).

18. Christensen R L, Strong W J and Palmer P E: 'A comparison of three methods of extracting resonance information from predictor-coefficient coded speech', IEEE Trans ASSP, ASSP – 24 , No 1, pp 8-14 (February 1976).

19. Yegnanarayana B: 'Formant extraction from linear-prediction phase spectra', JASA, 63 , No 5, pp 1638-1640 (May 1978).

20. Seeviour P M, Holmes J N and Judd M W: 'Automatic generation of control signals for a parallel formant speech synthesizer', Proc IEEE ICASSP'76, pp 690-693 (April 1976).

21. Lowry A, Hall M C and Hughes P M: 'Analysis and encoding of speech for a parallel formant synthesiser', Proc IEEE ICASSP'89, pp 492-492 (May 1989).

22. Lowry A, Hall M C and Hughes P M: 'Iterative parameter optimization techniques for parallel-formant encoding of speech', Proc IEE ECCTD, No 308, pp 537-541 (September 1989).

23. Hughes P M: 'Formant based speech synthesis', BT Technol J, 6 , No 2, pp 84-90 (April 1988).

24. Rabiner L R, Atal B S and Sambur M R: 'LPC prediction error — analysis of its variation with the position of the analysis frame', IEEE Trans ASSP, ASSP – 25 , No 5, pp 434-443 (October 1977).

25. Miyoshi Y, Yamato K, Mizoguchi R, Yanagida M and Kakusho O: 'Analysis of speech signals of short pitch period by a sample-selective linear prediction', IEEE Trans ASSP, ASSP – 35 , No 9, pp 1233-1240 (September 1987).

26. Wood L C and Pearce D J B: 'Excitation synchronous formant analysis', IEE Proc I, 136 , No 2, pp 110-118 (April 1989).

27. Hess W: 'Pitch determination of speech signals', Springer-Verlag, Berlin (1983).

28. Fourcin A J and Abberton E: 'First applications of a new laryngograph', Medical and Biological Illustration, 21 , pp 172-182 (1971).

29. Ananthapadmanabha T V and Yegnanarayana B: 'Epoch extraction of voiced speech', IEEE Trans ASSP, ASSP-23 , pp 562-569 (1975).

30. Ananthapadmanabha T V and Yegnanarayana B: 'Epoch extraction from linear prediction residual for identification of closed glottis interval', IEEE Trans ASSP, ASSP − 27 , pp 309-319 (1979).

31. Strube H W: 'Determination of the instant of glottal closure from the speech wave', JASA, 56 , pp 1625-1629 (1974).

32. Kadambe S and Bourdreaux-Bartels G F: 'A pitch detector based on event detection using the dyadic wavelet transform', Proc ICSLP '90, 1 , pp 469-472 (November 1990).

33. Hlawatsch F and Boudreaux-Bartels G F: 'Linear and quadratic time-frequency signal representations', IEEE Signal Processing Magazine, 9 , No 2, pp 21-67 (April 1992).

34. Rioul O and Vetterli M: 'Wavelets and signal processing', IEEE Signal Processing Magazine, pp 14-38 (October 1991).

35. Mendel J M: 'Tutorial on higher-order statistics (spectra) in signal processing and system theory: theoretical results and some applications', Proc IEEE, 79 , No 3, pp 278-305 (March 1991).

36. Paliwal K K and Sondhi M M: 'Recognition of noisy speech using cumulant-based linear prediction analysis', Proc IEEE ICASSP'91, pp 429-432 (May 1991).

37. Moreno A and Fonollosa A R: 'Pitch determination of noisy speech using higher order statistics', Proc IEEE ICASSP'92, 1 , pp 133-136 (1992).

38. Packard N H, Crutchfield J P, Farmer J D and Shaw R S: 'Geometry from a time series', Physical Review Letters, 45 , No 9, pp 712-716 (September 1980).

39. Wolf A, Swift J B, Swinney H L and Vastano J A: 'Determining Lyapunov exponents from a time series', Physica D, 16 , pp 285-317 (1985).

40. Grassberger P and Procaccia I: 'Characterization of strange attractors', Physical Review Letters, 50 , No 5, pp 346-349 (1983).

41. Tishby N: 'A dynamical systems approach to speech processing', Proc IEEE IC-ASSP '90, 1 , pp 365-368 (April 1990).

42. Herzel H and Wendler J: 'Evidence of chaos in phonatory samples', Proc Eurospeech '91, 1 , pp 263-266 (September 1991)

11

SPEECH CODING FOR TELECOMMUNICATIONS

I Boyd

11.1 INTRODUCTION

The function of a speech codec is to convert an analogue speech signal into a digital form for efficient transmission over a digital path, or storage on a digital storage medium, and to perform the complementary function of converting a received digital signal back to analogue. There are good grounds for claiming that the speech codec is the most important part of many speech systems. It is normally the speech codec which defines the basic speech quality for the whole system, since no amount of sophistication elsewhere in the system can compensate for degradation introduced in the codec. The users' perception of the system is strongly influenced by the quality of this one component.

The main thrust of recent work on speech coding has been to reduce the transmission bit rate. High quality (hi-fi) coding of a 20 kHz audio signal requires sampling at 40 ksamples/s and quantization to 16 bits resulting in a transmission rate of 640 kbit/s (40×16 kbit/s). This requirement is reduced in the case of a 3.4 kHz telephony signal to 96 kbit/s assuming 8 kHz sampling and 12-bit linear quantization. It is possible, however, to reduce the transmission rate of a telephony signal to 64 kbit/s by replacing the 12-bit linear quantization by 8-bit nonlinear quantization (using either an A-law

This chapter is based upon the paper which appeared in the IEE Electronics and Communications Journal, Vol 4, No 5 (October 1992), and is reproduced by kind permission.

or μ-law nonlinear quantization scheme). A transmission rate of 64 kbit/s is the standard rate of transmission for a telephony signal on the UK inland network; however, since speech is structured, its inherent redundancy can be exploited to reduce this rate still further (see sections 11.3-11.6).

Just a few years ago, the prospect of low bit rate speech coders being used for public telecommunications services seemed remote; there was little demand for them and the available speech quality was far too poor anyway. Since then, the growth of competition and the introduction of new services have led to a large number of important applications. Advances in speech coding techniques have been rapid too, leading to greatly improved speech quality at low bit rates.

By far the most important use of low bit rate speech coding is the creation of new mobile services by the efficient use of precious radio spectrum. The whole area of personal cordless and mobile communications is becoming increasingly important as users expect to have continuous access to communications services, and to be instantly contactable, without being tied to a particular location. New digital mobile radio (DMR) services are constantly being introduced to meet this market demand. For instance, BT introduced an aeronautical telephone service for passengers during 1990, based on a 9.6 kbit/s speech codec. Another example of a new DMR system is the full-rate pan-European digital cellular mobile telephone service, which employs a 13 kbit/s speech codec, and will be introduced in European countries throughout the early 1990s. Personal communications networks (PCN) in the UK, also targeted for introduction in the early 1990s, will use the same 13 kbit/s speech codec as the pan-European digital cellular system. Yet another indication of the popularity of mobile communications is the saturation in some locations of channel capacity for the current generation of cordless telephones. The next generation of cordless telephones (CT2) in the UK will employ 32 kbit/s speech codecs which will provide increased system capacity and improved quality and security of service. In all of these mobile services, the number of customers who can be served, and hence the revenue available to service providers, is crucially dependent on the bit rate at which the speech codecs can achieve the required performance. Lower bit rate codecs will be used as soon as possible, emphasising the importance of continued research and development in low bit rate speech coding.

Services using digital radio links are inevitably subject to transmission errors. Low bit rate speech codecs are used for these services not only to save bandwidth but also because they can be made very robust to digital errors. The capacity of digital cellular telephone systems, for example, is limited by radio interference from nearby cells using the same frequencies which causes digital errors in the received signal. The ability of the speech codecs to give good speech quality which is very little affected by these errors allows the radio frequencies to be reused much more intensively than in analogue systems; this is just as important in increasing the system capacity as that increase achieved by the reduction in bit rate.

Another important use for speech codecs is to reduce the unit cost of providing a service by improving the efficiency of usage of expensive capital equipment. For instance unit cost reduction is an important factor in the use of low bit rate codecs to multiplex several conversations on to a channel which would normally carry a single conversation (e.g. to replace one 64 kbit/s PCM channel with eight speech channels carrying speech coded at 8 kbit/s). Multiplexing is quite common on private communication circuits and on expensive international links in public networks (such as undersea cable links). It is rare, however, on the normal inland circuits of public networks, because the relatively low cost of providing short inland transmission links negates the possible cost benefits.

Cost reduction is also important in the use of low rate coding for speech storage. The amount of RAM or disk memory needed can be greatly reduced in voice mailbox or telephone answering machines, for example.

11.2 SPEECH QUALITY REQUIREMENTS AND ASSESSMENT

For all telecommunications services, quality of service is one of the critical elements determining the success of the service. One aspect of the quality of service is speech quality.

11.2.1 Speech quality requirements

The quality objective for speech codecs used in public telecommunications systems is usually to achieve the same speech quality as in a long distance telephone call on the analogue public switched telephone network (PSTN), often referred to as 'toll quality'. Since such calls are variable in quality, toll quality is a rather nebulous concept and some of the claims made for codecs are difficult to justify. A more precise target is to meet the transmission performance standards agreed by the International Telegraph and Telephone Consultative Committee (CCITT) for PSTNs (the CCITT is the international body concerned with the recommending of standards for telecommunications). However, there are also services where it would be uneconomic to meet the CCITT quality criteria with currently available technology and where lower speech quality is deemed acceptable. Current analogue cellular radio telephone services, for example, whilst much better than older mobile services when heavily loaded, give a quality below that of the fixed PSTN. The full-rate speech codec for the pan-European digital cellular system, which will also be used in the UK personal communications networks (PCN), was chosen to give quality which, on average, is at least as good as the analogue cellular systems. The speech quality achieved by cellular systems is below the CCITT quality standards. Similarly, lower speech quality is acceptable for maritime radio telephones and for aeronautical telephones. However, for cordless telephones it is considered necessary to achieve toll quality.

For speech codecs forming part of a transmission system the following factors affecting the speech quality or affecting the successful operation of the system must also be considered.

11.2.1.1 Digital transmission errors

Transmission systems using radio links may have very high error rates with the errors probably occurring in bursts. Channel coding, which provides error correction/detection, will usually be employed in these systems (possibly with interleaving to break up error bursts). However, this will still leave a residual average bit error rate at the input to the speech decoder, typically between 1 in 50 and 1 in 1000. The speech codec must therefore be designed to maintain speech quality when the decoder input data is corrupted.

11.2.1.2 Wide range of input speech

To be useful in a public service, speech codecs must operate well with a wide range of speakers, from adult males to young children, with a wide range of speech levels (a dynamic range of at least 25 dB is usually required), with distorted speech, with background acoustic noise, which in some applications can be very high, and with more than one person talking at a time. Each of these conditions is very difficult for certain types of speech codec to handle, yet each is very likely to arise if the speech codec input is from the analogue PSTN.

11.2.1.3 Tandem connections

Tandem connections of speech codecs will occur for mobile-to-mobile calls for DMR systems. Unfortunately, for certain types of speech codec, tandem connections with other speech codecs (identical or of a different type) can result in a small amount of distortion introduced by the first codec leading to a large increase in the distortion introduced by the second codec.

11.2.1.4 Delay

The delay of the system of which the speech codec is a part also affects the user's perception of the quality of the service. The mean one-way propagation time (often known as the delay) of a speech codec, may be measured by connecting the output of the encoder directly to the input of the decoder and measuring the time which elapses between speech entering the encoder and the reconstituted version of the same speech leaving the decoder.

For applications where the codecs are used purely for speech storage the delays introduced by speech codecs are likely to be unimportant. For transmission applications, though, delay is important for two reasons — firstly because excessive delay (more than about 400 ms) causes conversational difficulties and secondly because, if there are any sources of echo, either electrical or acoustic, in the transmission path, then propagation delay can cause the resulting echoes to become objectionable.

Both of these considerations can impose limits on the codec delay; the limit in any particular application will depend on the other delays in the connections involved. For example, if a speech codec is to be connected to the UK PSTN then its delay must be limited to 5 ms unless echo control devices are also provided.

11.2.1.5 Non-speech signal requirements

In addition to the speech quality requirements placed on speech codecs, for some systems they are also required to reproduce with low distortion modem data signals, multifrequency signalling tones and information tones, such as telephone ringing tones and engaged tones. Generally, the lower the bit rate of the speech codec and the more the encoding algorithm is 'speech specific' the more unlikely it is that the codec will pass these tones with sufficient fidelity.

11.2.2 Subjective assessment of speech quality

The assessment of speech quality from low bit rate speech codecs poses problems quite different from the assessment of the waveform codecs (such as A-law PCM) used at higher bit rates. The first reason for this is that the distortions produced by low bit rate codecs are so diverse in character; for instance they may be correlated with the speech signal or they may be entirely random. Moreover, different people's assessments of these diverse degradations can vary significantly; for example, some people find background noise or other non-speech sounds generated in the codecs more disturbing than distortion of the speech, whilst others are prepared to accept noise if the speech is clear.

The most straightforward way of assessing speech codecs is to make objective measurements, such as those specified in CCITT Recommendation G.712 for PCM systems. Unfortunately no objective tests have yet been devised which will reliably give the same results over a range of different coding methods as subjective assessment by potential users. Thus subjective testing [1] is used to check that speech codecs meet the requirements of their intended telecommunications applications. These subjective measurements have to be reliable, carried out in a way that takes account of the major interactions between the codec and the other parts of the transmission system, and calibrated in some way so that results from different laboratories can be compared. The 'informal' listening tests sometimes carried out on speech codecs can be unreliable and many of the claims for codec performance based upon them turn out to be incorrect or to relate only to a very restricted set of operating conditions.

The subjective testing of a speech codec should evaluate the effect of all the factors which are expected to influence its performance. For instance, in addition to the range of operating conditions discussed above (transmission errors, wide range of typical input conditions, tandem connections), a range of listening levels and environments may also need to be included in the assessment.

11.3   REVIEW OF SPEECH CODING TECHNIQUES

Speech coding algorithms can be classified into the following three types:

- waveform coding;
- vocoding;
- hybrid coding algorithms.

The most basic waveform coders do not attempt to exploit any knowledge of the speech production process in the encoding of the input signal. The aim of waveform coders, as the name implies, is to reproduce the original waveform as accurately as possible. As these coders are not speech specific they can cater for many non-speech signals, background noise and multiple speakers without difficulty. The penalty of a relatively high bit rate, however, must be paid for this 'acoustic robustness'.

In contrast, vocoders (voice + coders) make no attempt to reproduce the original waveform but instead derive a set of parameters at the encoder which can be used to control a speech production model at the decoder. The parameter set for the speech production model is relatively small and they can be efficiently quantized for transmission; hence vocoders operate at very low bit rates.

The most simple model of speech production used by vocoders is illustrated in Fig. 11.1. For voiced speech (such as the vowel sound 'a') the voiced excitation signal is modelled by a train of unipolar, unit amplitude impulses at the required fundamental frequency (the equivalent perceived frequency is known as the pitch frequency). For unvoiced sounds (such as the fricative 'f') the unvoiced excitation is modelled as the output from a pseudo-random noise generator. The voiced/unvoiced switch selects the appropriate excitation and the gain term controls the level of the excitation. The spectral shaping of the excitation signal by the vocal tract is modelled by a time-varying spectral shaper.

Linear predictive coding (LPC) analysis [2] may be used to derive the coefficients of a time-varying linear digital filter which models the spectral shaping of the vocal tract. For speech coding the parameters of this filter are updated typically at intervals of between 20 and 30 ms, with 20 ms being the most common update period and 30 ms normally reserved for codecs operating at bit rates of 4.8 kbit/s and below. The aim of LPC analysis is to extract the set of parameters from the speech signal which specifies the filter transfer function giving the best spectral match to the speech being encoded. An all-pole filter of order p (usually in the range 10 to 16) is used to model the spectral shaping of the vocal tract. This is a good model for reasons relating to the way in which speech is both produced and perceived. For example, the spectral envelope of the short-term speech signal contains a number of peaks at frequencies closely related to the formant frequencies, i.e. the resonant frequencies of the vocal tract. With regard to speech perception, it is the spectral peaks (formants) and not the spectral troughs which are the most significant. An all-pole filter model reproduces the spectral peaks much better than the troughs.

Hybrid coders combine features from both waveform coders and vocoders to form sophisticated coding schemes which provide good quality, efficient speech coding. In common with vocoders, hybrid coders make use of a speech production model. However, unlike the very simple representation of the excitation signal used by vocoders, a refined representation of the excitation is employed. Hybrid coders operate at a medium bit rate somewhere between vocoders and waveform coders.

11.4 WAVEFORM CODING

The simplest and best known waveform encoding technique is pulse code modulation (PCM). When PCM employs non-uniform 8-bit quantization (A-law or μ-law) with 8 kHz sampling, very good quality speech is achieved at 64 kbit/s.

The bit rate required by waveform coders for speech encoding can be reduced by exploiting the correlation between adjacent samples, for example by encoding the difference between successive samples rather than the samples themselves. One such scheme is known as differential pulse code modulation (DPCM). By adapting the quantizer step-size of a DPCM coder according to the short-term speech power, the speech coding technique known as adaptive differential pulse code modulation (ADPCM) is obtained [3].

ADPCM is an important speech coding technique and as a result several CCITT ADPCM speech coding standards exist. Perhaps the most important CCITT ADPCM standard is Recommendation G.721 (blue book), as this Recommendation also forms the core of other CCITT ADPCM standards. Recommendation G.721 fully defines the 32 kbit/s fixed bit rate CCITT speech coding algorithm.

ADPCM speech coding, in accordance with Recommendation G.721, has been specified in the common air interface (CAI) for the base stations used by the second generation cordless telephony (CT2) standard and it is proposed that CT2 handsets could use a 'cut-down' G.721 algorithm to reduce power consumption. The G.721 standard has also been adopted for the emerging digital European cordless technology (DECT) standard. DECT base stations will conform to G.721 and, as for CT2 standard, a reduced complexity G721 algorithm may be used for DECT handsets.

The CCITT Recommendation G.726 encompasses Recommendation G.721 and extends it to include operation of the algorithm at bit rates of 16, 24 and 40 kbit/s as well as 32 kbit/s. Recommendation G.726 is most commonly used for digital circuit multiplication equipment (DCME); this equipment is employed by network operators to increase the capacity of inter-continental connections by using the 32 kbit/s coding rate as the basic speech coding rate with 24 and 16 kbit/s employed on channels at peak overload periods. The 40 kbit/s coding rate is used for voiceband data transmission. The CCITT Recommendation G.727 is a variant of G.726 and defines an algorithm which uses an embedded structure [4] to provide speech encoding at the same four bit rates as G.726. Equipment conforming to G.727 is suitable for use on packetized networks.

11.5   VOCODING

The operation of vocoders is very closely based on the speech production model of Fig. 11.1. The vocal tract section of the model can be described in several ways; for example, it may be represented by the amplitude of the

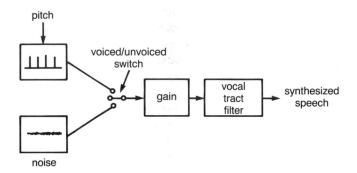

Fig. 11.1   Speech production model.

short-term frequency spectrum at specific frequencies (channel vocoder) [5], by the major spectral peaks (formant vocoder) [5], or by linear predictive coding (LPC) coefficients (LPC vocoder) [6].

The excitation is usually modelled in vocoders as shown in Fig. 11.1, i.e. a series of unipolar pulses spaced at the pitch frequency for voiced speech and noise for unvoiced speech. This excitation model is poor for several reasons, including the following:

- speech does not fall neatly into the two categories of voiced and unvoiced;

- pitch is not constant in voiced speech but subject to 'micro-variations' — the constant pitch of the excitation over several voiced excitation periods contributes to the synthetic sound of vocoders.

As vocoders are so strongly based on the simple speech production model of Fig. 11.1 they perform very poorly with high levels of background noise, multiple speakers and non-speech signals.

LPC vocoding is the most widely used vocoding technique today. LPC vocoders usually operate at around 2.4 kbit/s and give synthetic quality speech which is unacceptable for PSTN applications. It is, however, acceptable for military applications as a means of providing secure low bit rate communications — a US Federal standard algorithm is widely used for military communications [7].

11.6   HYBRID CODING

Hybrid coders combine features from both waveform coders and vocoders to form sophisticated coding schemes which provide good quality, efficient speech coding. The most common hybrid coders use the LPC synthesizer model as the speech production model. An understanding of the operation of LPC-based hybrid coders can be obtained by considering how the all-zero inverse LPC filter, i.e., the inverse of the all-pole LPC filter discussed in section 11.3, could be used to form a speech coder.

When a speech signal is filtered through the inverse LPC filter the short-term correlations of the speech signal are effectively removed leaving a noise-like waveform, known as the residual signal. If the speech is voiced the residual waveform will contain periodic spikes at the pitch frequency and if it is unvoiced the residual waveform has almost no discernible structure. If the residual signal is not quantized and is filtered by the unquantized all-pole LPC synthesis filter, the original speech signal is reproduced at the output of the synthesizer (Fig. 11.2). Even if the LPC filter is quantized, provided the quantization is relatively accurate, the reproduced speech is almost indistinguishable from the original. With this in mind, it is clear that provided the residual signal can be quantized accurately and efficiently, a high quality practical speech codec can be obtained.

The most obvious way to construct a codec based on this principle is to treat the residual as a waveform and quantize it directly — this form of coding is known as adaptive predictive coding (APC). However, as the residual is noise-like there is almost no correlation between adjacent samples, so none of the differential encoding schemes can be effectively employed and a relatively high bit rate is required.

A variant on APC is known as residual excited linear predictive coding (RELP) [8]. For a RELP coder the residual is low-pass filtered to band-limit the residual to around 1 kHz and this band-limited signal is then resampled at a lower rate. As the sampling rate of the residual signal has been reduced there are fewer samples to be transmitted and hence fewer bits are required to quantize this band-limited residual signal. A full-band signal

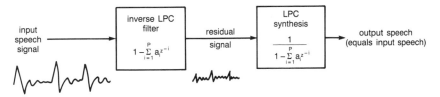

Fig. 11.2   The formation of the residual signal.

is obtained at the decoder by a nonlinear distortion technique which 'fills in' the missing high-frequency energy [8].

As the bit rate of both APC and RELP coders is reduced towards 8 kbit/s the number of bits available to represent the residual becomes insufficient and the speech quality deteriorates rapidly. At bit rates less than 8 kbit/s a very sophisticated and complex representation of the excitation is required for efficient and effective speech encoding. It is possible to view these more advanced coding techniques as 'residual substitution' techniques, the aim of which is to find a suitable replacement for the residual which can effectively substitute for the residual signal and act as the excitation for the LPC synthesizer. The substitute signal needs to:

- produce an output from the LPC synthesis filter similar to that obtained using the unquantized residual;

- be sufficiently accurately represented by much fewer bits than would be required for the residual.

To find a suitable substitution signal which meets these requirements a closed loop or analysis-by-synthesis technique is employed (Fig. 11.3).

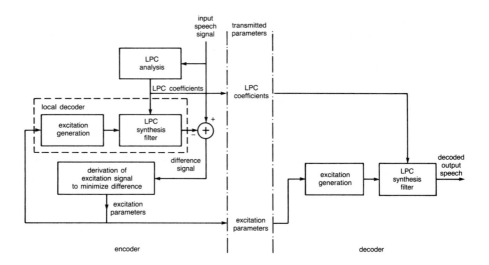

Fig. 11.3   Analysis-by-synthesis hybrid coding scheme.

For an LPC vocoder the LPC synthesizer model is used only at the decoder; however, for an analysis-by-synthesis LPC-based coder the LPC synthesizer model is employed at both the encoder and decoder. At the encoder the LPC coefficients of the synthesizer model are obtained directly from the input speech signal and the excitation signal is derived using a closed loop analysis-by-synthesis technique (Fig. 11.3). Analysis-by-synthesis, as the name implies, is the use of synthesis as an integral part of the analysis process. The objective of the encoder analysis-by-synthesis process is to derive an excitation signal, such that the difference between the input and synthesized signals is minimized according to some suitable criterion. To enable the difference signal to be obtained the synthesized signal has to be generated at the encoder using the LPC synthesizer/excitation generator combination (often referred to as the 'local decoder'). The LPC coefficients and the closed-loop derived excitation signal are transmitted as quantized values to the decoder. The decoded output signal is generated by passing the received excitation signal through the LPC synthesis model obtained using the received LPC coefficients (Fig. 11.3).

A brief description of the very popular low bit rate analysis-by-synthesis LPC coders is presented in the next few sub-sections.

11.6.1   Multipulse excited LPC

In 1982 Atal and Remde [9] presented a codec using a substitute excitation which they named multipulse excitation. A multipulse LPC (MPLPC) decoder is illustrated in Fig. 11.4. In MPLPC a series of non-uniformly spaced pulses with different amplitudes is used to excite the filter. Unlike LPC vocoding, no distinction is made between voiced and unvoiced speech: the same type of excitation waveform is used for all speech segments. For good quality speech, several pulses are required per pitch period. However, as all the pulse positions and amplitudes must be transmitted, a quality versus bit rate trade-off has to be made. The derivation of the appropriate pulse positions and amplitudes at the encoder is, of course, crucial to the coder performance.

An MPLPC encoder is illustrated in Fig. 11.5. This figure illustrates that the method of deriving the multipulse excitation involves an analysis-by-synthesis procedure, i.e. the input and synthesized speech are compared and the excitation is derived to minimize the error between the two signals. The excitation analysis procedure requires the partitioning of the input speech into small blocks (20-80 samples), and a search for the pulse positions and amplitudes which minimize the error, between the input and synthesized speech, over the block. A one-pass solution to finding the optimum positions

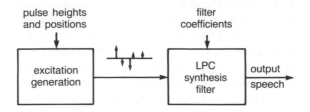

Fig. 11.4   Multipulse decoder.

and amplitudes is a highly nonlinear problem and is thus extremely complex. One possible method of finding the excitation is to split the problem into two parts, and try every possible combination of pulse positions. Given the pulse positions, the pulse amplitudes which minimize the error can be found relatively easily. The pulse positions and amplitudes which yield the lowest error over the block form the optimum excitation signal. Unfortunately, even for a small block size and only a few pulses per block, the number of possible combinations is relatively high, and results in an excessive computational load. Sub-optimal methods have thus been developed which find the pulse positions and amplitudes one at a time [9]. These sequential methods reduce the error minimization process to that of selecting a pulse position as the location at which a maximum occurs in a cross-correlation function. Once a pulse position has been found, the calculation of the corresponding pulse amplitude is straightforward. Given both the pulse position and amplitude, the cross-correlation function can be updated and the search for the next pulse can proceed.

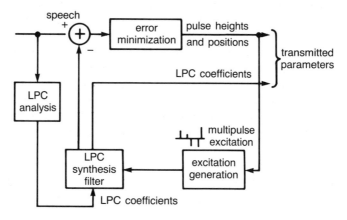

Fig. 11.5   Basic multipulse encoder.

The MPLPC coder described above can be extended [10] to include long-term prediction and perceptual weighting (Fig. 11.6). The long-term prediction included in the encoding process takes advantage of the long-term correlations in speech which arise primarily as a result of pitch related correlations in voiced speech. With the inclusion of long-term prediction, fewer pulses are required per pitch period to obtain the same speech quality. The long-term prediction parameters can be accurately quantized with relatively few bits.

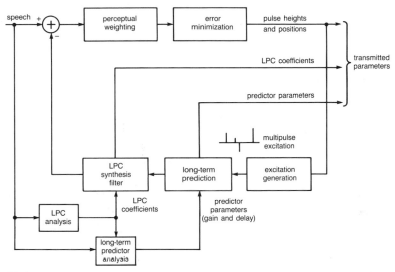

Fig. 11.6   Multipulse encoder including perceptual weighting and long-term prediction.

Perceptual weighting of the error is included at the encoder so that the pulse positions and amplitudes are chosen in a way which minimizes the perceived distortion rather than just minimizing the mean squared error. The approximately flat noise spectrum obtained at the subtractor output is shaped by trading-off increases in the noise power around spectral peaks (formants) in the speech, with decreases in the noise power between peaks. Due to auditory masking effects, this shaping of the noise spectrum allows the speech signal to more effectively mask the noise. Multipulse codecs can operate successfully over a wide range of bit rates — for example, from 8 kbit/s to 16 kbit/s.

A BT Laboratories-developed multipulse speech codec has been selected by INMARSAT and the Airlines Electronic Engineering Committee for a world-wide aeronautical telecommunications service. The codec operates at 9.6 kbit/s (8.9 kbit/s speech encoding rate plus 0.7 kbit/s for error correction/detection). This codec uses a 20 ms frame size for the LPC analysis and ten new LPC coefficients are generated every 20 ms. The long-term predictor parameters are also updated every 20 ms, whereas the analysis-by-synthesis based excitation analysis uses 4 ms sub-frames with three excitation pulses derived per sub-frame [11].

The BT codec was selected for this aeronautical application after extensive subjective testing of several speech codecs from the US and Japan showed the BT Laboratories-developed multipulse speech codec to provide the best speech quality under realistic operating conditions (transmission errors and acoustic background noise). BT, along with Norwegian and Singapore Telecom, have been offering a world-wide aeronautical satellite telecommunications service, known as Skyphone, using this codec since September 1990.

11.6.2   Regular pulse excited LPC

The regular pulse excited (RPE) LPC coder [12] is a variant of the multipulse coder. For a regular pulse excited coder, the excitation signal pulses are spaced uniformly, usually every 3-5 sample positions. With pulses spaced every four positions there are four possible candidate excitation vectors (Fig. 11.7). The encoding process involves finding the best candidate vector and the appropriate pulse amplitudes. An index defining the selected candidate vector and the quantized pulse amplitudes must be transmitted to the decoder.

Fig. 11.7   Four possible candidate vectors for an RPE coder — the ones indicate pulse positions.

The 13 kbit/s full-rate codec algorithm adopted as the standard for the forthcoming pan-European digital cellular mobile telephone service is essentially a regular pulse-excited LPC algorithm which includes long-term prediction [13]. The codec operates with an LPC frame size of 20 ms (160 samples) and employs an 8th order LPC analysis. A new set of long-term prediction parameters (gain and delay) and excitation parameters (RPE index and excitation pulse amplitudes) are derived every 5 ms using an analysis-by-synthesis procedure.

In addition to the 13 kbit/s speech coding rate, channel coding (using half-rate convolutional coding plus a cyclic redundancy check) is included, bringing the overall transmission bit rate of the full-rate pan-European digital cellular mobile telephony system to 22.8 kbit/s.

Although the full-rate system is now just starting to be introduced into service, the process to select a suitable speech and channel coding scheme for the half-rate system is already well under way. The half-rate system requires a combined speech and channel coding rate of 11.4 kbit/s; the selection of the split between speech and channel coding rates has been left to the codec designers. Most designers, however, have opted for a speech coding rate between 5 and 7 kbit/s.

In subjective tests completed during December 1991, six candidate half-rate codecs were compared with each other and the full-rate speech and channel coding scheme using extensive subjective testing. The subjective test results indicated that none of the candidates performed as well as the full-rate codec, and, as a result, the standardization of the half-rate codec has been delayed beyond the end of 1992. All six candidate speech codecs were based on the popular code-excited linear prediction algorithm, the basic principles of which are outlined below.

11.6.3   Code-excited linear prediction

The code-excited linear prediction (CELP), or stochastic-excited linear prediction (SELP) coder, is another coder based on an analysis-by-synthesis technique [14]. A block diagram of a CELP encoder is shown in Fig. 11.8. The LPC filter, long-term predictor and perceptual weighting filter are exactly as described for the multipulse excited LPC coder. The difference between multipulse and CELP coders is the excitation function — the pulses of multipulse are replaced by an 'innovation sequence'. At the decoder of a CELP codec each block of reconstructed speech samples is produced by filtering the selected innovation sequence through the long-term filter and then the LPC vocal tract filter. At the encoder there is a codebook which contains many different innovations. The encoder selects the 'optimum' innovation sequence by filtering each sequence in the codebook in turn; the sequence which results in the minimum weighted mean squared error between the input speech signal and the synthesized signal is chosen. A copy of the codebook is stored at the decoder and an index number is transmitted to identify the selected innovation sequence along with a gain term which sets the energy of the excitation signal. As with multipulse coders no distinction is made between voiced and unvoiced speech — the same method of analysis is used to determine the excitation waveform for all speech segments. The

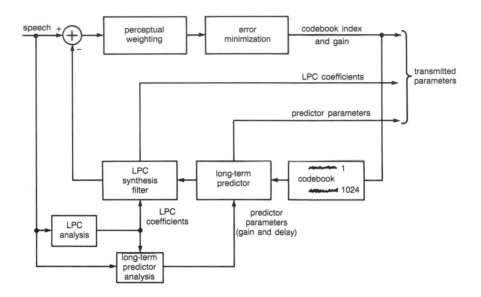

Fig. 11.8   Code-excited LPC encoder.

innovation sequence length is typically 40 samples and the long-term predictor analysis, or now more commonly, the adaptive codebook analysis [15], is also typically performed every 40 samples.

The innovation sequences can be either vectors of random numbers [14], sparse vectors (only a few sample positions have non-zero values) [16], sparse ternary vectors (all sample positions have pulses of amplitude 1, 0, or − 1) [16], or some form of structured codebook (e.g. a codebook where the next consecutive codebook entry is a shifted version of the previous entry [17]).

Figure 11.9 presents a spectral plot of a voiced segment of a speech signal and Fig. 11.10 presents the corresponding spectral plot of the decoded output speech from an 8 kbit/s CELP codec. Figure 11.9 clearly illustrates the subjectively important spectral peaks (formants) which are present in this voiced segment of speech, and Fig. 11.10 illustrates that the 8 kbit/s CELP codec reproduces the spectral peaks in the lower half of the frequency band quite accurately. However, from these figures it is also clear that the spectral peaks are not so well reproduced in the upper half of the frequency band. This lack of fidelity in the reproduction of the higher frequency information results in the decoded output speech sounding slightly 'muffled'.

CELP encoding is a very important coding technique — several digital cellular radio standards, the US Department of Defense (DOD) standard for secure telephony, and the emerging CCITT 16 kbit/s standard are all based on CELP coders.

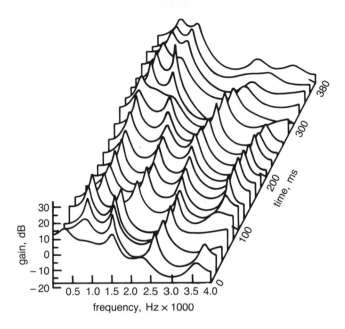

Fig. 11.9   Spectral plot of original speech signal.

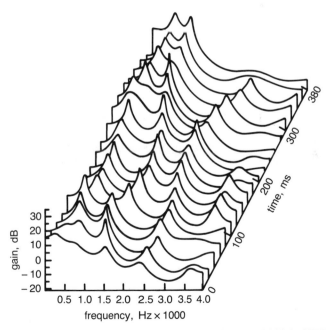

Fig. 11.10   Spectral plot of decoded output speech signal from an 8 kbit/s CELP codec.

The CELP codec specified in the US DOD 4.8 kbit/s standard [17] employs a tenth order LPC filter which is updated every 30 ms. The parameters of the long-term predictor, which is in the form of an adaptive codebook, and the fixed-excitation codebook are derived every 7.5 ms using an analysis-by-synthesis procedure. The codec has a structured fixed-excitation codebook with consecutive entries in the codebook shifted in time by two samples. The codec has been designed to operate at bit rates up to 4.8 kbit/s — at 4.8 kbit/s the full 512 entry fixed codebook is searched; for bit rates less than 4.8 kbit/s a subset of the codebook is searched. The complexity of the codec is substantially reduced when a subset codebook search is employed making a single digital signal processing device implementation possible. The codebook is structured so that codecs operating at different bit rates can intercommunicate, and the perceived quality will be equivalent to that of the lower bit rate codec.

The CELP codecs specified for the US and Japanese digital cellular telephony standards were both developed by Motorola and are known as vector-sum excited LPC (VSELP) codecs [18]. VSELP codecs employ the same LPC analysis and adaptive codebook structure as conventional CELP codecs — the major difference between VSELP and conventional CELP codecs is the structure of the fixed codebook(s) (two fixed codebooks are used for the 8 kbit/s US digital standard and one codebook for the 6.7 kbit/s Japanese standard). The fixed codebook(s) of VSELP codecs are formed from linear combinations of basis vectors which leads to structured codebook(s) being formed. The structure of each codebook is used to control the codebook search procedure such that a significant reduction in complexity is achieved [18].

One very important recent CELP codec development from AT&T Bell Laboratories is the low delay CELP (LD-CELP) speech coding algorithm which has been selected as the CCITT 16 kbit/s speech codec standard [19]. The most significant feature of this codec is that it has a delay of less than 2 ms. To achieve this low delay the traditional forward LPC analysis, which involves waiting until a block (20 ms) of speech samples is available before commencing the analysis, has had to be dispensed with in favour of a backward LPC analysis. In backward LPC analysis the block on which the LPC analysis is based is formed using the most recent past reconstructed speech signal. As the reconstructed speech is available at both the encoder and decoder, the LPC analysis is repeated at the decoder and hence there is no need to transmit the LPC coefficients.

To achieve low delay the excitation vector length has had to be limited to five samples compared with the more traditional 40 samples of low bit rate forward adaptive codecs. In a similar manner to the LPC analysis, the gain term associated with the selected innovation vector is also updated in a backward adaptive manner avoiding the need for the gain term to be transmitted. As no long-term predictor/adaptive codebook is used by this coder, the only parameter transmitted is the selected 10-bit codebook index every five input samples. The main disadvantage of this coder is the high complexity which results from both the large codebook (1024 entries) which has to be searched every five samples (0.625 ms) and the very high order (50th) backward adaptive LPC analysis which is employed. Unlike previous CCITT Recommendations for speech coding algorithms, the Recommendation for the 16 kbit/s algorithm (Recommendation G.728) is not a 'bit-exact' specification, i.e. every parameter and operation is not defined down to bit level. Thus to meet the requirements of the G.728 Recommendation, implementations of the algorithm may use either floating- or fixed-point arithmetic. Several implementations of the 16 kbit/s algorithm on different floating-point arithmetic digital signal processing devices already exist and implementations on fixed-point arithmetic digital signal processing devices are currently being developed. Extensive testing will be required to ensure that the different implementations of the algorithm interwork correctly with one another.

The CCITT 16 kbit/s standardization has progressed to the stage that a fixed-point version of the codec will soon be available and subjective tests will be performed to confirm that this codec performs at least as well as the 32 kbit/s G.721 ADPCM standard in all conditions tested. The CCITT 16 kbit/s standardization is due to be completed by mid-1993.

11.6.4   Frequency domain algorithms

Many of the time-domain coding algorithms outlined above have their counterparts in the frequency domain; for example sub-band coding (SBC) [20,21] and adaptive transform coding (ATC) [22] are two popular techniques. Frequency domain coders are most often used for the high quality encoding of wide bandwidth (approximately 7 kHz) audio signals. As for the low bit rate encoding of telephony signals, the standards for these audio codecs are important to ensure correct interworking and compatibility between different manufacturers' implementations. One of the most important standards is the CCITT Recommendation G.722 which specifies an algorithm for the encoding of 7 kHz audio signals within 64 kbit/s. This algorithm is based on the two-band coder which divides the incoming audio signal into two frequency bands, 0-4 kHz and 4-7 kHz — independent ADPCM encoders are employed to encode the two resulting signals. The encoder will operate at 64 kbit/s, 56 kbit/s and 48 kbit/s with a gradual decrease in the quality of the encoded signal as the bit rate is reduced. The availability of the 56 and 48 kbit/s bit rates enables a wideband audio signal and an 8 and 16 kbit/s data signal respectively to be multiplexed on to a 64 kbit/s carrier.

The International Organization for Standardization (ISO) is currently going through the process of ratifying the standardization of an audio codec capable of encoding audio signals of different bandwidth at correspondingly different bit rates. Probably the most important bandwidth/bit rate combination of this codec for wideband telephony is the encoding of 20 kHz audio signals at 128 kbit/s. In this mode of operation encoding of compact disc (CD) audio signal is reported as being transparent.

Recently a new frequency domain technique, known as multiband excitation (MBE) coding has been adopted for a satellite mobile communications application [23]. This coding technique is based on a more sophisticated model of the speech production process than that shown in Fig. 11.1. In the MBE model this speech signal is split into 20 ms speech frames and the spectral shaping of the vocal tract for each speech frame is modelled by means of a number of non-overlapping, variable width frequency bands. The number of frequency bands used for any particular speech frame is dependent on the fundamental frequency of voicing identified for that frame. The bandwidth of the frequency bands is set at three times the fundamental frequency of voicing. The required frequency band analysis is obtained using a discrete Fourier transform (DFT).

In the simple model of Fig. 11.1 each segment of speech is classified as either voiced or unvoiced; however, in the MBE model, voiced/unvoiced classifications are derived for each frequency band of each speech frame. If all the frequency bands are judged to be voiced, the excitation in all frequency bands will correspond to the voiced form of excitation (sinusoidal); if all the frequency bands are judged to be unvoiced, the excitation in all frequency bands will correspond to the unvoiced form of excitation (noise-like); if some bands are judged to be voiced and some unvoiced, then the excitation for each frame will be a combination of the voiced and unvoiced forms of excitation. The ability to have a mixed voiced and unvoiced form of excitation within the same speech frame overcomes one of the weaknesses of the more simple speech production model.

For each frequency band, three spectral amplitude values are derived for the three harmonics of the fundamental frequency of voicing which fall within that frequency band. These three amplitudes define the frequency response of the MBE model within a frequency band. The parameters transmitted per speech frame are the voiced/unvoiced decisions, the fundamental frequency of voicing identified, and the three spectral amplitude values per frequency band.

At the decoder, for each frequency band identified as voiced, the excitation is generated as a combination of three sinusoidal signals whose amplitudes are controlled by the transmitted spectral amplitudes and whose frequencies are at three harmonics of the fundamental frequency of voicing within that frequency band. For unvoiced frequency bands the excitation signal is derived from a white noise generator with the three spectral amplitude values controlling the energy of the excitation signal.

An improved multiband excitation (IMBE) coder has been adopted by INMARSAT for the INMARSAT-M land mobile satellite communications system [23]. The overall speech and channel coding rate for this system is 6.4 kbit/s with the speech codec operating at 4.15 kbit/s. The 2.25 kbit/s of channel coding is provided by a combination of Hamming and Golay codes.

11.7   COMPARISON OF THE SPEECH QUALITY AND COMPLEXITY OF DIFFERENT SPEECH CODING TECHNIQUES

There is no universally accepted measurement of complexity for speech coding algorithms, as neither the number of instructions per second (MIPS) figure nor the number of operations (MOPS) figure by themselves adequately define the complexity of an algorithm. In general, those who are concerned with the implementation of speech coding algorithms are actually concerned with how much power the implementation consumes, not how complex the algorithm is. While the two are clearly closely related, unfortunately the MIPS or MOPS complexity figures do not tell the whole story with regard to power consumption. For example, it is possible with many algorithms to lower the MIPS and MOPS figures of an algorithm implementation by increasing the amount of memory used to implement the algorithm. Thus to measure the complexity of an algorithm in such a way that it relates more directly to power consumption, the amount of memory required by an implementation must also be taken into account. While measurements of speech coding algorithm complexity have been employed based on *ad hoc* combinations of MIPS/MOPS figures and memory requirements, no universally accepted formula has been established. Without a meaningful widely accepted complexity measurement available, it is difficult to present precise complexity comparisons for different coding techniques. Therefore complexity comparisons in broad terms only are presented in Table 11.1 for various speech coding techniques. All comparisons are relative to PCM which has been assigned the arbitrary complexity figure of 1.

While the speech quality of a speech coder can be defined in terms of mean opinion scores (MOS), and, provided the same test conditions and references are used, meaningful comparisons of speech quality between coders can be obtained in terms of MOS, only a broad classification has been attempted in Table 11.1. The three speech quality categories used in Table 11.1 are:

● toll quality (quality of a long distance PSTN connection);

● communications quality (speaker identity maintained, good intelligibility but discernible loss in quality compared with toll quality);

● synthetic quality (speech sounds synthetic in nature, speaker identity information largely lost).

Table 11.1   Complexity and speech quality comparison for several speech coding techniques.

Speech coding technique	Bit rate (kbit/s)	Speech quality	Complexity
PCM (G711)	64	Toll	1
ADPCM (G721)	32	Toll	10
CELP (G728)	16	Toll	450
RPE-LPC (GSM)	13	Communications	100
VSELP (US cellular std)	8	Communications	250
CELP (US DOD)	4.8	Synthetic/ communications	400
IMBE (INMARSAT standard-M)	4.15	Synthetic/ communications	150
LPC10 (military std)	2.4	Synthetic	100

Table 11.1 illustrates, not surprisingly, that in general to increase speech quality or decrease coder bit rate results in the complexity of the coder increasing. However, while the complexity of new speech coding algorithms generally exceeds that of previous algorithms, the continued increase in the available processing power from new digital signal processing devices is keeping pace with the demand.

11.8   FUTURE CHALLENGES

The applications for speech coding over the past few years have presented many interesting and varied challenges for the speech codec designer, and the future looks to be just as full of challenges. In particular, focusing on the next five years, the major challenges facing the speech codec designer are the development of:

- a low delay CCITT standard 8 kbit/s speech coding algorithm offering equivalent quality to the 16 kbit/s standard now being set by the CCITT;

- a 4 kbit/s speech codec (but with no low delay constraint) providing equivalent quality to 32 kbit/s ADPCM;

- a 2.4 kbit/s codec achieving the quality of today's state-of-the-art 4.8 kbit/s algorithms.

No doubt other major challenges will also arise over this period and progress will not only need to be made in fundamental speech coding algorithm development, but also progress in the engineering of these basic algorithms to suitable forms for implementation may need to be even more rapid than it has been in the past.

11.9   CONCLUSIONS

The very high demand for mobile communications is currently the main driving force behind the development of new speech coding algorithms. Mobile communications applications are usually very demanding, requiring good quality speech transmission to be maintained in the presence of acoustic background noise and transmission errors, while the delay and the complexity of the codec must be kept to a minimum. As the demand for mobile communications continues to grow, speech codec designers are likely to be faced with very challenging problems for some years ahead.

REFERENCES

1.   CCITT: 'Handbook of Telephonometry, section 2 subjective tests', (Geneva 1987).

2.   Makhoul J: 'Linear prediction: a tutorial review', Proc of the IEEE, 63 , No 4, pp 561-580 (April 1975).

3.   Benvenuto N, Bertocci G and Daumer W R: 'The 32 kbit/s ADPCM coding standard', AT&T Techn J, 65 , No 5, pp 12-21 (September 1986).

4.   Goodman D J: 'Embedded DPCM for variable bit rate transmission', IEEE Trans on Commun, COM—28 , No 7, pp 1040-1046 (July 1980).

5.   Flanagan J L, Schroeder M R, Atal B S, Crochiere R E, Jayant N S and Tribolet J M: 'Speech coding', IEEE Trans Comm, COM—27 , No 4, pp 710-737 (April 1979).

6. Tremain T E: 'The Government standard linear predictive coding algorithm: LPC-10', Speech Technology, pp 40-49 (April 1982).

7. 'Telecommunications: Analog to digital conversion of voice by 2400 bit/sec. Linear Predictive Coding, FED-STD-1015', Office of Technology and Standards, National Communications System, Washington DC (March 1983).

8. Dankberg M D and Wong D Y: 'Development of a 4.8-9.6 kbit/s RELP vocoder', Proc IEEE Int Conf on Acoustics Speech and Signal Processing, pp 554-557 (1979).

9. Atal B S and Remde J R: 'A new model of LPC excitation for producing natural-sounding speech at low bit rates', Proc IEEE Int Conf on Acoustics Speech and Signal Processing, pp 614-617 (1982).

10. Singhal S and Atal B S: 'Improving the performance of multi-pulse LPC coders at low bit rates', Proc IEEE Int Conf on Acoustics Speech and Signal Processing, pp 1.3.1-1.3.4 (1984).

11. Boyd I and Southcott C B: 'A speech codec for the Skyphone service', BT Technol J, 6, No 2, pp 50-59 (April 1988).

12. Deprette E F and Kroon P: 'Regular excitation reduction for effective and efficient LP-coding of speech', Proc IEEE Int Conf on Acoustics Speech and Signal Processing', pp 25.8.1-25.8.4 (1985).

13. Vary P et al: 'Speech codec for the European mobile radio system', Proc IEEE Int Conf on Acoustics Speech and Signal Processing, pp 227-230 (1988).

14. Schroeder M R and Atal B S: 'Code-excited linear prediction (CELP): high quality speech at very low bit rates', Proc IEEE Int Conf on Acoustics Speech and Signal Processing, pp 937-940 (1985).

15. Kleijn W B, Krasinski D J and Ketchum R H: 'Improved speech quality and efficient vector quantization in SELP', Proc IEEE Int Conf on Acoustics Speech and Signal Processing, pp 155-158 (1988).

16. Xydeas C S, Ireton M A and Baghbadrani D K: 'Theory and real time implementation of a CELP coder at 4.8 and 6.0 kbit/s using ternary code excitation', Proc IERE Fifth Int Conf on Digital Processing of Signals in Communications, pp 167-174 (1988).

17. 'Telecommunications: Analog to digital conversion of radio voice by 4800 bit/sec. Code excited linear prediction (CELP), FED-STD-1016', second draft, Office of Technology and Standards, National Communications system, Washington DC (November 1989).

18. Gerson I and Jasuik M: 'Vector sum excited linear prediction (VSELP) speech coding at 8 kbit/s', Proc IEEE Int Conf on acoustics speech and signal processing, pp 461-464 (1990).

19. Chen I, Lin Y and Cox R V: 'A fixed point 16 kbit/s LD-CELP algorithm', Proc IEEE Int Conf on Acoustics Speech and Signal Processing, pp 21-24 (1991).

20. Crochiere R E, Webber S A and Flanagan J L: 'Digital coding of speech in sub-bands', Bell Syst Techn J, pp 1069-1085 (October 1976).

21. Kingsbury N G: 'Robust 8000 bit/s subband speech coder', IEE Proc F, $\underline{134}$, pp 352-366 (July 1987).

22. Zelinski R and Noll P: 'Adaptive transform coding of speech signals', IEEE Trans Acoustics, Speech and Signal Processing, $\underline{ASSP-25}$, No 4, pp 299-309 August 1977).

23. Lim J S and Hardwick J C: 'The application of the IMBE speech coder to mobile communications', Proc IEEE Int Conf on Acoustics Speech and Signal Processing, pp 249-252 (1991).

12

THE DSP DEVELOPMENT PLATFORM FOR THE SKYPHONE SPEECH CODEC

R M Mack, C D Gostling, I Boyd and C B Southcott

12.1   INTRODUCTION

Implementing speech codecs on digital signal processing (DSP) devices just after their introduction was a time-consuming and frustrating affair. It was time-consuming because the programmer had to learn all about the new device. It was frustrating because as soon as a particular device had been mastered, a new, faster device offering improved functionality would be introduced. Inevitably for the next speech codec implementation, the new device was required because the latest speech coding algorithm was much more complex than the previous one.

To avoid the unsatisfactory situation outlined above, it was concluded that a multi-DSP device computing unit was required for real time speech codec development. This unit became commonly known as the DSP computer. Its purpose was to enable the speech codec hardware designer to learn about one DSP device and to use it for some years to come. Change was to be avoided by choosing a DSP device with an upwardly compatible upgrade path and by using the multiprocessor capability of the unit to compensate for lack of speed or functionality in comparison to competing DSP devices from other vendors. The main objective of this development was to enable rapid transition from a high-level language description of an algorithm to a working DSP device implementation. This made the development of an efficient cross-compiler essential, and strongly suggested that the selected DSP device should have floating-point capability.

In the following sections, the development and functionality of the multi-DSP device computing unit is described. Its use is illustrated by a description of the development of a speech codec for the Skyphone aeronautical satellite telecommunications service [1]. The evolution of a second generation of the DSP computer is also outlined.

12.2   EXPERIENCES WITH THE EARLY DSP DEVICES

Prior to the emergence of floating-point DSP devices, BT Laboratories had set up a study group [2] to consider the case for design and production of an in-house DSP device satisfying all the requirements of a broad spectrum of telecommunications applications. Although the eventual decision was not to produce such a device, the participants in the study accumulated a good deal of knowledge relating to DSP devices available at that time. Over the same period, considerable practical experience had been gained with three integer DSP devices used in separate speech codec developments. The three devices were each of different manufacture, and each had their shortcomings for speech coding applications. An overview of the three devices is given below, highlighting their individual drawbacks. Table 12.1 compares the main features.

Table 12.1   Comparison of early devices.

Device	Fujitsu MB8764	NEC 7720	Texas Instruments TMS32010
Process	CMOS	CMOS	NMOS
Year	1983	1982	1980
Area (sq mm)	91	28	45
Power (W)	0.29	1.0	0.9
Precision (bits)	16	16	16
Multiplier	16*16: = 26 (integer)	16*16: = 32 (integer)	16*16: = 32 (integer)
Speed	100 ns	250 ns	200 ns
Program memory	1K*24 ROM/EXT	512*23 EPROM/ROM	4K*16 ROM/EXT
Data RAM	256*16	128*16	144*16
Data ROM	in program	512*13	in program

12.2.1   NEC 7720

This was the first DSP device with a hardware multiplier to become commercially available within the UK. Its most attractive features were its small size (28 pins) and the provision of on-chip erasable programmable read only memory (EPROM) for storing instructions and constants (in separate areas). There was also a version of the device which used masked ROM instead of EPROM, and allowed substantial cost savings for large quantity applications. The device had a serial input/output (I/O) capability and an 8-bit parallel port for communicating with standard general purpose microprocessors. A 32-bit result was available from the 16 * 16-bit multiplier, but at least three 250 ns instruction cycles were required before an accumulated product was available. The limitations of the device were its restricted on-chip random access memory (RAM) (128 * 16 bits), limited precision 13-bit constant ROM and virtually no external memory-addressing capability. The arithmetic logic unit (ALU) had an inadequate limiting facility, which meant that the programmer had to provide overflow protection.

12.2.2   Texas Instruments TMS32010

This device had the advantage of addressable external program memory at full speed. It had a 16 * 16-bit multiplier with 32-bit result, and had 32-bit ALU precision. A single multiply operation involved three 200 ns instruction cycles but, for consecutive multiplications, pipelined operation was available to reduce the effective multiply and accumulate time to 400 ns. A further two instructions were required to store the 32-bit result in memory. The architecture enabled program and constant data memory to be shared and hence traded. The device had only 144 words of on-chip RAM and could only access external RAM using multicycle fetch instructions. The lack of serial I/O pins was inconvenient as was the reduced accuracy of immediate data. Multiprocessor synchronization was difficult to achieve reliably and was not recommended.

12.2.3   Fujitsu MB8764

This device had two 128-word by 16-bit RAM areas. They could be used as two independent memories, with each RAM independently addressable using separate address arithmetic units, or as a single RAM having a continuous address space. In addition it could address up to 1024 words of external RAM at the full cycle speed of 100 ns. Instructions and constants could either be stored in internal ROM or in external memory. Concurrent operations within a single instruction could usually support a data move and an ALU function in the same cycle. It was possible to specify a multiply and add operation including a data move in one cycle. Within the device, however, there was two-stage pipelining of the multiply and accumulate operation. The instruction set also provided a divide instruction which took 17 cycles. The less attractive features of the device included the fact that only 26 bits of the 16 * 16-bit product were retained for processing by the ALU. This accuracy was inadequate for applications such as speech processing which requires a wide dynamic range. Also, the absence of any high speed serial I/O ports made it unsuitable for many practical applications.

12.3   THE MULTI-DSP DEVICE COMPUTING UNIT CONCEPT

While the DSP devices from different manufacturers each have their merits, the hardware designer must select the most suitable device for the particular application. In the rapidly changing realm of speech coding technology, there is a continuing search for better speech quality at ever decreasing bit rates. This progressive improvement results, almost invariably, in each new coding algorithm being more complex than its predecessor. In this situation, the hardware designer is forced to select the latest state-of-the-art DSP device in order to implement the algorithm. The penalty paid for adopting state-of-the-art technology is lack of experience with the new device, often coupled with incomplete documentation, inadequate development tools, and poor support from the device manufacturer. All of these factors combine to slow down the transfer of a high-level language simulation of a speech coding algorithm to a real time hardware realization.

The multi-DSP device computing unit, commonly known as the DSP computer, was conceived as a means of avoiding the above mentioned implementation delays, thereby allowing the most up-to-date speech coding technology to be evaluated for proposed new systems and services, and achieving a competitive edge in international codec selection and standardization activities. The DSP computer would offer the option to distribute the processing among several DSP devices in order that there would be sufficient processing power to implement the most complex algorithms for the foreseeable future. Ideally, the processing power would be expandable to any desired extent by adding DSP modules, and intercommunication between modules would be simple and easy to configure. A microprocessor for general housekeeping and control functions was envisaged, and would handle communication with the host computer, and download object code to the DSP device memory. An analogue interface would be developed as part of the overall system so that hardware aspects of the DSP device I/O facilities could be ignored for future speech codec implementations. With the objective of proceeding rapidly from a floating-point high-level language computer simulation to DSP device hardware, it was considered important that the chosen DSP device should also have floating-point capability so that dynamic range and scaling issues could be avoided in the code conversion. To further speed the transition to DSP device code, an efficient cross-compiler was to be developed for the selected DSP device, so removing most of the language translation task from the engineer.

If all of the above ideas could be realized in practice, then not only could the speech coder development team look forward to a long period of stability regarding the DSP device employed, but their knowledge of that device could, without consequence, be relatively superficial. A DSP device expert would no longer be essential to the production of speech codec demonstration hardware.

12.4 SELECTION OF THE DSP DEVICE

For the DSP computer, a high-level language programming capability was an important aspect of the overall concept. Accordingly, the chosen DSP device had to have an architecture and instruction set amenable to the eventual development of an efficient cross-compiler (or have a high-level language and optimizing compiler commercially available). The suitability of the device for multiprocessor applications was another factor and with these primary considerations in mind, a consultant from Queen's University, Belfast, was commissioned to make a study [3] of a short-list of DSP devices available at that time. As stated above, it was also desirable that the device should have the capability to directly perform floating-point arithmetic, but due to the limited number of such devices then available, two widely used fixed-point devices were included in the study on the grounds that the inconvenience and speed penalty of having to code floating-point operations might be tolerable if the device was sufficiently superior in other respects. A transputer device with floating-point capability was also included in the study.

12.4.1 Pros and cons of the various devices examined

The following devices were selected for study:

- INMOS F424 transputer;
- NEC μpD77230R;
- AT&T WE DSP32;

- Motorola MC56000;

- Texas Instruments TMS320C25;

- Texas Instruments TMS320C25 together with an associated 32-bit floating-point co-processor.

They were compared on the following six important aspects of performance:

- architecture — the complexity of the target machine is the most critical factor in the cost of developing an optimizing compiler;

- word size — since accurate representation of real numbers is required, 32-bit chips have a significant advantage over 16-bit devices;

- memory capacity — if the address range for external memory is limited, it may prohibit the implementation of complex algorithms;

- floating-point capability — this greatly simplifies the task of manipulating real numbers with maximum range and accuracy;

- speed — faster speed means more complex algorithms can be coded without resorting to parallel processing. Effective speed is dependent on not only cycle time, but also the design of the instruction set;

- multiprocessor capability — this may be necessary for the implementation of very complex algorithms, and some chips offer faster, more convenient communication between devices than others.

The relative performance of the devices is best illustrated by reproducing the tabulated rankings published in the study report [3] (Table 12.2). In the table, a minus suffix indicates a potentially prohibitive weakness with regard to the adoption of the device for the DSP computer.

As previously stated, the task of constructing an optimizing cross-compiler is critically dependent on the DSP device architecture. The criterion applied in the study report is that the assembly language programmer's model of the device should ideally be simple, and the fewer options available the better. The specific questions on which architectural complexity was assessed are reproduced here from the study report.

- Does the device have a reduced instruction set computer (RISC) architecture?

- How simple is the organization of the data memory space?

Table 12.2   Relative ranking of devices in critical areas of performance.

	F424	56000	TMS	TMS +	NEC	DSP32
Architecture	1	2	3	3	4	4
Word size	1	3	4	2	1	1
Memory capacity	1	2	3	3	5 –	4
Floating-point capability	3	4	5	3	2	1
Speed	5 –	1	2	1	3	4
Multi-processor capability	1	2	2	2	4	2

Note: TMS + = a TMS320C25 plus a 32-bit floating-point co-processor. This improves the TMS rankings on word size, floating-point capability and speed.

- How many different kinds of register are there?

- How convenient is floating-point arithmetic? For processors with a floating-point unit, is it pipelined? For fixed-point arithmetic, is there much programming effort required to manipulate real numbers with a wide dynamic range?

- Does the instruction set contain many special purpose instructions to increase efficiency (from the compiler writer's viewpoint, preferably not)?

The ideal device would be a RISC machine with a single uniform address space, a non-pipelined floating-point unit, and no registers at all (an unbounded evaluation stack being preferred).

Again from the study report, Table 12.3 compares the different devices on the above architectural complexity issues. In the table, a 'yes' statement is the preferred answer, while a 'yes/no' statement indicates that the device has some favourable aspects for the architectural issue in question.

The assessment categories for architectural complexity do not take into account the delayed branching and, for floating-point values, delayed memory write characteristics of the DSP32. As these would cause significant problems for compiler construction, the DSP32 was initially ranked lower than might be suggested by an inspection of Table 12.3.

Table 12.3   Architectural complexity comparison of the devices.

	F424	56000	TMS	TMS+	NEC	DSP32
RISC?	YES	YES	NO	NO	NO	YES
Simple data memory space?	YES	NO	NO	NO	NO	NO
Few kinds of register?	YES	YES/NO	NO	NO	NO	YES/NO
General purpose registers of single kind?	YES	YES/NO	NO	NO	YES	YES
Non-pipelined floating-point unit?	NO	—	—	?	NO	NO
Easy coding for fixed-point?	—	YES/NO	YES/NO	—	—	—
General-purpose instruction set?	YES	NO	NO	NO	NO	NO

Although featuring a 32-bit floating-point (pipelined) multiplier, the NEC device had fared poorly in other aspects of the study, particularly with regard to memory capacity, and was eliminated from further consideration. The F424 transputer was also eliminated since, despite achieving excellent rankings in nearly all aspects of the study, its slow speed for DSP applications was thought to be an overwhelming disadvantage. Of the remaining devices, the 56000 was ranked better than the TMS and approximately equal to the TMS with attached floating-point co-processor. The DSP32 was initially ranked below these, but the arrival of a faster (160 ns compared to 250 ns) version of this device removed one of its more major drawbacks. Its 32-bit floating-point multiplier could then be given the full merit it deserved. Doubts remained about the ease of developing an efficient cross-compiler for the DSP32, but a detailed reassessment showed that the difficulties were surmountable. Good support and interest in the project shown by AT&T representatives, coupled with the announcement of an enhanced upwardly compatible 80 ns device (the DSP32c) to be made available in the future, were sufficient to seal its selection for the DSP computer.

12.4.2   Description of the WE DSP32

The DSP32 [4] features 32-bit floating-point arithmetic and is available in both 250 ns and 160 ns instruction cycle versions. The on-chip memory consists of 2 kbytes of ROM and 4 kbytes of dynamic RAM with auto-refresh. The chip comes in either a 40-pin plastic dual in-line (DIP) or a 100-pin rectangular, ceramic pin-grid-array (PGA) package, but the PGA package must be used when external memory is to be accessed. The extra pins support the 32-bit data bus and the 16-bit address bus, and up to 56 kbytes of external memory can be accessed. As all instructions are 32-bit words, total program space is equivalent to about 16 kwords. The memory is divided into two banks but memory access can be made regardless of bank. However, as wait states are automatically inserted to resolve conflicts of memory access, for demanding applications maximum throughput is achieved by arranging for consecutive memory access to be to opposite memory banks.

The DSP32 features two execution units. The control arithmetic unit (CAU) performs 16-bit instructions for control and logical operations. It can execute 6.25 million instructions per second in the 160 ns device. The CAU contains a full-function ALU and 21 general purpose 16-bit registers.

The second execution unit, the data arithmetic unit (DAU), contains a full hardware floating-point multiplier and adder, and four accumulators. A 32-bit floating-point multiplier operand is comprised of 24 mantissa bits and 8 exponent bits, while the accumulators retain a further 8 mantissa bits for greater precision when input to the adder. The 160 ns device can perform 6.25 million multiply and accumulate operations per second; as this represents two floating-point operations, AT&T promote the 160 ns DSP32 as a 12.5 MFLOPS device. Although a succession of multiply-accumulate instructions can be performed at the full instruction cycle rate, this is achieved at the cost of a four stage pipeline (fetch/multiply/accumulate/store) within the DSP32. This means that up to four floating-point instructions can be in different stages of progress at any one time. Each individual multiply-accumulate instruction takes at least three cycles to complete, plus one additional cycle if the result of the accumulation is optionally written to memory. In addition to the multiply-accumulate operations provided by the DAU, special function instructions are provided for data type conversions such as floating-point to and from integer, and floating-point to and from μ-law or A-law.

The DSP32 features a serial I/O port and an 8-bit parallel I/O port, both of which support direct memory access (DMA) for transferring programs or data directly to or from the device memory without halting a program in execution. The parallel port is designed primarily for interfacing with an external microprocessor. The serial port allows a simple interface to PCM codec chips and time division multiplexed (TDM) data.

12.5   DESCRIPTION OF THE DSP COMPUTER

At the start of the development, the essential building blocks of the DSP computer had been identified as the following:

- a DSP32 device to perform the signal processing, with the capability to expand the processing power by the addition of further DSP32 modules;

- a microprocessor for system supervision and communication with a host computer (the host computer would be capable of running the AT&T software development tools for the DSP32, and would typically be an IBM personal computer (PC) or compatible computer);

- a high quality analogue interface;

- an A-law or μ-law PCM codec interface;

- an associated cross-compiler facility to allow DSP programs to be written in a high-level language.

In fact, the cross-compiler facility would be provided by the development of software to translate programs written in a subset of Pascal into AT&T source code for the DSP32, which could then be assembled in the normal way. The translation software became known as the Pascal compiler, and is described in some detail in section 12.7.

A block schematic of the hardware is shown in Fig. 12.1. Two RS232 serial ports handle external communications with the controlling microprocessor. This latter device was a Motorola MC68000, with provision for a maximum of 384 kbyte of EPROM and 128 kbyte of RAM on the board. A DSP32 device and its own 64 kbytes of RAM were also included on this board, together with a PCM codec device and all the necessary clock and timing logic. The board was, therefore, able to be used in stand-alone mode, with the PCM codec providing the analogue interface, and the DSP32 parallel port DMA facility being used to transfer the application program from the MC68000 EPROM to the DSP32 RAM.

The high quality analogue interface was designed as a separate module so that it could be isolated and screened from the digital noise prevalent on the main processor card. The analogue module featured 16-bit analogue-to-digital (A/D) and digital-to-analogue (D/A) converters, and a facility to change the anti-aliasing and reconstruction filter devices to cope with different

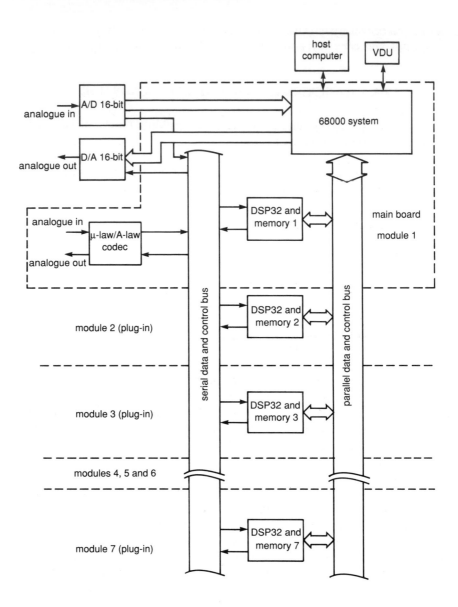

Fig. 12.1   DSP computer.

sampling rates. Two methods of connecting the analogue module to the main processor board were supported. The most commonly used of these was a serial interface linking with the serial port of the DSP32. The second method consisted of a 16-bit parallel interface to the MC68000 processor. This permitted a number of possible system configurations in which the MC68000 device could be used to perform part of the signal processing along with the DSP32 device(s).

To expand the processing power of the main board, individual modules containing a DSP32 device and associated memory could be linked to the main board. Each DSP32 device could be independently addressed by the MC68000, and associated data transfers took place via the DSP32 parallel port. Communication between DSP32 devices was via the serial port of the device. Although each DSP32 device has only one serial I/O port, data-multiplexer chips were used to allow up to seven DSP32 devices to be interconnected. An eighth input to the multiplexer devices allowed access to serial input data from the A/D converter or the PCM codec. The interconnection configuration of the DSP32 devices was controlled via the MC68000.

12.6 DEVELOPMENT OF THE SKYPHONE SPEECH CODEC ON THE DSP COMPUTER

12.6.1 Codec simulation and DSP computer implementation

An 8 kbit/s multipulse linear predictive coding (MLPC) [5] algorithm had been developed on a mini-computer and a real time implementation had subsequently been made using the Fujitsu MB8764 DSP device. The restricted dynamic range of this device made frequent scaling necessary, and its limited memory capacity prohibited any significant further development of the algorithm. However, further development was necessary to make the algorithm a viable contender for the Skyphone codec [1] selection; the line bit rate had to be increased to 9.6 kbit/s and more processing power and memory were required to enhance the algorithm. The prototype DSP computer had recently been completed and provided the necessary powerful real time hardware. As a first step towards the new implementation, the mini-computer simulation was expanded to 9.6 kbit/s by the inclusion of a pitch prediction routine, and by adding an extra excitation pulse for each subframe. (The Skyphone algorithm utilizes a 20 ms speech frame subdivided into five 4 ms excitation frames. Each excitation frame is served by a number of excitation pulses.) Using the simulation, the effect of transmission errors on certain parameters was investigated. In the light of the results it was considered necessary to provide a degree of error correction over the more important bits in the speech frame. Also, an error strategy was formulated for high error rates. When the number of errors in a speech frame exceeded a predefined threshold, the previous frame was repeated, and for a prolonged error burst the decoder output was muted.

At this point, the 9.6 kbit/s simulation was transferred to the DSP computer to run in real time. The transfer took place in a number of stages. Stage one implemented the encoding algorithm on the AT&T DSP32 development system, with the processed speech output derived from the 'local' decoder within the encoder. Stage two implemented an independent decoder receiving the transmitted parameters from the encoder. The third stage was the transfer of the DSP device code to the DSP computer. Finally, the parameter quantization was optimized, and the error correction strategy was tested and fine adjustments made.

12.6.2   Verification of the real time code

The real time implementation was verified by comparing the output from various stages of the real time processing with the output from the corresponding stages of the mini-computer simulation, when both versions were driven from an identical input stimulus. The algorithm stages, at which comparisons were made, were as follows:

Input sample block
Autocorrelation of the windowed samples
Durbin's recursion [6]
Quantization of reflection coefficients [1]
Filter tap derivation
Impulse response of filter

Five sub-blocks

Cross-correlation of filter impulse response and derived error
Pulse positions
Pulse heights
Quantization of pulse heights
Adaptation of pulse height quantizer
Output samples for each sub-block

As the floating-point formats for the mini-computer and the DSP32 were not identical, it proved necessary to test each stage of the algorithm in isolation, i.e. the input stimulus had to be injected directly into the input of the stage under test. If the individual stages were not isolated, and the input stimulus was applied to the first stage only, the different floating-point formats caused the outputs of the simulation and the real time implementation to diverge with progression through the algorithm, eventually preventing meaningful comparison.

12.7 THE PASCAL COMPILER

Developing DSP applications software can be a very slow process. This largely results from the complexity of the architecture of DSP devices and the low level at which the programmer usually has to operate. This factor slows down the development and the testing of new DSP algorithms, and makes it difficult to bring a product to market rapidly. The DSP32 has been designed to enable very high processing speeds to be obtained with careful programming. In the interest of performance, programming ease has often been sacrificed. For instance, pipelining of the floating-point unit is visible; there are multi-function instructions, and there are several awkward restrictions on when data can be accessed.

Using high-level languages for programming modern DSP devices is generally not practical. Traditional compilers [7], if they exist at all, are usually insufficiently sophisticated to exploit the complex architecture of a DSP device. For this reason a highly optimizing compiler was developed for the DSP32 processor which would enable algorithms to be coded in a subset of Pascal, while retaining a high degree of efficiency. The compiler implements a sophisticated set of optimizations that enable it to approach the efficiency of hand-coding. This opens the way to improved programming productivity, algorithm experimentation and product development. Care must be taken, however, in formulating the Pascal program, to enable the compiler to generate efficient code.

Although the compiler development was initially aimed at the DSP32 processor, at a later stage the necessary modifications were made to target the enhanced DSP32c device [8] of the second generation DSP computer (see section 12.8).

12.7.1 DSP32c architectural aspects

The DSP32c is a 32-bit floating-point DSP microprocessor device specifically designed for efficient execution of DSP programs. It is architecturally very similar to the DSP32, but with additional features. It processes data of two types:

- floating-point numbers (32 bits);
- integers or addresses (24 bits).

The addressing range is, therefore, vastly greater than for the DSP32 which has an integer number range of only 16 bits.

The DSP32c has 22 general purpose registers, r1, r2 ... r22, which can be used for addressing or integer calculations. Compared with the DSP32, r22 has been added to act as the interrupt vector table pointer when the interrupt handling facility of the DSP32c is utilized.

Like the DSP32, the DSP32c has four 40-bit floating-point accumulators, a0, a1, a2 and a3, which are used by the on-chip floating-point unit. The floating-point unit is pipelined; a full floating-point operation takes four instructions to complete. However, when processing arrays, it is possible to achieve a rate of one complete floating-point instruction per cycle. In general, however, care must be taken to avoid attempting to use the result of a floating-point operation before it has completed its passage through the pipeline. If no useful instructions can be performed while waiting for the result, then 'nop' instructions may have to be inserted. This is a source of inefficiency. Determining the minimum number of required 'nops' is a tricky task for a compiler.

12.7.2   Optimizations

In the following subsections, in all cases the Pascal compiler output code is presented exactly as it is generated by the compiler, including all comments associated with the code.

As an aid to understanding the DSP32c source code examples presented in the subsections, the meaning of a typical complex DSP32c multiply-accumulate instruction:

   *r1 = a0 = a1 + *r2+ + * *r3 + +

can be summarized as follows.

The product of two floating-point values stored in memory and indirectly addressed by registers r2 and r3 respectively, is added to the floating-point value stored in accumulator a1, and the result is stored in accumulator a0. The instruction has made use of the option to additionally store the result in a memory location indirectly addressed by register r1.

In the instruction, an asterisk prefixing a register identification indicates that register indirect addressing is being used, i.e. the register contains the address in memory of an operand to be used in the instruction or the memory address at which the result is to be stored. An asterisk also serves as a multiply symbol, but confusion between its two meanings is avoided because DAU instructions do not allow registers to be used as direct inputs to the multiplier, i.e. all DAU instruction register references will be prefixed by an asterisk indicating that the register is supplying a memory address.

A single plus symbol in the instruction indicates an addition following normal precedence rules, while two plus signs suffixing a register identification indicates that the address value stored in the register will be incremented following execution of the instruction (post-incremented). Similarly, a double minus sign would indicate that a post-decrement of the address is to occur.

In CAU instructions, if a reference to a register within the instruction has an 'e' suffix, this indicates that the instruction is performing 24-bit arithmetic as opposed to 16-bit. The option to perform 16-bit arithmetic is retained so that DSP32 programs are compatible with the DSP32c.

A typical 'branch on condition' instruction generated by the Pascal compiler would have the form:

$$\text{if (gt) goto pc} + (\text{labl} - . - 8)$$

This instruction can be described as follows.

If the result of the most recent instruction affecting CAU flags is greater than zero, then go to the instruction immediately following 'labl'. In the example instruction, the branch address is calculated relative to the program counter in order to avoid the jump address range limitation (16-bit) of using an immediate address. DAU flags, I/O flags and other CAU conditions can be tested by variations of this same branch instruction.

12.7.2.1 Optimal instruction selection

The DSP32c has a range of multiply-accumulate instructions, and several operations can be contained in the one instruction. An example is the multiply, accumulate and store instruction described above.

If the Pascal program contains an assignment statement such as:

```
sum := sum + a[i] * b[i]
```

a traditional compiler would generate a sequence of several single function instructions:

```
reg1 := a[i] * b[i];
reg2 := sum + reg1;
sum := reg2
```

For an assignment statement which has precisely the above form, it is much more efficient to evaluate the expression and store the result all in one step, using the above DSP32c instruction. To do this the compiler must match the syntactic form of assignments and expressions with the instruction set.

In general the compiler scans the instruction set for each expression evaluation or assignment, choosing the instruction which performs the most operations in one step.

12.7.2.2   Pipeline optimization

The compiler optimizes the use of the pipeline automatically. To do this, it tracks the contents of the pipeline at the end of each instruction, and keeps a record of which results are still in the pipeline (and cannot, therefore, be used). For each instruction, the compiler checks that operands which are required in the next cycle are not delayed in the pipe and inserts any necessary 'nop' instructions. The compiler also handles several other rather awkward restrictions and latency effects which would otherwise make life difficult for the assembly language programmer.

12.7.2.3   Standard optimizations

The compiler implements standard optimizations such as common sub-expressions elimination and loop invariant removal at various points. These are particularly important in generating efficient code for address calculations.

12.7.2.4   Autoincrement

One of the facilities of the DSP32c which is vital for efficient execution of loops is the ability to autoincrement (or decrement) addresses after use. It is a non-trivial task for a compiler to detect exactly when an address should be autoincremented. However, the Pascal cross-compiler developed for the DSP32/DSP32c is able to make good use of this facility. When the address of a variable is required, the compiler will first look to see if the address could have been available if a previous address had been incremented. The compiler can also cope in certain cases with incrementing by an address step greater than one unit of the data type. For instance, in compiling the loop:

```
for i := 1 to 256 do
    a[i] := b[3*i+1]
```

the compiler calculates that the address of b[3*i] has to be incremented by 12 bytes (3 floating-point locations) each time.

12.7.2.5   Special cases of the 'for' statement

Calculating variable addresses before entering a 'for' loop, and the use of autoincrement instead of explicit updating of the control variable, enable the compiler to rival the assembly language programmer for efficiency in several common special cases of 'for' loops.

The compiler can also spot automatically when it can make use of the DSP32c's special repetitive DO instruction, which enables certain 'for' loops to be implemented without looping overheads. Thus, the code fragment for calculating a summation:

```
sum : = 0.0;
for i : = 1 to 256 do
        sum : = sum + a[i] * b[i]
```

is compiled into the following assembly code:

```
/* sum : = 0.0 */
r1e = sum
r2e = flot1__0
*r1 = a0 = *r2
/* FOR i : = 1 to 256 */
r3e = a
r4e = b
do 0,255
```

```
for1:
    /* sum : = sum + a[i] * b[i] */
    *r1 = a0 = a0 + *r3+ + * *r4+ +
```

This code is quite close to optimal, and is comparable to what a reasonably proficient assembly language programmer might have produced. The one improvement which could be made is to avoid storing the summed result (a0) in memory (using *r1 every time round the loop). It would save one clock cycle each iteration to use the loop instruction:

```
a0 = a0 + *r3+ + * *r4+ +
```

and to store the result afterwards from a0 into memory. Specific improvements such as this are the subject of ongoing development of the compiler.

12.7.3   Other compiler features

To increase the usability of the compiler, and to enable programs to benefit from a certain amount of hand-coding for efficiency reasons, the compiler has been designed with several additional facilities:

- generating readable code — the compiler aims to generate very readable assembly language code, with the high-level Pascal format embedded as comments — this makes it much easier for a programmer who may wish to carry out some further hand-tuning of the generated code;

- external routines — by including a 'library' file which contains descriptions of existing routines, the programmer can make efficient use of external, previously hand-coded routines — the description of an external routine includes information for use by the compiler, such as the parameters to be passed, those registers that have their contents destroyed by the routine, the calling register, etc;

- interface to AT&T libraries — using the mechanism outlined above, an interface to the extensive AT&T libraries is available.

12.7.4   Compiler performance

To make it possible to use a high-level approach for DSP device programming, it is essential to have a highly optimizing compiler that can rival hand-coding for execution efficiency. The compiler presented here operates on a subset of Pascal and can approach the efficiency of hand-coding. A detailed analysis of compiler performance has still to be conducted, but for common operations, the current performance of the compiler is presented in Table 12.4. Three typical simple DSP algorithms have been selected and for each algorithm two DSP programs have been produced — a compiled Pascal version and a hand-coded assembly language version. These two versions were run on an AT&T simulator, and the resulting execution times compared. The three algorithms chosen were autocorrelation, Durbin's recursion and the conversion of the parcor coefficients to LPC 'a' coefficients. Table 12.4 shows the times obtained and, more importantly, the ratio of the execution times of the compiled and hand-coded versions. This ratio is the speed reduction factor obtained by using the compiler.

Table 12.4   Compiler performance — comparison with hand-coding.

Algorithm	Compiled	Hand-coded	Ratio: compile/hand
Autocorrelation	0.485 ms	0.3134 ms	1.5
Durbin	0.132 ms	0.1026 ms	1.3
Filter Taps	0.53 ms	0.294 ms	1.8

From this rather limited comparison, it can be seen that the performance of the compiled code appears to be within a factor of two of that of hand-crafted code. This makes it immediately suitable in an experimental or prototyping context and, given the readable nature of the compiled code, it is possible to hand tune the compiled code with only a small amount of effort to obtain a better performance. Using this approach, overheads of less than ten per cent are readily achieved. It is worth pointing out that the compiler is under constant development, and that the performance figures for the compiler are likely to improve with time. It is also worth noting that the figures obtained for hand-coding are not those of a first attempt. Developing a very efficient hand-coded DSP algorithm is an iterative process and takes considerable experience. Therefore, it is quite likely that the compiler will actually out-perform a relatively inexperienced DSP device assembly language programmer.

The advantage of automatic compilation is of course the productivity improvement which can be obtained, and means that the product and algorithm developers can spend more time on what they ought to be doing, rather than on detailed and tedious hand-coding in assembly language. In a DSP context the advantages of compilation are even greater, because of the very significant start-up time required to become familiar with the DSP architecture. It can take months before a good programmer can become sufficiently proficient in a particular architecture to be able to exploit its full performance potential. Using this compiler, a programmer new to the DSP32c can be producing acceptably efficient code in a very short time.

12.8   THE SECOND GENERATION DSP COMPUTER

The DSP computer had proved an immense aid to the rapid development of the Skyphone speech codec. However, the concentration of effort on the Skyphone development had prevented the DSP computer from being developed beyond the initial prototype design. While the functionality of the DSP computer was satisfactory, the main board was very large making it impractical to accommodate it in an equipment case. The expansion modules were designed to be piggy-back mounted on stand-off pillars on the main board, further complicating the housing problem and making it mechanically difficult to expand the system. It was clear that a physical redesign of the system was desirable, and since the first samples of the CMOS version of the DSP32 (the DSP32c) were just becoming available, any redesign should include an upgrade to this enhanced feature device. For the DSP computer, the most important performance enhancements offered by the DSP32c were a vast increase in memory capacity, and doubling of the processor speed.

At about the same time as the DSP computer redesign was contemplated, it became apparent that the DSP computer hardware satisfied many of the requirements of an equipment (see Chapter 13) to be developed by the transmission performance section at BT Laboratories. Their particular requirement was for signal processing hardware that could emulate any element of the telephone network, or combination of such elements, simply by running the appropriate software modules from a library of network element modules. A collaborative development led to a common specification for the new DSP computer/network emulator.

Figure 12.2 shows a diagrammatic representation of the new DSP computer/network emulator. In the new design, Eurocard sized DSP32c

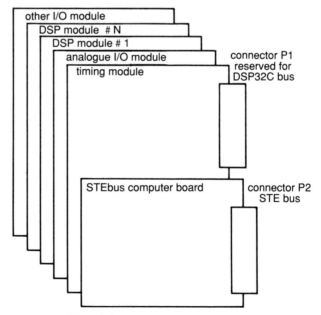

Fig. 12.2   Overview of the system.

based processor modules plug into a standard Eurocard (STE) bus backplane, via which they are controlled from an in-built IBM compatible PC. The PC can be used for DSP32c software development, and is used to download programs and data to the individual DSP modules. Additionally, it provides the user interface for interaction with the program. Part of the DSP32c device's 16-bit parallel I/O port is used to interface to the STE bus, permitting fast data transfers between the PC and the DSP module.

Despite the obvious similarities between the DSP computer and the network emulator, the multiprocessor requirement is fundamentally different for the two applications. For network emulation purposes each network element can be emulated as an independent process. Since no single element is likely to require the processing power of more than one DSP32c, multi-element networks can be emulated by linking DSP modules together in a pipeline fashion. In fact, two inputs to each module are required to model effects such as echo and sidetone. DSP modules can be linked by serial data connections, each operating at a common fixed data transfer rate. This contrasts to the multiprocessing facility envisaged for speech coder implementations on the second generation DSP computer. Here, the execution of a very complex algorithm could conceivably require the combined power of two or more DSP devices. The individual devices would need to interact frequently and must perform their respective assignments concurrently rather than in a pipelined sequential manner. Therefore, fast, flexible communications between devices is needed, and the fixed rate two-input port serial transfers might not be equal to the task. A more versatile means of inter-device communication was conceived, operating via the DSP32c device's external memory interface. This would allow very fast 32-bit parallel data transfers. However, to speed the realization of the complete equipment, the development of this parallel interface was deferred. To simplify its provision at a future date, the upper 96-way connector of the double Eurocard module is left totally uncommitted. In the short term, the serial communications facilities should prove adequate, since the high speed of the DSP32c will allow most speech coding algorithms to be implemented on no more than one or two DSP modules.

Figure 12.3 shows a block schematic of the DSP module. The serial interconnection of modules is set up by linking miniature coaxial sockets mounted on the front of the modules. Each module multiplexes two serial inputs and two serial outputs on to the DSP32c device's single serial I/O port. By the use of a 32-bit shift register in one of the channels, the transfers for the two channels are made to occur at the same time on the external interface. This allows greater freedom when interconnecting the modules. Clock rates for the transfers are generated in a separate timing module and routed via the backplane to the DSP modules. For this purpose, the STE

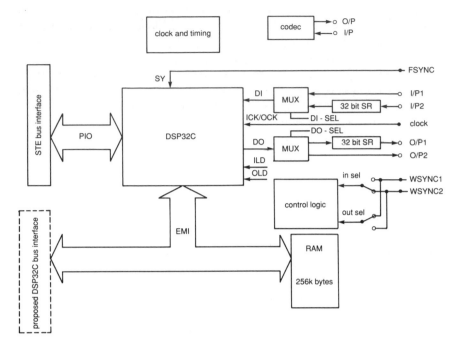

Fig. 12.3 DSP module.

bus backplane is customized to allow the use of the otherwise uncommitted row-B pins of the backplane connector. Two separate word synchronization rates are made available to the DSP modules to allow sample rate conversion to be implemented. All of the I/O timing and control logic is implemented in a logic cell array (LCA) programmable device.

The standard DSP module has provision for up to 256 kbytes of fast RAM with a facility to replace some of the RAM devices with fast EPROM devices. This enables the card to power-up into a stand-alone mode of operation without the need to download the program from the PC. An A-law or μ-law PCM codec is included on the card, and acts as the analogue interface in stand-alone mode.

A DSP module with a second memory area in addition to the 256 kbytes of fast RAM was also developed. This module is commonly called the error generator module as its primary function is to emulate the transmission errors that can occur in digital systems. However, it can also be configured to function as a standard DSP module within the system. This change of function is possible because, as with the standard DSP module, the I/O control logic is implemented in an LCA device, allowing the I/O logic to be reprogrammed for special applications. The second memory area of the error generator module can accommodate up to 512 kbytes of RAM or

2 Mbytes of EPROM, and is intended to be used to store transmission error data recorded from real systems. These error occurrences can then be generated in a subjective test environment, or used to demonstrate their effect on a transmission channel. Interactive software developed for use with this module allows either this stored error data or internally generated random errors to be selected as the error source. The error rate for random errors can be set to any value. If the transmitted data is composed of frames of bits, as in the case of low bit rate speech coders, then individual bits within the frame can be protected from or exposed to corruption by the errors by masking the appropriate bits on a monitor display of the frame bits. This has proved an extremely powerful tool in the development of robust low bit rate speech coders.

Compared with the original DSP computer, the high quality analogue interface module has been completely redesigned. The A/D and D/A converters are 16-bit hybrid circuit modules which plug into the mother-card. Different sampling rates can be accommodated by changing the anti-aliasing and reconstruction analogue filter circuits which are also implemented as small plug-in modules to the mother-card. Data transfer to the DSP modules is serial via the front mounted coaxial sockets of the modules. To prevent digital noise disturbing the accuracy of the A/D and D/A conversions, all digital input and output signals to the card are transmitted via opto-isolator devices, thus ensuring thorough separation of the analogue and digital power and ground lines.

An advantage of the modular concept of the DSP computer is that interface modules can be developed as the need arises. A recently developed module for network emulator applications provides a bidirectional interface to a TDM 2048 kbit/s PCM signal. The interface conforms to the relevant CCITT G.700 series Recommendations [9]. The same module also provides a bidirectional link conforming to the 64 kbit/s co-directional standard outlined in CCITT Recommendation G.703 [10]. A dual channel high quality analogue interface module has also been added to the range, and a module permitting parallel I/O data transfers has just been developed. Furthermore, DSP modules can be developed for different DSP devices. At the time of writing, the latest addition to the range of modules available for the DSP computer is a DSP module designed around the Analog Devices fixed-point ADSP2101 device. This module was produced because Analog Devices had already developed software to implement the full-rate groupe speciale mobile (GSM) speech coding algorithm for mobile communications [11,12]. As software development and verification of precisely specified algorithms can absorb much time and effort, the development of the new module offered an attractive route to implementing the GSM algorithm, required as a speech quality reference for comparison with half-rate GSM algorithm proposals and for certain network emulation scenarios. Being based on a fixed-point device, the new module may also be appropriate to the implementation of other precisely specified fixed-point speech coding algorithms in the future.

12.9   CONCLUSIONS

The multi-DSP device processing unit known as the DSP computer was conceived as a means of progressing rapidly from a computer simulation of a speech coder to a real time hardware implementation. When only at the prototype stage, the DSP computer was put to good use in the development of 9.6 kbit/s multipulse speech codec hardware for the Skyphone aeronautical satellite telecommunications service. With very tight timescales and against strong competition, the DSP computer proved a prodigious aid to this development, and the eventual selection of the BT codec as the standard for the service stands as testimony to the significance of the DSP computer concept.

The success of the Skyphone speech codec spurred the progression to a second generation of the DSP computer, building on the experience gained with the first. This latest equipment was developed jointly with the transmission performance section at BT Laboratories, and includes features to make it suitable for network emulation purposes. In its original role, the DSP computer continues to provide telling aid to the implementation and refinement of speech and channel codecs.

REFERENCES

1.   Boyd I and Southcott C B: 'A speech codec for the Skyphone service', BT Technol J, 6, No 2, pp 50-59 (April 1988).

2.   BT Laboratories: 'Report of the DSP architecture study group', (October 1984).

3.   Crookes D: 'Proposals for a DSP computer — a software perspective', Study report (1986).

4.   Bodie J R, Gadenz R N, Kershaw R N, Hayes W P and Tow J: 'The DSP32 digital signal processor and its technical application development tools', AT&T Technical Journal, 65 , No 5, pp 89-104 (Sept/Oct 1986).

5.   Atal B S and Remde J R: 'A new model of LPC excitation for producing natural-sounding speech at low bit rates', Proc IEEE Int Conf on Acoustics Speech and Signal Processing', pp 614-617 (1982).

6.   Makhoul J: 'Linear prediction: a tutorial review', Proc of the IEEE, 63 , No 4, pp 516-580 (April 1975).

7. Hartung J, Gay S L and Haigh S: 'A practical C language compiler/optimizer for real-time implementations of a family of floating-point DSPs', AT&T Bell Laboratories.

8. AT&T: 'WE DSP32c digital signal processor information manual', (1988).

9. CCITT: 'General aspects of digital transmission systems; Terminal equipments', G Series Recommendations, Blue Book III (1989).

10. CCITT: 'Physical/electrical characteristics of hierarchical digital interfaces', Recommendation G.703, Blue Book III (1989).

11. Balston D M: 'Pan-European cellular radio: or 1991 and all that', Electronics & Communications Engineering Journal, 1 , No 1, pp 7-13 (January/February 1989).

12. Vary P, Hellwig K, Hofmann R, Sluyter R J, Galand C and Rosso M: 'Speech codec for the European mobile radio system', Proc IEEE Int Conf on Acoustics Speech and Signal Processing, pp 227-230 (1988).

13

THE DSP NETWORK EMULATOR FOR SUBJECTIVE ASSESSMENT

D R Guard and I Goetz

13.1   INTRODUCTION

New and prospective products, systems and services can be demonstrated in typical network configurations in real time. With the network emulator two people can talk over an emulated call from, for example, London to New York simply and quickly, without using additional equipment. It emulates the speech degradations caused by all parts of the telephone connection (for which software modules have been written). The facility of emulating calls 'on the spot' has generated much interest.

The network emulator allows managers to hear for themselves any connections causing concern, on any proposed system or service. Further, it facilitates formal subjective testing to give reliable answers more quickly, before commitment is given to expensive in-service trials.

Historically the first emulator was produced using the now ubiquitous TMS32010 digital signal processing device. This emulator was used in the subjective testing area at BT Laboratories to emulate circuit elements, such as ADPCM (adaptive differential pulse code modulation), filters and delay. These emulations allowed the subjective tests to be performed without recourse to buying or borrowing and interfacing to real equipments. However it soon became apparent that the fixed-point arithmetic TMS32010 was not powerful enough to emulate complete network connections, especially when compared with the new digital signal processors then becoming available. The present network emulator (hereafter called 'the emulator') contains cards of AT&T DSP32Cs, being one of the few floating-point 32-bit digital signal processors then available. This particular processor was chosen because the signal processing and coding section at BT Laboratories was already designing their digital signal processing (DSP) development platform (see Chapter 12) upon the AT&T DSP32C device. The emulator hardware is the DSP development platform, which is now being jointly developed by the signal processing and coding section and the transmission performance section at BT Laboratories. It contains an integral computer to control the software running on the digital signal processing cards as illustrated in Fig. 13.1.

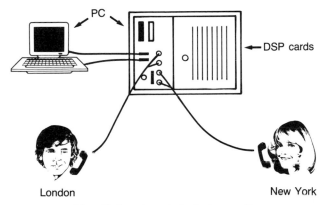

Fig. 13.1   A typical telephone call.

13.2   NETWORK EMULATOR DEMONSTRATORS

The emulator is the ideal vehicle for exhibiting new and potential systems and services and, in the longer term, it may be a useful tool to demonstrate to prospective customers the various qualities of service available from BT.

13.2.1   Groupe speciale mobile demonstrator

Several demonstrations have been made to various groups illustrating the pan-European GSM (groupe speciale mobile) network connected through national and international circuits.

For example, at a recent BT Laboratories demonstration the emulator was used for a complete connection emulating the GSM coder, the radio path fading errors and the fixed network (Fig. 13.2).

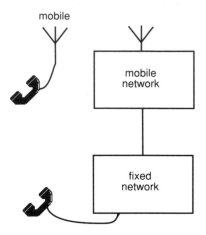

Fig. 13.2   A typical mobile call.

13.2.2   Facility demonstrator

The emulator is incorporated into Marlin, a hardware platform used to demonstrate potential intelligent network services. With these services, calls could be diverted internationally or forwarded nationally through various networks. To demonstrate these facilities at BT Laboratories a public branch exchange (PBX) is used to manage the call routeing, an emulator to appropriately degrade the connections, and a computer workstation to control the system (Fig. 13.3).

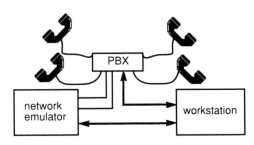

Fig. 13.3   The Marlin system.

13.2.3   Echo demonstrator

The above two examples concentrate upon using the emulator to demonstrate potential facilities and services. The emulator also highlights any degradations that new systems may unintentionally generate when interconnected to various other systems.

The example shown in Fig. 13.4 is a scenario where the time delays of various services build up to an unacceptable total delay. A Glasgow number is dialled from Exeter and passed through a PBX and an MPR (multipoint radio) link. The call is then diverted in Glasgow (tromboned — by a call forwarding facility) back to Taunton through a DECT PBX. Although the delay incurred by each individual service is reasonable, the total delay associated with tandemed connections such as this will be unacceptable unless due care is taken. Fortunately in this case, echo control has been addressed in DECT standards.

There are many such potential scenarios for which BT now has a tool which can demonstrate various alternative configurations of systems and services, and then eliminate those which give a poor performance. Strategic decisions can then be made based upon professional subjective and objective assessment as described in section 13.3. Before strategic decisions are made, due care and attention must be given to such tandemed connection scenarios.

13.3   SUBJECTIVE ASSESSMENT USING THE NETWORK EMULATOR

Rigorous professional subjective assessment is the only sure way of predicting the reactions of customers to new systems and services. Since 'one man's meat is another man's poison', the demonstration of the degradation of new systems and services frequently generates totally unrepresentative impressions. Considerable care must be taken when demonstrating to people involved in the launch of a new product or service, who often believe that their responses are representative. But such responses are probably highly biased by their own personal involvement. Consequently all decisions of strategic importance must be based upon scientifically run subjective tests with suitable statistical analyses performed upon the results, to give a representative opinion of what the average customer would think of a call. For example, only a slight imperceptible increase in speech degradation could encourage the average user to clear down after say five minutes rather than persist for quarter of an hour. This would result in a 66% loss in call revenue!

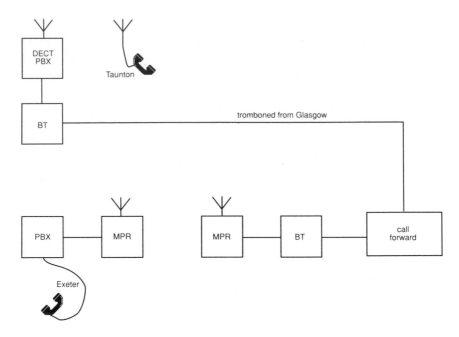

Fig. 13.4   The buildup of unacceptable time delays.

13.3.1   Formal subjective testing

In BT, the formal assessment of speech systems is done by the subjective testing group at BT Laboratories using tools such as the emulator.

Depending upon the system being considered, the subjective assessment is performed by either a conversation or listening test. In a listening test, people (who are called subjects) are placed alone in a controlled environment and asked to listen to speech transmitted through a series of randomly presented network connections (which are called conditions). At the end of each condition the subject is asked to vote on their opinion of the speech heard. During a conversation test two subjects are placed in separate rooms and asked to converse over randomly presented conditions. (As an aid to conversation they will be given a simple, irrelevant task to perform for each condition.) At the end of each condition they each vote on the condition without discussing it with each other. To improve the statistical reliability of the results, this is repeated several times with different subjects having the conditions presented in a different order.

To reliably assess a new piece of equipment or service, many network configurations must be considered. The total number of variations can be of the order of five hundred. However subjective tests generally cope with less than twenty conditions. Consequently, the actual conditions tested must be reduced to a representative set. Thus, considerable skill must be exercised to produce a minimal set of conditions. Furthermore, the results from the tests must be carefully applied to avoid misinterpretation.

13.3.2 Using the emulator in subjective tests

Before the existence of the emulator, every subjective test had to organise the loan of real equipments, or the set-up of unusual real network connections to BT Laboratories. This severely reduced the flexibility of the subjective testing facility. There are several advantages of the emulator.

- Previously if a slight change was thought to be desirable at the last moment, a new network connection had to be re-negotiated. For example, if a satellite hop was originally requested and then altered to a grazing satellite link, a total re-routeing would be required. In contrast, the emulator only requires a single run-time parameter to be changed.

- In the past, some tests would not have been contemplated, since the logistics of setting up the complete system would have been impractical. With the emulator the majority of the system is inside the DSP environment and can be changed quickly, efficiently and repeatably, ensuring that a reliable system is functioning throughout the test.

- It is now possible to set up theoretical connections — maybe potential products and services, or new equipments — which are not currently realisable, by selecting various unusual combinations of emulations.

When a new piece of equipment is to be assessed, three methods of testing are possible.

- If the equipment is available, easily transportable and has appropriate interfaces, then the surest way of testing it is to connect it into the emulator. The emulator provides the remainder of the network connections required for the subjective test. (The emulator has several types of interface enabling the interconnection to external equipment as many times as is necessary — see Fig. 13.5.)

- It may be possible to adequately emulate the new equipment by using a collation of existing software modules. For example under very specific circumstances a DCME (digital circuit multiplication equipment) could be emulated by an ADPCM module.

- Sometimes it may be necessary to build an emulation quickly using high-level software tools. Although this approach often generates inefficient code requiring many digital signal processing devices (as discussed below), it is a very quick way of generating a 'one-off' pseudo-emulation of an equipment or system.

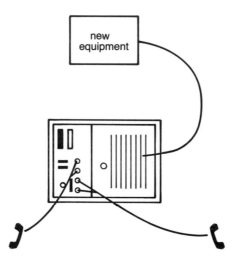

Fig. 13.5   Connecting external equipment.

To integrate the emulator into the subjective testing environment at BT Laboratories, two interfaces are required:

- a method of controlling the emulator from the subjective test computer, to give each subject the required ordering of conditions;

- a channel for the subjective test computer to provide the emulator with source speech for processing during listening tests (see Fig. 13.6).

Thus the emulator is a very powerful addition to the subjective test facility at BT Laboratories, enabling tests to be performed without the actual equipment being available, or, under extreme circumstances, even designed.

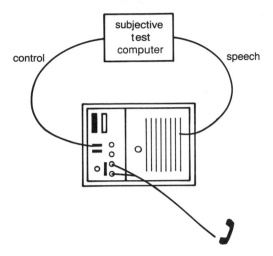

Fig. 13.6   Controlled by the subjective test computer.

13.3.3   Expert assessment

Before a formal subjective test is designed the number of conditions to be tested must be reduced to a practical figure. This is done by assessing past work, using modelling tools and trying out some conditions. This last process, which is often called a 'golden ear test', requires a number of truly expert assessors from the transmission performance area to decide upon the conditions to be formally tested. The emulator allows various conditions to be set up simply and quickly upon demand. It should be noted that a 'golden ear test' must not be confused with the beliefs of other people as described below.

13.3.4   Unprofessional assessment

Practically all developers of equipments, systems and demonstrators have a 'quick listen' to what their development sounds like. The emulator allows this process to be simply extended, so that it can be crudely assessed in the context of some typical network connections or network error conditions. However most people believe that they are an expert and can too easily take this process too far. Several people have made expensive erroneous judgements because they had presumed that their opinions represented the perception of the end user, and had based their judgements upon insufficient material. Consequently no apology is made for again emphasising the fact that decisions must not be made which are not based on professional subjective advice.

13.3.5   Subjective modelling

The subjective assessment of speech systems requires modelling tools. In addition to emulating circuit conditions, the emulator may also collect conversation parameters during subjective tests (Fig. 13.7). The collation of parameters, such as speech levels and speech duration, will facilitate the development of further network models.

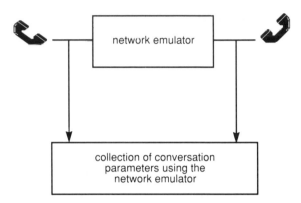

Fig. 13.7   Collection of parameters by the emulator.

An objective measurement tool [1, 2] is being developed which will predict subjective opinion of a coder. Ideally this tool could be used to give a subjective opinion of a complete emulated connection (Fig. 13.8).

13.3.6   Objective assessment

The above example shows the emulator being used for objective assessment. The signal processing and coding section has developed and objectively tested several speech coders, such as Skyphone (see Chapter 12), using the DSP development platform, alias the network emulator. The following section includes objective assessment of mobile systems using the emulator.

It is also probable that the emulator will be used on behalf of the speech language technology unit to examine speaker verification function, when receiving speech over various network connections. The emulator may also be used to validate a tester being developed within the network interface testing area.

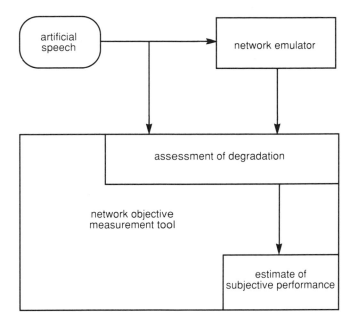

Fig. 13.8 Using the emulator in a subjective model.

13.4 EMULATION OF MOBILE COMMUNICATIONS NETWORKS

One of the major growth areas in telecommunications today is mobile networks. The nature of their design is to support the maximum number of subscribers on the allocated radio spectrum. This, coupled with the nature of the radio propagation, has led to the incorporation of low bit rate speech coders, long transmission delays and echo control equipment in the system design, many of which are being reproduced on the emulator. It is important to understand the interactions of these systems with the network, to ensure acceptable speech quality for its customers.

Fixed network interactions with mobile systems (and other transmission aspects) are assessed using the emulator as part of the mobile transmission test system (MTTS). The results of this work are used to optimize transmission plans and influence emerging mobile standards, to ensure quality communications on the move.

13.4.1   Digital cellular radio

The pan-European digital cellular communications system, GSM, was introduced in 1992. The system takes its name from the ETSI technical committee who have standardized the technology. The system incorporates low bit rate speech coding and echo control. A real time GSM speech coder has been developed which uses a 13 kbit/s residually excited linear predictive (RELP) algorithm and incorporates a voice activity detector (VAD) to further optimize the use of the channel. The VAD is used to control a discontinuous transmission function (DTX) in the radio circuitry. DTX switches the radio transmitter off when no speech is present. To prevent the far user thinking the line is dead, comfort noise is injected on the other side of the disabled radio link. The injected comfort noise is of the same amplitude and spectral content as the ambient noise surrounding the users. Combined with the radio interface channel coder and error correction, the GSM system has a mean one-way transmission delay of 95 ms.

The real time speech coder was designed to run on four AT&T DSP32C processor cards in the emulator. In this way a mobile terminal and a fixed radio base station can be emulated. Either an encoder or a decoder is implemented on each card (see Fig. 13.9).

The GSM encoder has a bit-exact definition within the GSM 06 series specifications and is a fixed-point arithmetic algorithm. The original implementation was developed on AT&T floating-point processors, as these were all that was available on the emulator. The floating-point implementation was tested to ensure speech quality and VAD activity were the same as with a fixed-point implementation. Once a fixed-point processor became available, it was incorporated into the emulator. The fixed-point cards use the Analog Devices ADSP2101 processor and have successfully passed the digital test vectors used for GSM coder verification.

To allow the effect of radio path degradations to be taken into account, error files can be run on a DSP32C card. These error files are generated from computer simulations of GSM, produced for cellular network planning purposes. An emulation of a vehicle driving from one base station to another is produced with two files, because separate error generator cards are needed to emulate the up-link and the down-link.

To prevent the far end user hearing echo, the GSM system relies on an adequate acoustic loss of the GSM terminal. The GSM specification originally stated that the weighted terminal coupling loss (TCL_W) of the terminal should be 56 dB. Measurements, carried out by the transmission performance unit at BT Laboratories on several commercially available analogue mobile terminals, showed that additional echo control would be needed in the terminal to achieve the 56 dB figure. Several acoustic echo control (AEC) options were developed on the emulator to enable this aspect of the GSM network to be examined by cellular radio systems. These options included echo suppressors, echo cancellers and a version that controlled the DTX VAD in the coder to prevent poor comfort noise performance, as shown in Fig. 13.9.

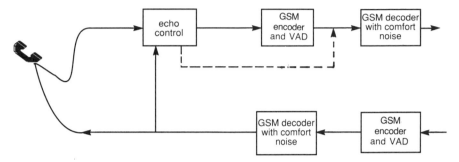

Fig. 13.9   Emulation of GSM.

The emulator implementation of the GSM speech coder forms part of the mobile transmission test system (MTTS). This system uses the emulator and real network components to carry out objective testing on real connection scenarios. The system uses proprietary echo cancellers and suppressors, adhering to CCITT Recommendations G.165 and G.164 respectively, to simulate the various echo control components in the connections. The system has been used in GSM/PSTN interaction testing, including testing at exchange sites. The results of these tests have been used to help formulate the technical content of interconnect agreements between BT and the GSM cellular operators (Cellnet and Vodafone).

For example the MTTS was used to evaluate several G.165 echo cancellers for network interactions, for use in the Cellnet mobile switching centres. The cancellers were tested with both white noise, as specified in CCITT G.165, and speech. It was found that some echo cancellers operated effectively with a white noise signal source but were ineffective when tested with a speech source. CCITT Recommendation G.165 uses white noise as its source and hence manufacturers need only optimize their cancellers to pass the white noise test. This testing also showed that tandemed echo cancellers produce less effective echo control than single cancellers. The results of these tests have been sent to CCITT for use in amending the G.165 Recommendation. In this way the emulator, as part of this test suite, has already influenced international telecommunications standards.

The MTTS was also used to objectively assess the different acoustic echo control (AEC) options available on the emulator itself, to make recommendations on the best system to use.

13.4.2   Digital cordless telephony

The number of mobile communications systems is steadily increasing. Not only are there several analogue cellular networks in Europe and the GSM digital system, there are now two digital cordless communications systems. The CT2 system was originally developed in the United Kingdom as a British Standard and has recently been accepted by ETSI as an Interim European Telecommunications Standard (I-ETS); it is currently available as a residential cordless telephone. The digital European cordless telephone (DECT) standard is nearing completion. The first systems should be available towards the end of 1992.

Both systems incorporate 32 kbit/s ADPCM for speech coding. The DECT system uses network echo control and has several handset TCL_W options. CT2 has a similar specification but has no echo control due to its shorter transmission delay. The various options make it difficult to simulate the DECT and CT2 standards for transmission testing. BT Laboratories are developing a DECT model to run in real time on the emulator where all the options will be modelled. The operator will be capable of selecting the desired test option. The CT2 model will be a subset of the DECT model. These models will be used in conjunction with the mobile transmission test system to provide information on the interoperability with other networks.

Future mobile systems will be standardized on a global basis. These networks, conceived as advanced replacements for European digital mobile systems, will have to work with telecommunications networks on a global basis. The standards are starting to emerge for this universal mobile telecommunications system (UMTS) and already experience drawn from testing using the emulator is ensuring a system that will provide quality mobile communications around the world.

13.5   THE NETWORK EMULATOR HARDWARE

To achieve the emulator objectives (see Chapter 1), a very powerful DSP engine was required to emulate the degradations inflicted upon speech as it is passed through a complete network connection. To be future proof, flexibility has been the keyword in its development. It is modular in all respects — hardware, software and interconnectability. Nevertheless, to keep to manageable objectives, the emulator only mimics speech degradations. However the emulator's modularity does enable it to be upgraded in the future to take other degradations (for example signalling, ringing or modem transmission) into account.

13.5.1   The DSP engine hardware

As discussed in the introduction, the heart of the engine is a set of AT&T DSP32C digital signal processing cards which are controlled by an IBM PC (or clone such as the BT M5000). The DSP software is set up on the IBM PC and downloaded to the processors via an STE bus. This bus is extended to contain timing synchronization. The speech data flow between the processors is achieved via two physically separated channels individually socketed on the front of the cards, as shown in Fig. 13.10. Generally, the two channels are daisy-chained so that each card has both the send and receive telephony channels. The first and last cards are interfaces to the outside world. The data at each socket is transmitted at 5 Mbit/s in bursts at the sampling rate of 8 kHz (there are hardware switches for 8/16 kHz sampling). This is the general basis of the design, but individual cards can have different facilities. The basic set of cards consists of:

- the DSP32C card (the DSP speech encoder development platform also has cards with different I/O configurations for the frame clocks);

- the Analog Devices fixed-point ADSP2101 processor card, to facilitate the fixed-point GSM coder;

- the analogue conversion cards which have 16-bit linear inputs and outputs;

- the PCM (pulse code modulation) card which has a 2 Mbit/s interface and a separate 64 kbit/s interface for both input and output (within the 2 Mbit/s interface two 64 kbit/s channels may be selected for both input and output);

- a card to connect a four-wire handset to the analogue card;

- an EBU (European Broadcasting Union) interface (primarily for communication with digital audio tape-recorders);

- a 16-bit parallel interface for downloading speech in real time for the listening tests as described in section 13.3.2.

A parallel interface card for GSM half-rate testing is also available.

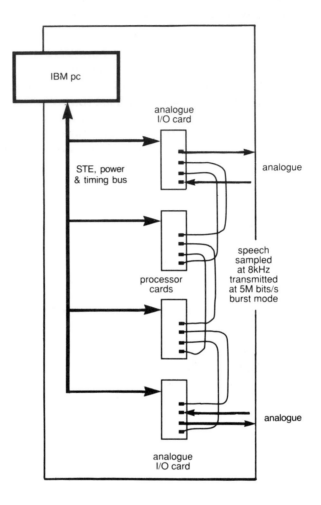

Fig. 13.10   Hardware block diagram.

13.5.2   The network emulator interfaces

As can be seen from the above list, the majority of the cards are interfaces, enabling speech transmission between other systems as often as required. External control of the PC in the emulator has been achieved via an RS232 interface, however if a separate PC with an ethernet card is used with the emulator, then the PC could be controlled via an ethernet. External control is provided by the subjective testing area and several other work areas in BT Laboratories, as mentioned previously.

13.5.3   Sizes of network emulator

The emulator is currently available within BT in a number of sizes:

- the standard integral PC version with fourteen card slots as shown in Fig. 13.1 — the PC was an XT, but 286s and 386s are now being supplied;

- the simple rack version which has an integral power supply and eighteen card slots — an external PC controls the STE bus;

- as the first version, but with only seven card slots.

In the near future the emulator will meet the BT electromagnetic compatibility and safety standards to allow it to be used in operational centres.

13.6   DSP SOFTWARE MODULES

Network connections are made up of separate network component elements. The modular block diagram structure of a real network connection lends itself to being split up on to separate digital signal processing cards, and thus separate software programs. However there can be a prohibitively large number of elements in a connection for a separate processor to be allocated to each element; consequently several elements must be allocated to each processor. Each element is emulated by software modules. As illustrated in Fig. 13.11, processors can have one module, several modules in one direction, many modules in both directions, or, exceptionally, a module may require more than one processor.

13.6.1   Emulation modules

The following facilities are currently available from numerous software modules:

- variable gain and delay;

- local two- and four-wire telephone lines, including sidetone and hybrids;

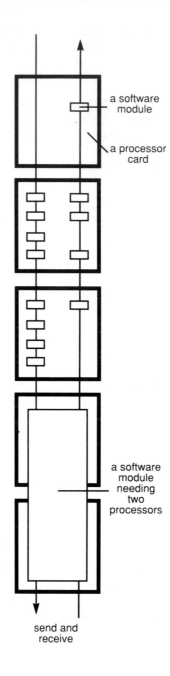

Fig. 13.11   Software modules and DSP cards.

- echo suppression, cancellation and control, for network and acoustic echo paths;

- full rate GSM, Skyphone, PCM and ADPCM coders;

- noise sources;

- random, burst and fading error generators;

- VADs (voice activity detectors) with comfort noise;

- interfaces (see section 13.6.2).

13.6.2   Software interface modules

There are five classes of interfaces which must be supported by the emulator software:

- a set of interface standards (and conversion modules) for communication between modules residing on the same processor;

- a set of interface standards (and conversion modules) for communication between processor cards;

- a set of standards for external interfaces, e.g. the PCM card requires a suite of modules to run on its dedicated processor for channel selection and synchronization;

- a set of up-sample and down-sample standards to convert between 44 kHz on the EBU interface, 8 kHz on the emulator and 16 kHz on the subjective test computer (the emulator may also later use 16 kHz),

- a set of modules which enable the PC to monitor, for example, the speech levels.

13.6.3   Writing modules

The majority of these modules have been written in assembler, as they had to be handcrafted to obtain the required efficiency in both code size and speed. In the future it is anticipated that a recently obtained block diagram tool — signal processing worksystem (SPW^{TM}) [3] (see Chapter 5), which uses assembler and C code submodules — will generate code prior to handcrafting. Although this tool may not be efficient at present (see Chapter 1), it should enable the fast development of 'one off' pseudo-emulations (see Chapter 17) which may use several cards. A Pascal compiler has also been used (see

Chapter 12). All modules must comply with a Pascal calling method to allow them to be collated into programs to be run on several processors. In addition the modules allow some parameters to be altered while the programs are running, as mentioned in section 13.7.2.

13.7   USER INTERFACE

The emulator consists of DSP hardware and network element software modules. However the user wants his choice of network connection to be run automatically and transparently. To achieve this a program running on the PC must:

● collate the appropriate software modules into groups, forming programs of suitable size and duration for the individual processor cards available, including the required interface modules as described in section 13.6.2;

● compile, link and download these programs into the processor cards;

● synchronize and run the programs.

13.7.1   Codegen

A program has been written called Codegen which runs on the PC. The user creates a text file listing the network elements in the order that is required, together with appropriate parameters, such as the desired gain of an amplifier. Codegen takes this list, consults a file containing the details of the cards residing in the particular emulator, ascertains the number of modules that can be placed on each processor card (in both directions of transmission, taking account of cross-coupling) and generates a top-level program for each processor. The top-level program for each processor (containing the call statements to the network element and interface modules) is compiled and then linked to a software library of the modules. The complete programs are then downloaded to the processor cards ready for running. Any unused cards are loaded with a software short circuit program so that the signals reach the I/O cards (Fig. 13.12).

13.7.2   Graphical network emulator front end (GNET)

The PC program GNET runs Codegen and then displays a graphical representation of the complete connection on the screen as illustrated in Fig. 13.13. The user may then use a mouse to click on each module to display and alter run-time parameters. A single click in the filled-in corner of each module will toggle the module bypass on and off. These two features make GNET an invaluable aid in demonstrations.

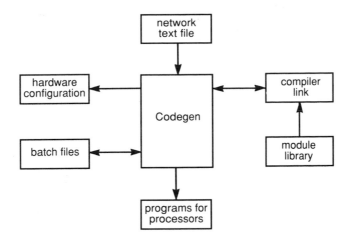

Fig. 13.12 The Codegen program.

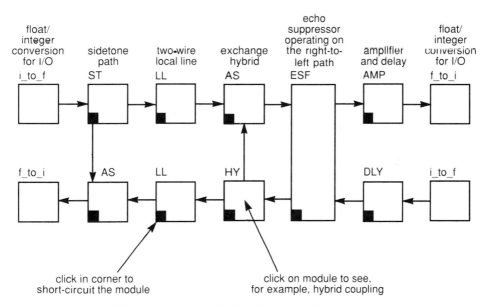

Fig. 13.13 The screen displayed by the GNET program.

13.7.3   Remote control

Remote control of the emulator conditions is required during subjective tests and similar situations, as discussed previously. There is a program (also called up by both Codegen and GNET) which facilitates batch files and the diversion of the keyboard to other ports for remote control.

13.8   FUTURE DEVELOPMENTS

In the short term, the time is ripe for the development of a complete suite of software modules to represent all current and new network elements, making efficient use of SPW^{TM} [3]. To support the collation of a network connection, a user-friendly front end would be a boon, even for the expert user. It is very probable that this tool will become part of the repertoire of network planners, sellers of systems and the subjective test facility, where a simple-to-use windows front end will be essential. An increase in processing power may also be required. In the short term a multiprocessor card containing more powerful processors is envisaged. However, the latter may be delayed until the next generation of emulator which might be based upon a computer workstation and a VME bus.

13.9   CONCLUSIONS

The network emulator is widely used at BT Laboratories for demonstrating the speech path of new and prospective network scenarios. Strategically, it is essential to assess all new services with the evolving total network to ensure any speech degradations are acceptable to the customer. Currently this activity constitutes the major use of the emulator, where it saves time and money by emulating network equipments quickly, efficiently and repeatably. The emulator has flexible hardware, software and interfaces, so that it can be expanded internally, connected externally to other network equipments, and controlled by other computers.

REFERENCES

1.   CCITT Study Group XII Question 18, Contribution XII-8-E, Nippon Telegraph and Telephone Public Corporation (April 1985).

2.   Hollier M P et al: 'Non-R2 pseudo-noise sequences for transfer function measurement', Proc of Inst of Acoustics, <u>13</u> , Pt 7, pp 219-226 (1991).

3.   Signal Processing WorkSystem User's Guide, COMDISCO Systems Inc (1989).

14

A HYBRID BANDSPLITTING ACOUSTIC NOISE CANCELLER

J M Connell, C S Xydeas and D K Anthony

14.1   INTRODUCTION

The problem of signals being contaminated by unwanted interference is a universal one. In many instances, the spectrum of the interference allows for some form of fixed filtering to enhance the signal-to-noise ratio (SNR). If the signal of interest and the noise are spectrally overlapped then the enhancement process must take a more sophisticated approach. Where the signal of interest is speech and the interference is broadband acoustic noise, the range of options is severely restricted.

14.1.1   Publications on active acoustic noise reduction

Published material on the enhancement of speech in the presence of acoustic noise is dominated by military funded research into the effects of background noise in fighter aircraft cockpits [1,2,3], where sound pressure levels can be in excess of 90 dBA. Treatment of speech communications in helicopters has also received attention [4]. More recently, the proliferation of mobile communications networks has led to acoustic noise reduction techniques being applied to telephone speech in cars [5,6,7].

The noise reduction processes most common in the referenced publications are adaptive noise cancellation (ANC), using a two-microphone input system, and single microphone spectral subtraction (SS). Broadcasting stationary white noise using a single loudspeaker in a partially soundproofed room, to simulate the aircraft cockpit [1,2,3], achieved noise attenuation levels in excess of 10 dB using ANC. This performance is found to deteriorate rapidly as the sound field is made more diffuse by introducing more loudspeakers. Cancellation of complex noise sources with a linear adaptive filter is feasible only in very specific instances.

The SS method has been applied in Boll [4] using actual helicopter cockpit data. In this process, the spectral envelope of the noisy speech is reduced in amplitude, at frequency bin level, by an amount determined by the noise estimate. The advantages of SS over ANC include the degree of selectivity provided by resolving a signal into its frequency components and also the elimination of noise estimate phase considerations. In Boll [4], noise reduction effectiveness is measured subjectively with improvement in output speech quality and intelligibility being recorded.

14.1.2   Acoustic noise in telephone kiosks

The provision of a public communications service very often requires siting telephone kiosks in locations as diverse as roadsides, public concourses and railway stations in which the user is often subject to substantial levels of background acoustic interference. The variety of interference is extensive and may include traffic, footfall and even external speech signals, which collectively encompass a wide range of sound power levels and sound spectra. These can have adverse effects on the quality of the service for both the calling and the called parties and thus some facility for acoustic noise reduction is desirable.

There are several other contributing factors to the problem of eliminating background noise in telephone speech. The choice of location for a telephone kiosk is influenced by a number of factors and, thus, positioning for the best available acoustic protection may not always be possible. In addition, payphone housings can vary from the fully enclosed form to hood coverings, common in railway stations.

More practical influences on the problem are the developments in kiosk design. The need for anti-vandalism precautions and ground level ventilation has lessened the potential for acoustic insulation. The transparent panelling used in the construction has low absorption, which makes the kiosk interior reverberant, and limited damping, which results in resonant behaviour. These kiosk attributes will be seen to influence the degree of success found in all applied signal processing techniques. More suitable absorption materials and panel shaping designs are available but are currently limited in application due to the threat of vandalism.

This chapter describes a successful acoustic noise cancellation scheme which is based on a hybrid of time and frequency domain techniques. The chapter is divided into six sections. In section 14.2, an outline of the data collection procedure is given. In addition, this section discusses the application of existing noise cancellation techniques to the recorded kiosk data and their associated limitations. In section 14.3, the basis for the proposed noise cancellation technique is discussed, whereas in section 14.4 the operation of the emergent hybrid bandsplitting scheme (HBS) is fully described. A real time implementation of the scheme is outlined in section 14.5. Conclusions are drawn in section 14.6 on the effectiveness and limitations of the scheme's performance.

14.2   CONVENTIONAL NOISE CANCELLATION TECHNIQUES APPLIED TO TELEPHONE KIOSK DATA

14.2.1   Collection of data

In the search for a practical solution to the acoustic noise reduction problem it is imperative to experiment with real environment data. This provides a realistic basis for analysis and proposal evaluation. For this purpose, research data was collected from two principal experiments. These included recordings in telephone kiosks in a dual carriageway lay-by, where passing traffic was accelerating or had achieved cruising speed, and secondly, in a more controlled environment, where the noise source was a motorcycle positioned on the roadside approximately 1.5 m from the kiosk. Four-band synchronized analogue recordings were made in each experiment.

The microphones used in the experiments were omnidirectional miniature electrets which had similar sensitivity profiles and were essentially flat over the range 0-4 kHz. They have similar characteristics to those used in standard kiosk handsets, except that the latter are normally band-limited. There is an inherent difficulty in using directional microphones inside the kiosk due to the reflective nature of the environment. Their usage outside the kiosk relies heavily on noise sources being fixed and reflectivity levels being low. This condition is not met by public telephone kiosks and thus omnidirectional microphones are most appropriate.

A dummy telephone handset was used with one microphone mounted in the mouthpiece and a second microphone mounted in a small casing behind the earpiece. The handset was then held and used as in a normal conversation. The remaining two microphones were encased and mounted on tripods outside the kiosk at a distance of 1 m and at a height of 1.5 m. Encasing the microphones modifies the sensitivity profile slightly. Plots of the free-standing microphone sensitivity and the encased microphone sensitivity are shown in Fig. 14.1.

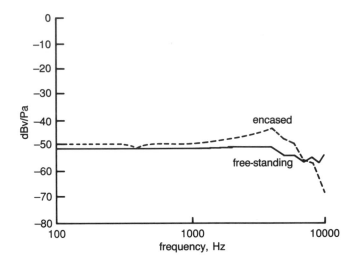

Fig. 14.1   Microphone sensitivity.

During digitization, the recorded signals were lowpass filtered to 3.4 kHz, using fifth-order, switched capacitor, elliptic filters, and sampled at 8 kHz using a synchronized, four-channel, 12-bit, analogue to digital converter module.

Henceforth, the mouthpiece microphone signal will be referred to as the primary input. The earpiece and outside microphone signals will be collectively referred to as the reference inputs. These primary/reference input signals were used in evaluating the performance of conventional noise cancellation systems when applied to the telephone kiosk environment. These systems are discussed in the remaining parts of this section.

14.2.2   Adaptive noise cancellation

The basic concept of the two-microphone ANC system, shown in Fig. 14.2, is the elimination of the primary noise component using the signal from a separate noise collecting reference microphone. In series with the reference input is an adaptive filter which attempts to model the transfer function (TF) corresponding to the primary/reference microphone separation. Processing the reference noise signal thus produces an approximation of the primary noise component. Time domain subtraction now improves the SNR of the primary channel.

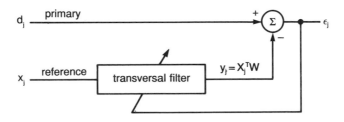

Fig. 14.2 Basic adaptive noise canceller.

The system equation, derived in Widrow and Stearns [8], is given by:

$$E[\epsilon_j^2] \;=\; E[d_j^2] \;+\; W^T\, E[X_j X_j^T]\; W \;-\; 2E[d_j X_j^T]\; W \qquad \ldots (14.1)$$

where X is the filter memory matrix,
  W is the filter coefficient matrix and
  j is the adaptation iteration.

The adaptive filter coefficients are updated using the gradient descent based LMS algorithm [8], which aims to minimize the power of the error signal, ϵ, generated in modelling the transfer function. This is consistent with maximizing the output SNR if the speech signal is uncorrelated with the noise signal. The LMS algorithm is given as

$$w_k^{j+1} \;=\; w_k^j \;+\; 2\,\mu\epsilon^j\, x_k^j \qquad\qquad \ldots (14.2)$$

where μ regulates the rate of convergence,
  x_k^j is the kth memory sample,
  w_k^j is the kth coefficient at the jth adaptation iteration.

ANC was applied to recorded telephone kiosk data from the motorcycle experiment using the mouthpiece signal as the primary input and an outside tripod-mounted microphone as the reference input. This arrangement deviates from the original assumptions of ANC theory in several key areas which will therefore constitute limitations on the system. The first limitation is that in the kiosk environment, a unique solution to the required filter impulse response, W, does not exist. Continuously changing conditions both inside and outside the kiosk, due to speaker movement, noise source movement and indeed air movement, produce an unstable reverberance regime between the microphones. This results in an ever-variant microphone separation TF which cannot be combatted by simply increasing the filter length. The limiting aspect of this regime for adaptive processes is that the filter coefficients may now be required to change at a rate which is in excess of the system convergence time. This will be manifested in poor ANC performance and is a feature of the kiosk data processing. Similar findings have also been reported [1, 2, 3] where stationary white noise is broadcast in a partially soundproofed room using distributed sound sources.

A multi-input, lattice filter based system using a least squares adaptation criterion, as investigated by Proudler et al [9], yields an improved noise attenuation performance through faster filter convergence. However, the computational complexity is increased and the processed speech contains more distortion and echo than that produced by the transversal filter ANC approach. The dynamic nature of the microphone separation TF is undoubtedly the most crucial factor in the failure of adaptive filtering in the kiosk environment.

The second limitation concerns the restraints placed on filter convergence speed by the nature of the input signals. The ANC system equation, given in equation (14.1), describes a hyperparabolic error surface of fixed geometry and position but only for fixed primary and reference signal statistics. In consideration of the received kiosk signals, this implies that the point of minimum mean square error (MMSE) in this instance is 'mobile'. The lag in tracking the MMSE point may be reduced by increasing μ but this is offset by an increase in the noise level within coefficient values which can lead to instability.

The shape of the noise spectrum also influences the speed of convergence. The concentration of energy at a fundamental frequency and at several harmonics is indicative of a large eigenvalue spread within the input correlation matrix [8]. The extent of the spread is inversely proportional to the speed of the convergence process. An optimum condition exists when the eigenvalue spread is unity. In this respect, ANC is more suited to spectrally flat, stationary white noise. In comparison, performing ANC on kiosk data will be more restrictive in terms of adaptation speed.

A final limitation concerns causality in the adaptive system. If the reference noise signal is a delayed version of the primary noise signal, then the adaptation process will be unable to minimize the power of the output signal. This inability will induce oscillation into the weight magnitudes. It is therefore common practice to include a delay block in the primary channel to allow the system 'room for manoeuvre'. Using kiosk recorded data, the inclusion of a fixed digital delay, of any length, on the primary channel has no effect on ANC performance. This implies that weight magnitude behaviour will have an ever-present oscillatory component. Plots of the filter impulse response with time, for kiosk data, contribute to this notion.

14.2.3 Spectral subtraction

Spectral subtraction using dual input microphone systems can be an effective means of achieving noise reduction. Its operation involves subtracting the reference spectral envelope from the primary spectral envelope [4] using:

$$|\epsilon(e^{jk\omega})| = |d(e^{jk\omega})| - |x(e^{jk\omega})| \; ; \; 0 \le k < N \qquad \qquad \text{... (14.3)}$$

where ϵ, d and x are the output, primary and reference signals respectively and N is the block size, i.e. the spectral resolution. Negative magnitudes which result from the subtraction process may be set to zero, a process referred to as halfwave rectification, or alternatively to a small fraction of the primary input. The output time domain data is reconstructed using the resultant magnitude data and the original primary phase information. The use of primary phase data in the output reconstruction is an approximation which is not normally noticeable due to the ear's relative insensitivity to phase.

The use of halfwave rectification can have a serious perceptual effect on the quality of the recovered signal in that the 'enhanced' spectrum now contains 'spectral holes' between the magnitude peaks. These holes, whose position varies randomly with time, are perceived as musical tones embedded in the output signal and may be subjectively irritating. Using a minimum threshold, i.e. replacing negative or zero magnitudes with a fraction of the primary value, is usually ineffective. Substantial thresholds would be needed to generate the required envelope energy around the magnitude peaks in order to recreate a recognizable noise residual. The noise reduction achieved by the subtraction process is now largely reversed.

The limitation on the SS process centres on one fundamental issue. For single input systems [11], where the primary noise estimate derives from averaging past noise-only frames, it is the accuracy of using past bin magnitude values to estimate the current values, and, in dual input systems, it is the lack of correlation between primary bin magnitude values and their corresponding reference values. Using blocks of 256 samples, overlapped by 128 samples, from noise-only segments of the data used in section 14.2.2, plots of bin magnitude behaviour with time, over the whole spectrum, yield very random waveforms. Successive magnitude values can increase or decrease by as much as fiftyfold on the previous value. This behaviour is more severe for recorded signals inside the kiosk than for those outside, due to the reverberant nature of the kiosk.

Based on viewing these random bin magnitudes as functions of time, an autocorrelation function was calculated using data from the 500 Hz component of the mouthpiece primary signal as shown in Fig. 14.3. Clearly, estimates based on past bin magnitude values in the same channel will not be an accurate indication of the current bin magnitude value. The cross-correlation function, for the same 500 Hz component, using primary and corresponding reference data, is shown in Fig. 14.4. This plot suggests that more accurate primary bin estimates could be derived from processing current and past values of the corresponding reference bin. In practice, this is found to be the case.

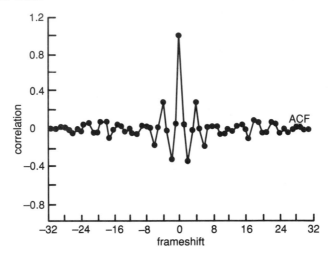

Fig. 14.3   ACF of primary bin (500 Hz) magnitude.

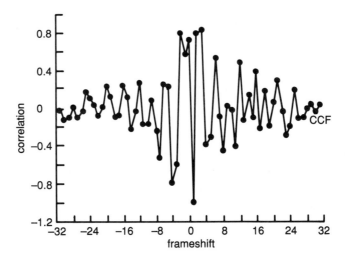

Fig. 14.4   CCF of primary and reference bin (500 Hz) magnitudes.

The level of correlation between the primary and reference noise signals has proved the dominant factor in the relative lack of success of existing noise cancellation techniques in both the time and frequency domains. A definitive way of classifying the nature of the kiosk environment, from a signal processing perspective, is in the calculation of the coherence function. Denoted as $\gamma(j\omega)$, it is usually expressed in its magnitude-squared form (MSC) by:

$$\gamma^2(k\omega) = \frac{|P_{xy}(jk\omega)|^2}{P_{xx}^2(k\omega).P_{yy}^2(k\omega)} \qquad \text{... (14.4)}$$

where $P_{xy}(jk\omega)$, $P_{xx}(k\omega)$ and $P_{yy}(k\omega)$ are the power spectral densities relating to the signals x and y. The coherence function may be used to indicate the degree to which two discrete time signals, $x(nT)$ and $y(nT)$, are linearly related. In this instance, where the mouthpiece-reference TF is essentially linear, the evaluation of the coherence function will be of more use in indicating the degree to which this transfer function is stable with time.

To estimate the coherence function, a set of N samples, from both the primary and reference channels, is divided into p segments of length m. The autocorrelation and cross-correlation values are then evaluated for each segment and averaged over p. The resultant correlation functions are Fourier transformed to yield the power spectral densities.

Since, for any segment, p, the primary-reference TF may be evaluated using:

$$h^p(jk\omega) = P_{xy}^p(jk\omega)/P_{xx}^p(k\omega),$$

the evaluation of equation (14.4) inherently implies a summation of p transfer functions. If successive complex bin values of $h^p(jk\omega)$ are not closely matched, i.e. they are inconsistent, then cancellation will result during the summation thus yielding low values of coherence. The inconsistency in $h^p(jk\omega)$ may be due to the effect of finite data sets, a high incidence of uncorrelated noise energy in the measurements of $x(nT)$ and $y(nT)$, or statistical bias and variance in the spectral estimates [12]. In this case however, the existence of a highly diffuse and dynamic ambient noise field is most likely to be the predominant influence.

A coherence estimate (see equation (14.5)), based on identical noise-only data to that used in section 14.2.2, is shown in Fig. 14.5. The segment, one second in length, was partitioned into frames of 256 samples, overlapped by 128 samples. Clearly, from the waveform, the effectiveness of ANC will be essentially confined to low frequencies, less than 1 kHz. This is borne out in practice where spectral plots of the ANC output show attenuation in the low frequency band with the remainder of the spectrum virtually unaffected. The shape of Fig. 14.5 is in close agreement with the theoretical analysis of diffuse noise fields presented by Bendat and Piersol [12]. Coherence plots produced for the aircraft cockpit [13] and the car interior [6] are also similar. The major acoustical factors which contribute to this plot are those outlined in section 14.2.2, i.e. noise source movement, handset movement, multipath reverberance and air movement. The signal processing characterization of the kiosk environment using the coherence estimate is one of the most significant results produced by this research.

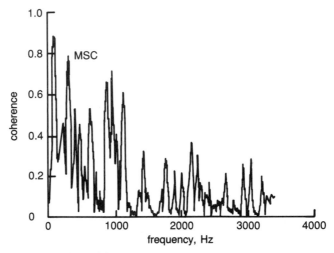

Fig. 14.5   Coherence function for kiosk data.

A final insight into the limitations of spectral subtraction may be gained by analysing the addition of noise to speech in time. This is of use in highlighting the potential, from a speech signal prerspective, for destructive interference at source. Considering a vector model of the addition process, shown in Fig. 14.6, a speech frequency bin component, $a+jb$, is chosen to have arbitrary phase and magnitude. Assume that the corresponding noise frequency bin component, $c+jd$, can have any phase value, relative to the speech vector, with equal probability. The speech vector magnitude is given by $(a^2+b^2)^{1/2}$. Following the addition of the noise and speech vectors, the composite vector magnitude is $((a+c)^2 + (b+d)^2)^{1/2}$. If this is smaller than the speech magnitude, i.e.

$$((a+c)^2 + (b+d)^2)^{1/2} \le (a^2 + b^2)^{1/2},$$

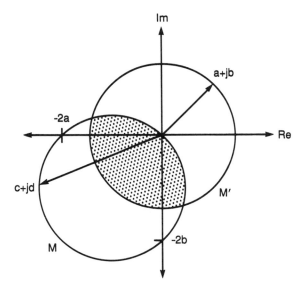

Fig. 14.6   Addition of vectors.

then destructive interference has taken place. This equation may be expressed as:

$$((a+c)^2/(a^2+b^2)) + ((b+d)^2/(a^2+b^2)) \leq 1 \qquad \ldots (14.5)$$

which is the equation of a circle, M. If the noise vector lies within M, then the speech vector is attenuated at source. If it is also assumed that $(c^2+d^2)^{1/2} \leq (a^2+b^2)^{1/2}$, then this has a theoretical probability given by the ratio of the shaded area of M' and yields a value of 0.38. Clearly, under these circumstances, further attenuation by spectral subtraction will compound the destructive interference.

14.3   EVOLUTION OF THE BANDSPLITTING APPROACH

In general, the implementation of dual input spectral subtraction can result in large reference bin magnitudes being used to cancel quite small primary bin magnitudes. This is an ineffective use of reference energy. A more successful approach is to first reshape the reference envelope to resemble the primary envelope shape so that:

$$\frac{|x(jk\omega)|^2}{Pow_{\text{ref}}} = \frac{|d(jk\omega)|^2}{Pow_{\text{prim}}}$$

where *Pow* is the frame energy.

This transformation is therefore achieved by:

$$|x'(jk\omega)| = |d(jk\omega)| \cdot \left(\frac{Pow_{\text{ref}}}{Pow_{\text{prim}}}\right)^{1/2} \qquad \qquad \ldots (14.6)$$

for $0 \le k < N$. Subtracting the reshaped reference magnitude spectrum, $x'(jk\omega)$, from the primary spectrum now yields an output:

$$|\epsilon(jk\omega)| = |d(jk\omega)| \cdot \left(1 - \left(\frac{Pow_{\text{ref}}}{Pow_{\text{prim}}}\right)^{1/2}\right) \qquad \qquad \ldots (14.7)$$

where $\epsilon(jk\omega)$ is the process output.

This is equivalent to reducing the primary frame energy by an amount equal to the reference frame energy and constitutes a power scaling operation. Since time energy and spectral energy are related by the factor N, this operation can similarly be effected in the time domain by scaling each primary sample, $d(nT)$, by:

$$\left(1 - \left(\frac{Pow_{\text{ref}}}{Pow_{\text{prim}}}\right)^{1/2}\right) \qquad \qquad \ldots (14.8)$$

Perceptually, this produces superior noise attenuation results to conventional spectral subtraction for considerably less processing. The process also proves advantageous in ensuring that the noise residual is recognizable due to the preservation of the primary spectral shape. When the reference energy is greater than the primary energy, during noise-only frames, the minimum threshold concept may now be implemented to eliminate the annoying random silent gaps in the output signal.

One drawback of the power scaling approach is directly attributable to the nature of speech and its spectral energy profile. The process of power scaling can occasionally cause clipping in low power speech, such as word endings and unvoiced sounds. High-power voiced sounds emerge from the noise attenuation process perceptually unscathed. Since the quality and intelligibility of the output speech relies on both low power and high power sounds, some allowance must be made during the reduction process for low power speech. Examination of the spectra associated with low power speech yields a concentration of energy in the 2.9-3.4 kHz band. Great caution must therefore be exercised here during noise suppression. It is complicated however by the presence of a significant noise energy content which also resides in this band.

The exercise of restraint when suppressing noise in the 2.9-3.4 kHz band will be considerably aided by improving on the accuracy of the primary noise energy estimate. This prompts an investigation into the levels of correlation between successive frame energies within the same channel and between primary and reference frame energies.

Using identical data to that in section 14.2.2, with frame sizes of 256 samples, overlapped by 128 samples, plots were made of successive primary and reference frame energies during noise only. These both prove to be of a similar random nature to the plots of successive bin magnitudes but they exhibit less variance. The primary frame energy behaviour is less restrained than that of the reference due to the reverberant nature of the environment. The primary-reference frame energy ratio also shows a variant profile with time but it operates between well-defined limits.

As in the bin magnitude case, viewing the behaviour of primary and reference frame energies as functions of time, allows the generation of correlation functions. A plot of the primary energy autocorrelation function is shown in Fig. 14.7 whilst the cross-correlation between primary and reference energies is shown in Fig. 14.8. Processing current and past reference energy values could clearly yield a more accurate primary noise power estimate. This is found to be the case. The introduction of an eighth order moving average (MA) adaptive filter, using the LMS algorithm in an ANC format to process reference frame energy values, dramatically increases the level of noise reduction. However, the suppression of low-power speech has only marginally increased in comparison with the earlier power scaling scheme using the current reference energy value.

Combining the strategy of power tracking and the need for separate noise reduction mechanisms for the high and low spectral bands provides the basis for an emergent scheme. The primary and reference signals are split into two bands of interest — the low band, d.c.-2.9 kHz, and the high band, 2.9-4 kHz, which is of much lower energy and requires a greater degree of precision during cancellation. The energy contribution of the 3.4-4 kHz region will be negligible due to anti-aliasing filtering. Separate MA adaptive filters process the reference band energies to estimate the primary noise band energies. Using the filter outputs, a potential noise cancellation algorithm must strike a self-regulating balance between maximizing noise reduction and maintaining output speech quality. The following proposed scheme purports to meet this requirement.

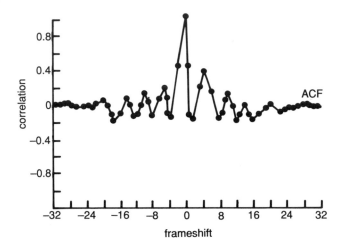

Fig. 14.7   Autocorrelation of primary energy values.

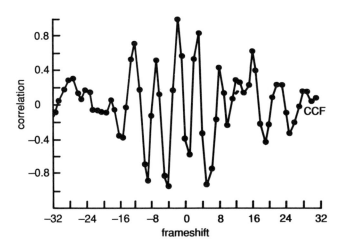

Fig. 14.8   Cross-correlation between primary and reference energy values.

14.4   PROPOSED HYBRID BANDSPLITTING SCHEME (HBS) METHOD OF NOISE REDUCTION

A block diagram of the HBS is shown in Fig 14.9. The operating frame size at a sampling frequency of 8 kHz is 256 samples. Successive frames in both channels are overlapped by 128 samples. The delay between input and output is therefore 16 ms. Both frames are weighted by a Hanning window for spectral smoothing and to minimize leakage. The Hanning window also proves to be most accurate when producing output time samples by the overlap-add process. The data in both channels is then fast Fourier transformed (FFT) to yield a 256 point spectrum. Real time implementations can utilize a simultaneous transformation of the primary and reference time data by using a single real-imaginary array pair.

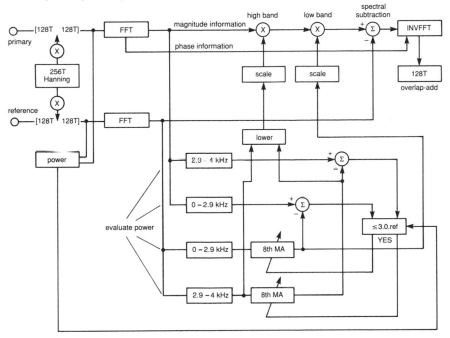

Fig. 14.9   HBS block diagram.

The low band and high band energies are evaluated by summing the bin magnitude-squared values over the range d.c.-2.9 kHz and 2.9-3.4 kHz respectively. The reference energies form the inputs to the separate energy tracking filters. The filter outputs are used as estimates of the corresponding primary noise band energies. The errors produced by subtracting the estimates from the actual primary energies, are used to update the filter coefficients when the current primary frame contains noise only.

The task of marking frames as containing noise only or speech plus noise is normally performed by a voice activity detector (VAD). Their usual mode of operation involves comparing the current primary frame energy with fixed thresholds. Using kiosk recorded data, accurate frame discrimination based on in-built, fixed threshold values proved impossible. The result of this simple VAD strategy was an increase in speech clipping due to the reference filters being allowed to track incorrectly marked noise-only primary frames.

A computationally simpler yet more accurate alternative exists however. Based on the analysis of all available sets of experimental data, it is found that the primary/reference frame energy ratio, during noise only, is confined almost exclusively to within a 3 dB limit. Thus, a very simple yet effective criterion exists for weight updating in the kiosks studied. With provision for headroom, the primary frame energy is allowed to exceed the reference energy by a factor of 3 (4.77 dB), within which weight updating takes place using the returned error signals. Primary energy above this level is assumed to imply the presence of speech. Instances in which the primary energy falls within this limit, but contains speech, are very infrequent and are not sustained. Thus, whilst the filters may instantaneously be caused to track erroneously, filter convergence time acts as a restraining element so that the update trend is dominated by primary noise energy behaviour. Thus, a simple threshold criterion obviates the need for a VAD and thus combines minimum processing with optimum performance.

The filter weights are updated using the LMS algorithm, equation (14.2). The convergence factor expression is given as:

$$\mu = \frac{misadjustment}{\sum\limits_{n=0}^{L-1} x^2(n)} \qquad \dots (14.9)$$

where $x(n)$ are the filter memory samples and L is the filter length.

A value of 0.1 is used for the misadjustment. This strikes an acceptable balance between tracking speed and an in-built inertia to tracking incorrectly identified noise-only frames.

At the expense of increased computation, it seems likely that FFTs could be performed at more regular intervals to achieve greater filter tracking accuracy. This proves to be a retrograde step. As a consequence of the now increased overlap between successive frames, filters now track more deeply into erroneously marked noise-only primary frames before the primary energy level crosses the ratio threshold of 3. In practice, this causes more frequent attenuation of low power speech — thus, minimum computation and optimal filter tracking are also consistent.

The method of noise reduction is different for both bands. The primary low band is scaled by a value determined by the low-band filter output. The primary high band is scaled by the lower of the current reference high-band energy and the high-band filter output. This forms the essential part of the conservative treatment of the high frequency part of the spectrum. Noise reduction is effected but not at the expense of output speech quality.

Due to the different treatment of the spectral bands, the output noise residual from this reduction stage contains a low frequency component, which sounds like the noise source, and a high pitched component which resembles a bird chirrup. The speech is perceptually unharmed. It is subsequently found that the reduction process may be further enhanced by spectrally subtracting the current reference envelope from the scaled primary envelope, over the entire frequency range. The net effect, in addition to further reducing the noise residual energy, is that the low-frequency component of the residual no longer indicates the noise source, owing to the break-up of the primary envelope, and the high pitched chirping element is now reduced to an occasional 'tweet'. The overall perception of the noise residual is that of a very low energy musicality. Importantly, the noise residual exhibits no noticeable energy changes during transitions between speech-on and speech-off.

The resultant spectral envelope is then inverse transformed, using the input primary phase information, to yield a weighted output frame which is overlap-added to the previous frame to produce 128 output time samples.

Using identical data to that in section 14.2.2 three dimensional spectral plots of the original primary data and the processed output are shown in Figs. 14.10 and 14.11. The plots have been constructed from frames of 256 samples, overlapped by 176 samples, which are Hanning windowed. Identical vertical magnitude linear scaling has been used in both plots. Speech begins at the 1.5 s point.

The scheme was subsequently applied to data from the dual carriageway experiment with the background noise being almost completely eliminated, except for an occasional 'tweet' which arises from residual noise energy in the high band. An interesting comparison may be made between these results and those from the lower average primary SNR motorcycle experiment, where the engine speed was continuous. Incidences of word ending suppression occur in the dual carriageway processed output but not in the motorcycle output. These are due to sharp peaks in background noise energy caused by heavy vehicles passing the kiosk. Under these circumstances, the disparity between the energy of noise-only frames and those which contain speech becomes less. This results in more incorrectly marked frames, as dictated by the ratio threshold, which leads to incorrect filter tracking. In addition, the tolerance on the filters' outputs decreases so that the effect of inaccurate band energy

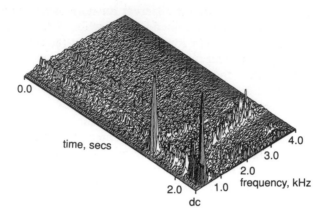

Fig. 14.10   Primary microphone signal for motorcycle experiment.

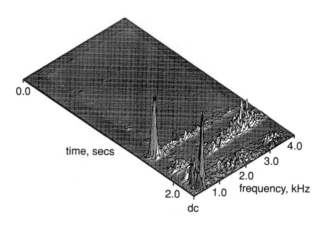

Fig. 14.11   Processed primary microphone signal for motorcycle experiment.

estimates becomes more perceptually noticeable. This phenomenon does not arise in the motorcycle experiment due to the constant nature of the background noise energy.

A final test on the ability of the proposed HBS scheme to generalize used recorded data from a KX100, fully enclosed telephone kiosk in which the ground level ventilation facility was covered over. This alters the acoustic insulation characteristics of the kiosk. Analysis of the recordings showed that the noise attenuation due to the kiosk structure has now increased from essentially zero attenuation, to providing 10 dB of attenuation. The reference data samples were therefore scaled by a factor of 0.33 prior to processing.

Speech data corrupted by background traffic noise, railway station noise, which included external speech, and stationary coloured noise was used as the target. Both male and female subjects were used. Results showed excellent noise cancellation performance but with similar incidences of low-power speech suppression when noise surges caused the primary SNR to drop below the threshold level.

Based on data from the motorcycle experiment, the noise attenuation performance in each experimental case (summarized in Table 14.1) will be expressed in terms of dB. The attenuation figure derives from an average of output-primary frame energy ratios over a one second period, during noise only. The frame size is 256 time samples. The mouthpiece signal is used as the primary input while an outside noise-collecting microphone is used as the reference input.

Table 14.1   Performance of noise suppression algorithms as applied to telephone kiosk data.

Noise suppression algorithm	Performance (dB)
ANC	1.4
Spectral subtraction	4.3
Multi-input lattice with LS	10.2
Bin proc. using MA filter	10.4
Power scaling	13.7
Power scaling using MA filter	29.2
HBS	39.5

Using the ANC format of Fig. 14.2, with an optimum LMS convergence factor expression, developed by Connell and Xydeas [10], and a filter length of 200, the processed output yielded a noise attenuation level of 1.4 dB. Introducing a multi-input lattice filter system, which uses a faster least squares algorithm, increases the noise attenuation level to 10.2 dB.

Transforming the primary and reference data into the frequency domain using block sizes of 256 samples, overlapped by 128 samples, and performing spectral subtraction, yielded a noise attenuation level of 4.3 dB. Clearly the lack of phase considerations has contributed favourably. Introducing an eighth-order adaptive filter on each reference spectral bin, to track the corresponding primary bin magnitude value in ANC format, increases the noise attenuation performance to 10.4 dB.

Reshaping the reference spectral envelope, to resemble that of the primary, prior to subtraction, increases the noise attenuation level from 4.3 dB to 13.7 dB. Making this power-scaling operation more effective by processing successive reference energies with an eighth-order adaptive filter, in ANC format, yielded a noise reduction level of 29.2 dB.

Over the same data, the proposed HBS reduces the level of noise by 39.5 dB.

14.5 IMPLEMENTATION OF HBS AS A REAL TIME SYSTEM

The hybrid band-splitting scheme described was implemented using the Signal WorkSystem (SPW) [14] at BT Laboratories. The application was run on a Sun SparcStation 1 +. SPW is a block diagram simulator which enables systems to be built up using either standard library or custom made blocks. The latter can either be constructed using the library blocks or implemented as C code. Overall, this approach facilitates rapid prototyping and simulation of systems to be achieved. SPW also allows the direct generation of DSP device code for a number of supported devices. This allows the simulations to be run in non-real and, if achieveable, in real time on the target DSP device. For the work described here the Texas Instruments TMS320C30 DSP device [15] was used.

14.5.1 Description of the simulation system

The HBS was implemented and the top level system block diagram is shown in Fig. 14.12. The processing was performed using 256 point frequency domain records with 50% record overlap, and ultimately was to run with a sampling rate of 8000 samples-per-second. The noise reduction processing was housed within the 'noise_reduct' block. This block only processes the magnitude of the frequency bins from both the primary and reference inputs. The phase information from the primary input was passed directly to the inverse FFT to construct the enhanced output. The vector sizes of the data used in the system (i.e. the data size that is processed in one simulation instance) is shown at pertinent points within the system. For the top level system, the vector sizes correspond to the record sizes for the Fourier transform operations.

The noise reduction processing was achieved separately for the two frequency bands 0-2.9 kHz and 2.9-4 kHz. These are depicted by reference to the frequency bin numbers 0-93 and 94-128 respectively.

The detail of the 'noise_reduct' block is shown in Fig. 14.13. Both inputs are frequency domain 'double-sided spectrum' transforms and, to avoid the unnecessary overhead of processing the same data twice, the processing was performed on 'single-sided spectrum' transforms (frames). A subsequent mirroring operation after the processing converts back to the 'double-sided spectrum' transform necessary to recover the output in the time domain. These processes are achieved by the 'vector split' blocks and the 'single_double' block respectively. The spectral shaping and subtraction was performed by the 'scale_subtract' block. The two scaling factors required

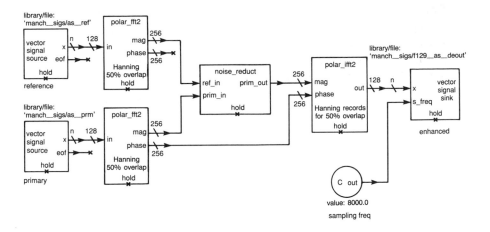

Fig. 14.12 SPW implementation of hybrid band-splitting scheme.

are calculated by the 'power_scale' block. The 'vrectifier' block prevents an output being produced should a negative magnitude result from the spectral subtraction.

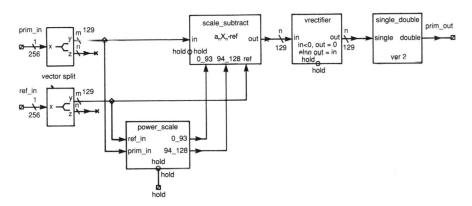

Fig. 14.13 Detail of 'noise_reduct' block.

The detail of the 'power_scale' block is shown in Fig. 14.14. Using the magnitude of the primary and reference bins this block generates the two scaling factors required to pre-shape the magnitude bins prior to spectral subtraction. These factors are calculated using the total frame and split frame powers in both inputs, generated by the 'power_estim' blocks. Each factor was calculated using an adaptive filter block 'adaptive_FIR' which adapts in such a manner as to estimate the power in the primary signal from the reference signal during non-speech periods. For this the speech detection was performed by a comparison of the total frame powers by the 'comp__xy' block. For the higher band the lowest signal of the primary noise estimate and the reference signal was taken; this was achieved by the 'lowest__xy' block. Both noise band estimates, 0__93 and 94__128, are then scaled with respect to the total primary power as detailed in equation (14.7).

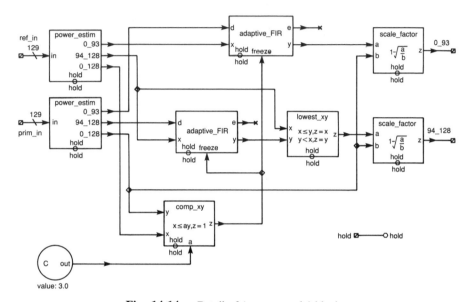

Fig. 14.14   Detail of 'power__scale' block.

The 'adaptive__FIR' detail is shown in Fig. 14.15. The filter block was implemented around the SPW adaptive filter block 'LMS ALC' with additional elements to control the rate of adaptation and block initialization. The 'T__OFF' block connects the output of the block to the input 'x' for the first eight simulation instances, after which output of the 'LMS ALC' block was used. The convergence factor μ was determined by the sum of the power estimates of the input values held in the filter, R.

A misadjustment factor (*misadj*) of 0.1 was used; μ is given as:

$$\mu \;=\; \frac{misadj}{\displaystyle\sum_{n=0}^{L-1} x^2(n)} \;=\; \frac{0.1}{\displaystyle\sum_{n=0}^{L-1} x^2(n)} \qquad\qquad \dots\,(14.10)$$

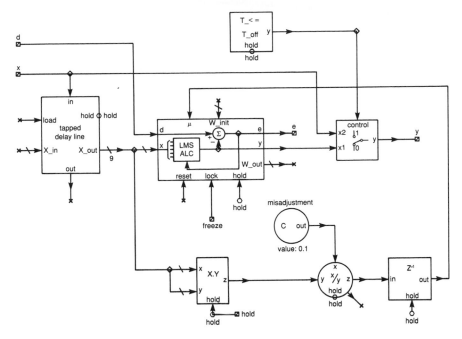

Fig. 14.15 Detail of 'adaptive_FIR' block.

14.5.2 Operation of the HBS on a DSP device

The HBS was implemented and simulated using SPW. The resultant speech files were found to be consistent with those obtained in section 14.4. Then, using the code generation system (CGS), C code was produced for the system from which compiled down-assembler code for the target DSP (TMS320C30) was produced. The DSP device was located on a PC system board (Loughborough Sound Images Ltd [16]) which resided in a personal computer connected to the workstation via a network server. The system was run, firstly, in non-real time using the target DSP device. The results were again found to be consistent. Using the analogue interface on the system board the HBS can be run in real time as a stand-alone process (i.e. running independently of the workstation).

The real time HBS is asynchronous in that data sampled at a constant rate is assembled in vectors of 128 and then passed on to the frequency domain processing part of the scheme. At the time this chapter was prepared CGS did not allow a real time implementation of this system to be achieved.

A fair estimate of the processing frame time of the HBS can still be achieved using CGS and implementing a system which consists of a signal source block connected directly to a signal output block with the HBS scheme included but left unconnected. The code for all the items included in the system was compiled, executable code produced for, and executed, despite the HBS being left floating. The signal source/output arrangement is required in order to demonstrate the whole system is running successfully at the sampling frequency requested by the user. It should be noted that as the inputs and outputs to the 'noise__reduct' block are of the form of size 128 vectors, the system sampling frequency corresponds to the frame rate of the HBS.

14.5.3   SPW implementation of HBS

The highest frame rate of the HBS which allowed the system to be successfully run was 55 frames per second. With the frame size of 128, this then corresponds to a sampling frequency of around 7000 samples per second. The required frame rate was 62.5 per second. Arguments to support that the HBS is achievable at this rate on the target DSP device are threefold.

Firstly, redundancy exists in two areas of the system implementation. Both magnitude and phase values are calculated for all the bins within each frame, whereas only values for 129 of the bins are used in the frequency domain processing. (It is estimated that the polar-magnitude conversions account for about 12 ms of the frame time and the removal of this redundancy alone should recover enough processing time to achieve the desired sampling rate.) Also, the squared magnitudes of the bins are calculated within the 'polar__fft2' blocks for each channel, and then again in the 'power__estim' blocks. (Investigation showed that the removal of this particular redundancy alone allowed an increased frame rate of 57 per second.)

Secondly, and most significantly, the polar-rectangular conversions (which take place in the 'polar__fft2' and 'polar__ifft2' blocks) are calculated for each set of values as opposed to using a 'look-up' table. The use of the latter can also be expected to decrease the computational overhead significantly. In addition, due to the ear's relative insensitivity to phase, the accuracy of this parameter within the look-up tables may be relaxed, allowing a further overhead reduction.

Thirdly, with a view to achieving a practically efficient algorithm, practical redundancies are also sought. An example of this would be the processing of the very high frequency bins whose amplitudes are severely attenuated due to filtering, and whose contribution to the frame power is minimal. Hence it is proposed, with confidence, that the desired sampling frequency could be achieved, after such optimizations, for the HBS running on the target DSP device.

14.6 CONCLUSIONS

The effectiveness of this scheme lies in the level of noise reduction attainable with minimum computational complexity. Its construction is based on an in-depth study of the success and limitations of existing techniques and an acknowledgement of the extent to which linear processing can make use of the limited correlation which exists in a time, spectral or power context. Filter lengths and algorithm complexity have been honed to combine an efficient use of memory with minimal mathematical overheads. The system components reflect precise requirements and combine an in-built consideration of the kiosk environment and the spectral nature of both speech and kiosk background noise. As discussed in section 14.4, its noise attenuation performance over a wide range of data has proved consistent. Within acceptable primary SNR limits, it maintains speech quality and intelligibility whilst providing a noise attenuation level of 39.5 dB. The computational costs in achieving this are approximately 25% more than that for SS. Consequently, HBS is a novel signal processing technique applied to acoustic noise reduction.

The essence of the scheme's contribution to the active acoustic noise reduction area is its solid foundation in an extensive variety of real environment data. This separates it from much of the published work which is based on either artificial data or on artificial recording environments. It does not offer the limited degree of primary/reference correlation as a concluding reason for failure but rather uses it as a starting point for analysis. Subsequent development then produces a scheme which optimizes the use of this level of correlation. It is this philosophy which defines the HBS as a major progression in the area of active noise reduction.

The limitations of the system are twofold. Low primary SNR values cause suppression of low-power speech sounds since the power ratio criterion for weight adaptation is violated by an increasing number of speech frames. The filter tracking trend is now unduly influenced by speech energy which leads to its cancellation. For most telephone kiosks however, this is not a serious problem. The SNR limitation is more likely in incidence of transient, high-powered, acoustic noise occurring simultaneously with low-power speech. Such occurrences are rare.

A second aspect of the system's limitation is the noise residual. This becomes more noticeable with a decreasing primary SNR. Perceptually, it appears as a chirruping effect which, although low level, could be subjectively irritating. The availability of an accurate VAD would allow reshaping of the residual into a more 'recognizable' form, such as the original primary or reference spectral envelope. However, trials with VADs invariably result in reshaping incorrectly marked frames which leads to speech clipping.

The HBS was successfully implemented on a DSP device, using a development path via SPW/CGS processing tools. The complete scheme was not able to be fully run in real time due to problems which currently exist with one of these tools. However, valid measurements enabled the speed of the HBS implemented on the DSP device to be measured. The scheme was found to be able to be run in real time up to sampling rates of 7300 per second. Failure to operate at the required sampling rate of 8000 samples per second is mainly due to overheads and inefficiencies in SPW block processing code, as outlined in section 13.5.3. A direct assembler-based approach to the real time implementation of HBS, coupled with the discussed reductions in bin magnitude calculations and the introduction of a look-up table, should render it entirely operational within the available processing time for a current generation DSP.

REFERENCES

1.   Harrison W A, Lim J S and Singer E: 'Adaptive noise cancellation in a fighter cockpit environment', Proc Intern Conf on ASSP, pp 18A.4.1-18A.4.4 (1984).

2.   Rodriguez J J and Lim J S: 'Adaptive noise reduction in aircraft communication systems', Proc Intern Conf on ASSP, pp 169-173 (1987).

3.   Harrison W A, Lim J S and Singer E: 'A new application of adaptive noise cancellation', IEEE Trans on ASSP, ASSP—34, No 1, pp 21-27 (February 1986).

4.   Boll S F: 'Suppression of acoustic noise in speech using spectral subtraction', IEEE Trans on ASSP, ASSP—27, No 2, pp 113-120 (April 1979)

5.   Dal Dagen N and Prati C: 'Acoustic noise analysis and speech enhancement techniques for mobile radio applications', Signal Processing, No 15, pp 43-56 (1988).

6.   Armbruster W, Czarnach T and Vary P: 'Adaptive noise cancellation with reference input — possible applications and theoretical limits', Proc EUSIPCO-86, pp 391-394 (September 1986).

7.   Erwood A F and Xydeas C S: 'A multiframe spectral weighting system for the enhancement of speech signals corrupted by acoustic noise', Proc EUSIPCO-90, pp L1-4 (1990).

8.   Widrow B and Stearns D S: 'Adaptive signal processing', Prentice-Hall, Englewood Cliffs, NJ (1985).

9.   Proudler I K, McWhirter J G and Shepherd T J: 'Fast QRD based algorithms for least squares linear prediction', in J G McWhirter (Ed): 'Mathematics in Signal Processing II', Oxford Press, pp 465-488 (1990).

10. Connell J M and Xydeas C S: 'A comparison of acoustic noise cancellation techniques for telephone speech', Proc Loughborough Intern Conf, pp 320-325 (1991).

11. Munday E: 'Noise reduction using frequency-domain nonlinear processing for the enhancement of speech', BT Technol J, 6, No 2, pp 71-83 (April 1988).

12. Bendat J S and Piersol J S: 'Engineering applications of correlation and spectral analysis', Wiley and Sons (1980).

13. Powell F A, Darlington P and Wheeler P D: 'Practical adaptive noise reduction in the aircraft cockpit environment', Proc Intern Conf on ASSP, pp 173-177 (1987).

14. 'SPW system, installation and tutorial users guide', COMDISCO Systems Inc.

15. 'Third generation TMS320 user's guide, SPRU031', Texas Instruments.

16. 'TMS320C30 PC system board user's manual, Version 1.0', Loughborough Sound Images Ltd.

15

VIDEO CODEC DESIGN USING DSP DEVICES

M W Whybray

15.1   INTRODUCTION

Data compression algorithms now exist which are capable of compressing a television picture from a raw bit rate of over 100 Mbit/s down to 64 kbit/s, for applications such as integrated services digital network (ISDN) videophones. One such example is the CCITT Recommendation H.261 algorithm [1] which can be used to transmit compressed video for videophone and videoconference at bit rates from 64 kbit/s to 2 Mbit/s.

Such algorithms have usually been implemented in dedicated hardware because of the high volume of data that must be processed in real time. For example, the mean input data rate of a television picture sampled at a resolution suitable for H.261 encoding is 40 Mbit/s, compared to a normal pulse code modulated speech data rate of 64 kbit/s. Digital signal processing devices (DSP devices) have been used for some years for real time speech processing, but have only recently been applied to the problem of encoding a video signal with almost three orders of magnitude greater sampling rate, and implementing an algorithm as complex as H.261.

There have been several reasons for this. Firstly, the sheer volume of data to be handled is such that even the data input, local storage, and data output processes were a very significant load. Secondly, the processing power of a single DSP device was hitherto insufficient to perform an algorithm of significant complexity in real time on the amount of data required. This task can be tackled by parallel processing, but this gives rise to a third problem, or set of problems:

- the complexity involved in the hardware for supporting the parallelism;

- the cost of multiple DSP devices;

- the inefficiencies in the use of processor time due to the overheads of parallelism and of imperfect load sharing.

The result is that, once more than a few DSP devices are involved, the solution quickly becomes more costly and complex than a dedicated hardware implementation would have been in the first place.

Nevertheless, there are still many advantages to a DSP device approach to video codec design, if it can be made to work. These advantages all stem from the programmability of the devices, which confer the following benefits on systems using them:

- algorithm complexity is enshrined in the software rather than the hardware, resulting in potentially simple (hence cheap and compact) hardware;

- hardware development can proceed before the algorithm is fully defined, reducing time to market and avoiding costly reworks;

- the same codec hardware design may be adapted to fulfil several different functional requirements, merely by a change of software;

- the use of standard rather than specialised components is likely to reduce costs.

The rest of this chapter describes the H.261 algorithm, and derives from it estimates of the processing power apparently required to implement it. These estimates are then revised downwards in the light of various mitigating circumstances and it is shown how a single DSP device can perform a video encoding or decoding function complying with Recommendation H.261.

15.2   VIDEO ENCODING ALGORITHM

During the five years of CCITT international collaboration to form the new Recommendation H.261, many alternative picture encoding algorithms were explored. After much debate, an encoding scheme based on the discrete cosine transform (DCT) with motion-compensated interframe prediction and variable length coding was selected. The following sections describe the picture format which is an integral part of the Recommendation, and the details of the compression algorithm itself.

15.2.1   Picture format

A common intermediate format (CIF) was adopted for H.261 by the CCITT specialists group, based on 288 non-interlaced lines per picture at 30 pictures per second[1].

Since there are 288 active lines per field (576 active lines per frame) in standard 625-line television, 625-line/25 Hz frame rate codecs have in principle only to perform a picture rate conversion to meet the 30 Hz picture rate requirement. 525-line/30 Hz frame rate codecs already have the correct frame frequency but instead have to convert the number of active lines from 480. Picture impairments due to this pre- and post-processing are usually negligible when compared with those introduced by the compression encoding. In the horizontal direction, CIF is formed by sampling the incoming television signal at 6.75 MHz which results in 352 luminance (Y) samples to be encoded. The two chrominance or colour difference signals (Cr and Cb) in CIF are sampled at half the luminance resolution in both directions, giving a chrominance resolution of 176 pixels (picture elements) horizontally by 144 vertically.

CIF is an appropriate choice for many applications, including video-conferencing. However, for some applications (e.g. face-to-face videophone) a lower resolution would suffice. For this reason a second picture format was included in H.261, having 176 horizontal luminance samples per line and 144 lines — half the resolution of CIF in each dimension, with corresponding reductions for chrominance. This format is termed quarter CIF (QCIF).

Pictures are divided into blocks for subsequent processing. The smallest block size is an 8×8 pixel block, but a group of four such luminance blocks, and the two corresponding chrominance blocks that cover the same area at half the luminance resolution, are collectively called a macro-block (see Fig. 15.1). Further, 33 macro-blocks, grouped as shown in Fig. 15.2, are known as a group of blocks (GOB), and these are in turn used to build up a full CIF or QCIF picture (Fig. 15.3).

---

[1] In fact the precise picture rate is 30 000/1001, or approximately 29.97 picture/s.

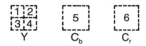

Fig. 15.1 Arrangement of blocks in a macroblock.

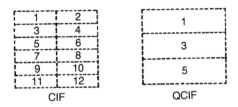

Fig. 15.2 Arrangement of macroblocks in a GOB.

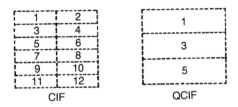

Fig. 15.3 Arrangement of GOBs in a picture.

The basic picture rate of CIF and QCIF is 30 picture/s, but a video codec is not constrained to encode every picture. Particularly at the lower bit rates it is usual to omit one or more pictures between each encoded one, as a means of helping to reduce the number of bits generated.

15.2.2 The Recommendation H.261 encoding algorithm

A block diagram of the H.261 encoding algorithm is shown in Fig. 15.4. Having obtained an input picture in CIF or QCIF format, a predicted image is then subtracted from the input picture and the resultant 'difference-picture' (point {A} in Fig. 15.4) passed to the discrete cosine transform (DCT) unit. The prediction image is derived from the previously encoded image in a manner to be described later. Taking the 'difference-picture' substantially reduces the amount of information that needs to be transmitted, since most scenes (especially those of videoconferences) usually only contain fairly small regions of change and all other information in the picture remains approximately constant. After 'differencing', the picture is divided into 8×8 blocks and then subjected to the DCT process. The transfer function for the DCT is given by:

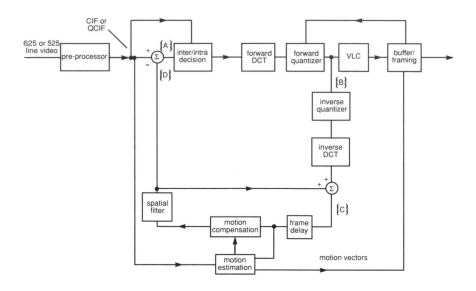

Fig. 15.4   CCITT Recommendation H.261 encoding loop.

$$F(u,v) = \frac{1}{4} \sum_{x=0}^{7} \sum_{y=0}^{7} f(x,y)\cos[(2x+1)u/16]\cos[(2y+1)v/16]$$

with $u,v = 0, 1, 2, \ldots 7$

where x,y = spatial co-ordinates in the pixel domain

u,v = co-ordinates in the transform domain.

A formal description of the DCT process can be found in Clarke [2]. In brief, the DCT produces a series of coefficient values which relate to the spatial frequency content of each 8×8 block. Before transformation, each block is made up of 64×8-bit values which represent the brightness of each point within the 8×8 space. After transformation, 64 coefficient values result which represent the magnitudes of the various spatial frequency components present at the input to the transform. Each coefficient is then processed in a sequence determined by a zig-zag scanning path shown in Fig. 15.5. The first coefficient value (known as the DC coefficient) represents the mean value (grey level) of the 8×8 input block. The second value represents the magnitude of the lowest frequency in the horizontal direction, the third value gives the magnitude of the lowest frequency in the vertical direction, and so on up to the 64th value which gives the magnitude of the very highest frequencies in both the horizontal and vertical planes.

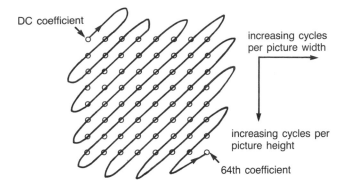

DC coefficient

increasing cycles
per picture width

increasing cycles per
picture height

64th coefficient

Fig. 15.5   Coefficient scanning sequence for each 8×8 block.

The DCT process in itself does not achieve any image compression — 64 values at the input result in 64 values at the output. The compression comes from further processing which takes advantage of redundancy in image statistics. Pictures, and also differences between pictures, tend to be made up of regions of similar values. Most 8×8 blocks can be represented by only a few transform coefficient values — usually only the DC coefficient and a few low-frequency coefficients have significant magnitude. The subsequent quantization process sets all small values to zero and quantizes all non-zero transformed values to a set of nearest preferred magnitudes ready for transmission (point {B} in Fig. 15.4). Advantage is taken of the fact that for most of the time only a few coefficient values need to be transmitted for each block. Even in the case where a few tens of coefficients need to be transmitted, these values most often occur consecutively within the coefficient sequence and so by using a form of run-length encoding [3], which exploits the statistics of the occurrence of non-zero coefficient values, relatively efficient transmission is ensured.

Further gain is obtained by the use of variable-length encoding [4] which exploits residual redundancy in the coefficient magnitude statistics. The probability distributions of coefficient values have a large peak at zero, and decay away rapidly for progressively larger positive or negative values. Short codewords are therefore used for the commonly occurring values near zero, and longer ones for the relatively rare occurrences of the larger coefficient values. The variable-length codewords (VLC), plus other control information are multiplexed together, and placed in a data buffer for transmission.

The quantizer is split into a forward and an inverse part. The forward quantizer maps each input coefficient into one of a finite set of notional levels, which are used to look up the appropriate variable length code in a table.

The inverse quantizer converts the notional levels back into real (but quantized) numerical values again for subsequent processing within the prediction loop.

These quantized transform coefficients are subsequently inverse transformed and then added to the previously encoded image. This results in 8×8 blocks of image data (point {C} in Fig. 15.4) which are very similar to the video data at the input, though not exactly the same due to the quantization process. The error appears as noise distributed throughout the block, usually with some spatial frequency structure evident due to the nature of the transform. However, if the encoding parameters are chosen carefully, an acceptable image results. This data is then delayed for approximately one picture period ready to be used in the prediction process for the next encoded picture.

The motion compensation unit serves to minimize the inter-frame differences which have to be encoded. Objects which move simply by translation between sequential images (i.e. no rotation, zooming, or deformation) can be most efficiently encoded by remapping the object's position from the previously stored image. For example, a person's hand located at a certain point in the previous image can simply be moved to the new position in the current image. Only the displacement vector need be transmitted, resulting in significantly less information than that required to completely reconstruct the hand itself in the new position. Unfortunately, in many situations motion is represented not only by simple translation but also by some distortion of the image content by rotation, object deformation, etc; so merely moving the previously encoded image is rarely more than approximately correct. However, motion compensation does significantly reduce the magnitude of the picture difference signal and thus the bit count. In practice the motion estimation process works by taking a macro-block (a 16×16 luminance pixel area) from the current input video image and searching in the previous encoded image over a region of up to ± 15 pixels in the horizontal and vertical directions to find a best match. In Fig. 15.4, the motion compensation unit is effectively a variable length store, whose length is dynamically adjusted so that the best-matched position is used for the subtraction (at point {D}) from the input video image and the values for the horizontal and vertical translations are included as additional data in the output data buffer. Motion is usually only estimated from luminance information, but compensation is applied both to luminance, and, at half the pixel displacement, to the two chrominance components of the macro-block as well.

Since the motion compensation process is only approximate, the higher frequency components in particular may be bad matches to, and hence predictions of, the new image data to be encoded. The compensation unit is therefore followed by a spatial filter which low-pass filters or blurs the prediction image data when required, thereby removing the poor prediction components. The filter may be switched in or out on a macro-block by macro-block basis, and is typically only switched in if a non-zero motion vector is detected.

The output of the motion compensator and filter processes is the prediction image used by the subtractor at the start of the encoding loop. For some macro-blocks where a large change in image content has occurred, even the motion compensated prediction may be quite poor, in which case the subtraction process can be disabled and the input picture data for that macro-block simply encoded directly (intraframe mode). Since this mode encodes a macro-block without danger of corruption from any previously received data that could contain errors, it is also used to progressively update each block in the picture over a period of time, to mop up any errors that may have accumulated due to transmission errors or small differences in arithmetic rounding within the DCT process. Recommendation H.261 sets specific limits on the DCT calculation error and on the minimum intra-frame mode refresh period for macro-blocks.

The different sources of data within the encoder are assembled into an orderly bit stream according to a video multiplex defined within H.261. The multiplex is a hierarchically nested structure, starting with information about the complete frame, then each GOB, then each macro-block, and finally the transform coefficient data. The start of each encoded picture is marked by a unique 20-bit code-word that cannot be mimicked by any other video data, and allows the decoder to synchronize to the start of each picture. The following data is then uniquely decodable to identify which macro-blocks within each GOB have been transmitted, and the values of all the associated coefficient values, motion vectors and so on.

Unfortunately, variable length codes are very sensitive to errors, and a single bit error can scramble all the data in a GOB before resynchronization is obtained. To counteract this, an additional level of multiplex is added which incorporates a BCH (Bose-Chaudhuri-Hocquenghem) error correction encoding scheme.

A simplified block diagram of the H.261 decoder is shown in Fig. 15.6. It matches the final part of the encoding loop in the encoder, so displacement vectors and coefficient values are input to the decoder and processed in a similar way to the data at point {C} in the encoder (see Fig. 15.4). If there are no transmission errors then the resulting video data at point {E} in the decoder will be an exact replica of that at point {C} in Fig. 15.4.

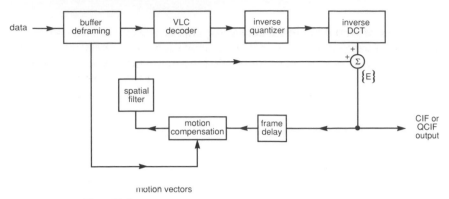

Fig. 15.6   CCITT Recommendation H.261 decoding loop.

15.3   WORST-CASE PROCESSING POWER REQUIREMENTS FOR RECOMMENDATION H.261

Given the description of the H.261 algorithm above, it is possible to derive estimates of the processing power apparently required to implement it, at least for the major functions. These estimates are for the encoder and assume it has to handle worst-case conditions which are:

- spatial resolution — full CIF (352 by 288 luminance pixels, 176 by 144 by 2 chrominance pixels);

- temporal resolution — full CIF rate (30 CIF picture/s);

- output bit rate — the H.261 maximum of 2 Mbit/s;

- motion estimation — every macro-block in the picture, over a full search area of ±15 pixels;

- forward and inverse transform and loop filter — every block in the picture;

- zig-zag scan, quantize and variable length code — every coefficient in each block.

These worst case processing power estimates are derived below.

15.3.1   Motion estimation

The number of macro-blocks processed per second is the number in a CIF picture (396) times the picture rate (30 Hz), giving 11 880. If a full search algorithm is used the number of block-compare operations is 31×31 or 961. Each block-compare typically requires summing the squares of the differences of the 256 pixel pairs. Multiplying this out, and counting the difference, square and accumulate as three separate operations, gives a total of 8768 MOPS (million operations per second).

15.3.2   Discrete cosine transform

The number of blocks processed per second is the number of macro-blocks per second (396×30) times the number of blocks in a macro-block (6), giving 71 280. If the DCT is implemented by a pure matrix multiply method, then the number of multiplies and accumulates required is the number for a 1-dimensional DCT (64 of each) times the 8 rows plus 8 columns in an 8×8 block, giving 146 MOPS. Each block has to be forward and inverse transformed, doubling this to 292 MOPS.

15.3.3   Loop filter

The loop filter requires convolution of a block with a 3×3 kernel, so each of the 64 pixels requires nine multiply and accumulate operations to produce the output pixel. Multiplying this by the number of blocks processed per second (71 280) gives 82 MOPS.

15.3.4   Zig-zag scan, quantize and variable length code

These items are harder to quantify as the operations required are not easily translatable to simple multiply or accumulate operations. However, if it is assumed that each operation will be performed by a table look-up process, the maximum number of these table look-ups may be calculated as 71 280 times the number of pixels per block (64) times three (one for each table look-up), and it is reasonable to perhaps treble this to allow for other overheads, giving 41 MOPS.

These estimates cover the main encoder processing loads in decreasing order of size. Other encoder functions are of lower order than those of section 15.3.4, and are therefore negligible in comparison to those of sections 15.3.1 to 15.3.3 above. The above processing power requirements, of the order of 9000 MOPS, are well above those offered by currently available general purpose DSP devices, which are typically rated at a few tens of MOPS. In particular it is clear that motion estimation is by far the most dominant processing load. A hardware codec usually has to cater for all the above worst-case conditions occurring, and therefore is normally grossly under-used when typical or average requirements are considered.

15.4   REDUCED PROCESSING POWER REQUIREMENTS FOR RECOMMENDATION H.261

In this section, it will be shown how the various constraints and options in H.261 mean that the worst case conditions described in section 15.3 can be scaled down considerably, particularly in a DSP device implementation.

15.4.1   Limitation imposed by the output bit rate

Because Recommendation H.261 defines the multiplex of the bit stream, the worst case bit stream can be calculated in terms of the maximum number of blocks requiring processing to generate a given bit rate.

In considering, firstly, the case where all blocks being transmitted contain just one non-zero transform coefficient (the lowest number possible still requiring the DCT calculation to be invoked) detailed examination of Recommendation H.261 shows that the minimum number of bits per block required to signal this condition is at least 1 bit for macro-block overheads, at least 2 bits for the shortest coefficient data code-word, and a further 2 bits for the end of block code, giving a total of 5 bits.

If the codec is working to a 64 kbit/s line for example, this means that the maximum number of such blocks that can be transmitted per second is only 12 800 compared to the 71 280 blocks per second in uncoded CIF. (In fact this is an over-estimate as no allowance has been made for the other parts of the video multiplex, which reduce the bit rate available for block data.) Thus the peak computational power for encoder DCT requirements is already reduced from the 292 MOPS of section 15.3.2 to around 52 MOPS, if operation at the 64 kbit/s rate is required. Corresponding reduction for the filter gives 15 MOPS, and, for the zig-zag scan, quantizer and VLC, 8 MOPS.

Even the above figure is still very much a worst case, as in practice most blocks will contain several active DCT coefficients, all of which use up more bits on the channel and reduce the actual number of blocks transmitted per second several times more. Similarly, for blocks which are part of a motion compensated macro-block, or have the loop filter activated, several more bits are used up to active these functions.

15.4.2 Efficient implementation of the DCT and loop filter

The formal definition of the DCT in section 15.2.2 cannot be implemented exactly, as the cosine values are indefinite numbers, i.e. they would require infinite precision arithmetic, whereas in practice some form of finite calculation precision must be used. For an encoder and decoder loop to track exactly they must use exactly the same calculation precision. However, this would require defining the DCT calculation down to the level of the precise sequence of operations and roundings to be used.

There are, in fact, many different ways of structuring the DCT calculation, and, to avoid ruling out some of these more efficient implementations, the CCITT decided to avoid specifying only one method of calculation, but instead to impose various limits on the average and peak errors allowed when processing a large predefined set of pseudo-random test data. This opens the way for so-called fast DCT algorithms, similar in nature to the fast Fourier transform in that they rely on butterfly rather than matrix multiply operations. An example is that due to Chen, Smith, and Fralick [5], for

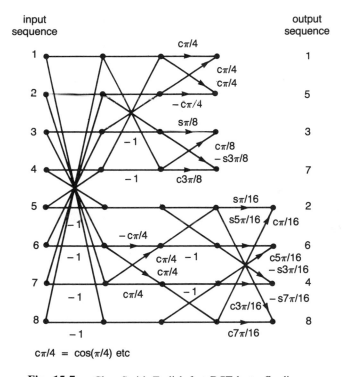

$c\pi/4 = \cos(\pi/4)$ etc

Fig. 15.7 Chen-Smith-Fralick fast DCT butterfly diagram.

which the butterfly diagram for a one-dimensional forward transform is shown in Fig. 15.7. The number of multiplies and additions required is reduced from 64 each per transform to 20 and 26 respectively. Most DSP devices are able to perform a multiply and accumulate as a combined single operation, but, in practice, because of the algorithm structure, not all of the operations can be done in parallel. Also the increased addressing complexity may cause additional overheads, but a speed increase of up to three times above the matrix method can be achieved.

Combining this improvement with that indicated in section 15.4.1 gives a maximum encoder processing load for DCT calculations of only 17 MOPS in the encoder, compared to the 292 MOPS of the superficial estimate in section 15.3.2. The decoder only has to do the inverse transform and so requires only 9 MOPS.

The loop filter calculation may similarly be speeded up by making use of the fact that the 3×3 FIR filter kernel is separable into two 1×3 kernels. This reduces the number of multiply and accumulate operations from nine to six each, and again since most DSP devices can perform these in parallel the processing requirement is reduced by an overall factor of about three times. Combining this with the bit rate limitation factor indicates a maximum processor power requirement for filter operations of 5 MOPS, which is the same for encoder and decoder.

15.4.3   Decoder options in Recommendation H.261

Recommendation H.261 specifies the structure and meaning of bits in the video multiplex, and how a decoder must respond to these bits. In addition, to ensure compatibility between all H.261 codecs, there is provision for a negotiation phase when an encoder and decoder are connected at the start of a call, during which the decoder signals to the encoder what its maximum decoding picture rate and decodable resolution are.

The two resolutions available are CIF and QCIF, but all CIF-capable codecs must also handle QCIF, so any decoder may opt for the QCIF choice. The main operations performed in a decoder are VLC decoding, zig-zag scan, inverse quantization, the inverse DCT, the filter and motion compensation. However this last item is only a matter of off-setting the address for a block of data according to the received motion vector. VLC decoding, zig-zag scan and inverse quantization are quite straightforward, and also have the property that for every extra VLC decoded or coefficient inverse quantized, an extra increment of processor time has been made available due to the time taken to transmit those bits, so these can never become major overheads, and amount to under 1 MOP. The DCT and filter are the remaining items, and as noted in section 15.4.2 these require about 9 and 5 MOPS processor power respectively at 64 kbit/s.

The maximum decoder picture rate capability is actually expressed in H.261 in terms of the number of dropped pictures between active ones sent by the encoder. This number may be from 0 to 4, giving maximum picture rates of 30 Hz down to 7.5 Hz. By opting for a resolution of QCIF and, say, a 15 Hz maximum picture rate, the number of blocks required to be processed per second is reduced to 15 times the number in a QCIF picture (594) giving 8910, which is actually below the limit set by the capacity of a 64 kbit/s channel. This option would reduce the peak decoder processing power requirements for DCT and filter operations to six and under 4 MOPS respectively.

Perhaps a more important reason for limiting the decode frame rate is to reduce the amount of data-moving required to reconstruct and display complete frames, which at a 30 Hz rate can become a more serious problem for a DSP device than actual calculations.

15.4.4   Encoder options in Recommendation H.261

Although a decoder can specify certain options as above, it must nevertheless be able to cope with the worst-case bit stream that H.261 can impose upon it. An encoder however has more options. In particular, as long as the bit stream it generates conforms to the H.261 syntax, and the inverse quantizer, inverse transform, motion compensator and filter (which together comprise the part of the coding loop that matches the decoder) conform to the Recommendation, the rest of the encoding loop can be quite freely modified. This is to allow different codec manufacturers the freedom to choose different optimizations for features such as the criterion on which to base the inter/intra decision, the choice of quantizer step size, the method of motion estimation and so on. For instance it would also be possible to use a forward transform with very poor accuracy and still conform to H.261, since the rest of the encoder loop would still accurately track any decoder loop, and the fact that the encoder would be wasting bits correcting its own errors all the time is immaterial as far as the standard goes, though the resulting codec would probably be unmarketable.

As was evident in section 15.3, one major area where a DSP device-based encoder needs to save processing power is the motion estimator — a combination of several techniques may be used to do this. The first technique is to restrict the search area to something below the maximum allowed range of ±15, for example ±7.

The second technique is to use a sparse search pattern in multiple stages to home in on the 'best' match position, rather than searching every possible position. An example of a multistage search is illustrated in Fig. 15.8. The first stage searches at the zero position and ±4 pixel displacements, finds the best match and then uses this position as the start point for the next stage. The second stage then searches at relative offsets of ±2 pixels and similarly locates the best start position for the third stage which searches at ±1 pixel displacement. The procedure thus attempts to home in on the minimum error position, in the search area that extends up to a maximum range of ±7 pixels, using only 25 macro-block compare operations as opposed to 225 for a full search on the same area.

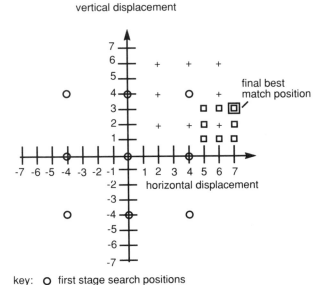

key: O first stage search positions
      + example second stage search positions
      □ example third stage search positions

Fig. 15.8   Example three-stage motion estimation search strategy finding the best vector at (7,3).

The danger is, of course, that if the wrong direction is chosen at the first or second stages the procedure may not find the best match at all, but only second or third best, but because images, or sub-images of this size, are generally fairly smooth the error surfaces are similarly smooth and the search usually proceeds in the right direction. Also, it does not matter if some 'wrong' motion vectors are chosen, because the predictive coding loop corrects the prediction error anyway, and the picture quality is only affected indirectly by the small reduction in efficiency. Simulations have shown only a small loss of picture quality through using partial rather than full searches. The range of the search could be of greater importance, but for some applications, such as videophony where the picture content is usually a fairly static person, a reduced range is often quite adequate to track motion. This can be seen from monitoring the motion vector statistics, which show very low usage of large displacement vectors in such scenes. It can also be noted that at QCIF resolution, ±7 pixels corresponds to the same range of motion in the original scene as ±14 pixels does in a CIF resolution picture.

A third technique for reducing the number of operations required for motion estimation is to compare only a subset of the pixels in the macro-block during each test. Again, because natural images do not usually contain a lot of high spatial frequency energy, this subsampling operation can have a fairly small effect on final picture quality, and a saving factor of two is quite reasonable here.

A fourth technique is to monitor the error value as it is accumulated during the macro-block match test at each position, and abort the test if the error exceeds the minimum error found in previous tests. The saving from this method is indeterminate as it depends on the statistics of the error surfaces, but is usually quite significant — at least a factor of two.

Combining all the above savings, noting that a standard difference, square (or absolute value) and add operation per pixel for the block match can be performed in two cycles on most DSP devices, and encoding only a QCIF picture, the processing power estimate for motion estimation falls from the 8768 MOPS of section 15.3.1 to a much more reasonable 10 MOPS. Of course, these estimates ignore various overheads incurred in controlling the whole operation, and make various assumptions about picture statistics, but even so it can be seen that very significant savings can be made.

15.4.5   Change detection in the encoder

The above calculations have assumed that the worst case condition is when each processed block corresponds to only 5 bits being transmitted on the channel. This is indeed the worst case for the decoder, but for an encoder there is the possibility that it performs motion estimation, the DCT, quantization, etc, on a block, only to find that all coefficients are zero valued and no data is generated. This will in fact be quite normal for static areas of the picture such as the background and so is not a minor problem. However, it can be addressed by first performing some form of change detection process on all the blocks in a picture.

A typical process would be to compare each incoming block with the one in the encoding loop prediction store, and threshold some block error functions. Static areas can be identified at once and only changed areas, which will actually generate some useful bits, will subsequently use up valuable processing power. As with motion estimation, the processing power saved here is dependent upon picture statistics, but in a videophone scene only around half of the picture might change on average between encoded pictures, reducing the motion estimation task still further to around 5 MOPS. Unfortunately this does not necessarily save any DCT, filter bits, etc, as the number of these operations required is already limited by the channel capacity to a level even below that of a picture sequence with only 50% change per picture. The processing power required for the change detection is under 3 MOPS.

As a final point, it should be noted that the DCT, filter, zig-zag, quantizer and VLC calculations have all assumed the worst case bit stream of only 5 bits per block, which corresponds to only the DC coefficient being non-zero, and with a quantizer index value of ± 1. If any other coefficients are active, or higher quantization values are used, the number of bits used will rise considerably, and this will be the usual situation. The mean processing power requirements for the above functions will therefore be on average a factor of two or more below their worst-case values.

15.4.6   Final estimates of processing power requirements

All the preceding calculations are summarized in Table 15.1.

Table 15.1   Estimates of processing power required for a QCIF 64 kbit/s H.261 compatible encoder and decoder.

	MOPS	
Encoder tasks	Peak	Mean
Change detection	3	3
Discrete cosine transform	17	9
Loop filter	5	2
Motion estimation	10	5
Zig-zag scan, quantize, VLC	8	4
Encoder total	43	23
Decoder tasks	Peak	Mean
Discrete cosine transform	9	5
Loop filter	5	2
Zig-zag scan, inverse quantize, VLC	1	1
Decoder total	15	8

As noted before, a decoder has to cope with the worst case bit stream that H.261 allows, and hence the peak figure of 15 MOPS is the critical one. However, the encoder is essentially in charge of the bit stream, and as long as it is up to the mean processing requirement, momentary overloads due to worst case conditions can usually be tolerated. If these overloads are short, then they will be catered for by the output data buffer anyway. If they persist, the encoder has two main options for generating data fast enough to prevent a buffer underflow.

The first is to reduce the quantizer value such that it is more likely to pick up non-zero coefficients after quantization. This is an appropriate thing to do because a low buffer level is usually associated with a reduction in the number of changed blocks in the picture, so the codec can take the opportunity to use a finer quantizer to improve the picture quality.

If this does not work, an encoder has one last stratagem that it can employ if it is proving difficult to generate enough bits to keep the channel occupied. It can insert certain bit-stuffing codes defined within H.261 into the channel, to buy itself more processing time. This last option will prevent it violating H.261 by avoiding random data being sent to line, but of course it is really wasting valuable capacity to the detriment of picture quality, and so should not be used except *in extremis*. Thus the mean encoder processor power of 23 MOPS is the more appropriate value to use when designing an encoder.

15.5   THE VC9000-001 CODEC

This section describes a DSP device-based codec that makes use of the above considerations to allow an H.261 compatible codec, the VC9000-001, to be realized using only a single DSP device for the encoder, and another for the decoder.

15.5.1   Choice of DSP device

The processing power estimates, derived in section 15.4 for a QCIF codec capable of working to a 64 kbit/s transmission channel, were 23 MOPS for the encoder and 15 MOPS for the decoder. Although these are based on a certain number of reasonable assumptions, they do not properly account for all the extra overhead involved in practice. Also, it is impossible to account accurately for the processing requirements without considering the software implementation in the assembly language of a specific DSP device. The figures given assume that an operation (OP) equates to a single cycle instruction that includes a multiply and add (or equivalent), plus sufficient data moves to keep the central processor unit (CPU) fed with data. This contrasts with a typical manufacturer's use of the term where every extra add, multiply, or data move performed in the same instruction counts as an extra OP. Thus a 10 million instructions per second (MIPS) DSP device is termed a 30 or more MOPS device.

Manufacturers' specifications in terms of MOPS, or benchmarks for standard operations, can be quite misleading; therefore it is usually worthwhile to attempt to write the key parts of the most processor-intensive components of an algorithm in DSP device assembly language, and to optimize the code to make the most efficient use of each DSP device's different architecture and instruction set. Only in this way is a proper feel for a particular DSP device's strengths and weaknesses gained. For example, a processor may have a CPU capable of doing a multiply-and-accumulate in one cycle, but have restrictions on the data source and destination address generation which make it impossible to keep the CPU fully loaded on anything except a simple finite impulse response (FIR) filter calculation.

Another factor ignored in the processing power estimates is the problem of moving around the considerable amounts of data required for video processing. A processor with inadequate input and output capabilities can run into problems if data moving cannot be done in parallel with CPU calculations, or if there is bus contention with the program bus.

At the time the VC9000-001 was specified, the processor which most nearly fitted the requirements was the TMS320C30 (C30) from Texas Instruments. It had a 60 ns cycle time, giving it an instruction execution rate of 17 MIPS, equating to 17 MOPS in the terminology of this chapter. It was therefore capable of performing the peak decoding operation and, although not quite meeting the mean encoding requirement of 23 MOPS, the extra flexibility open to the encoder for reducing the processor load was felt to make it still a viable option.

Other useful features of the C30 were:

- on-chip direct memory access (DMA) controller capable of transferring data on or off chip with zero CPU overhead;

- two serial data ports;

- two on-chip timers;

- two external buses to avoid program/data bus conflicts;

- 2 k word of on-chip RAM, able to support up to four accesses per instruction cycle;

- mask programmable for future cost/size reduction in volume quantities.

However, it should be mentioned that the C30 also has some features which are less desirable; in particular the dual 32-bit wide buses make for a very high pin count (181), the large die and pinout currently require the use of a pin grid array package which is both large and expensive, and the part cost is significantly higher than many 16- or 24-bit DSP devices.

15.5.2   Overall codec architecture

The main architectural considerations for a video codec concern the design of the frame-stores necessary to capture, store and display pictures. To avoid wasted CPU time accessing external memory directly, it was decided to use the C30's DMA controller as the primary means of moving data on- and off-chip for processing. It was known that the coding algorithm to be implemented was based on blocks of pixels, and that the DMA controller would not be able to address these blocks directly if the frame-store memory appeared simply as contiguous addresses in the C30's memory space. Also, only the program bus had sufficient address lines to directly address the frame-store memory, but using this bus would cause program/data bus conflicts.

The design of the codec architecture also had to meet certain basic requirements, which were either specific to the encoder or decoder parts, or common requirements for both:

- encoder requirements — to capture QCIF pictures from a 525-line/ 30 Hz or 625-line/25 Hz colour video source with appropriate spatial prefiltering and subsampling;

- decoder requirements — to output 525-line/30 Hz or 625-line/25 Hz colour video by spatial interpolation of QCIF pictures;

- common requirements:

  — to provide up to four QCIF framestore areas to allow for buffering of input or output pictures;

  — to provide C30 access to the four frame-stores without interfering with the picture input or output processes, and without causing the C30 to generate bus wait states whilst performing DMA transfers;

  — to provide framestore address generation to allow a C30 DMA controller to transfer two-dimensional blocks of Y, Cr, or Cb data of arbitrary size;

  — to allow display of uncoded incoming video in a mirror-image (laterally reversed) format for self-view purposes.

The final solution to these requirements is the architecture shown in Fig. 15.9. The codec contains two C30s — one for the encoder and a second for the decoder. Each of these is linked to its own frame store sub-system which provides the four QCIF sized stores that each C30 requires.

The frame-store sub-systems allow the C30s free access to the stored pictures, in parallel with pictures being either captured or displayed. To allow the C30s' DMA controllers to be used to read and write blocks of data, each frame-store contains its own DMA controller that can be programmed to access any arbitrarily sized block of image data, with any required X and Y origin, and from any of the three colour planes, Y, Cr and Cb.

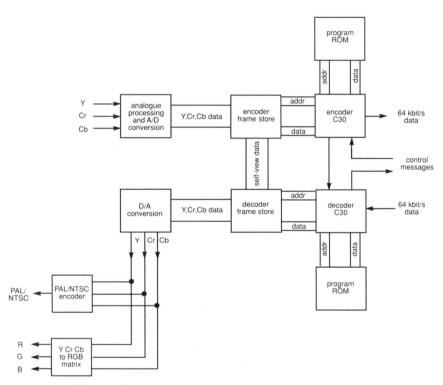

Fig. 15.9   Codec block diagram.

The analogue Y, Cr and Cb input signals are prefiltered, clamped and analogue to digital converted, before being vertically filtered and subsampled to the required QCIF resolution, to be packed into the encoder frame store. Output data from the decoder frame store is vertically interpolated to provide the required number of output lines for a 525- or 625-line television system, then digital to analogue converted and passed through both a PAL/NTSC encoder and a Y,Cr,Cb to R,G,B matrix to provide outputs capable of directly driving displays.

In addition, there is a direct data path between the two frame stores to allow digitized input pictures to be passed directly to the display to provide a self-view function. The advantage of going through the digitizing process rather than simply switching over the analogue inputs to the outputs is that the self-view display appears at the same QCIF spatial resolution as incoming pictures do, hence avoiding highlighting the fact that the pictures are at reduced resolution compared to normal television. Also, a digital line store can be used to laterally reverse the picture, giving a mirror image type display which is more suitable for self-view purposes.

One set of C30 serial ports is used to interface to the 64 kbit/s data line, and the other set provides a simple serial message interface to allow the codec operating modes to be modified by an external controller. Facilities provided include test pattern generation, freeze-frame, self-view, and in-view for which a small self-view picture insert is overlaid in the corner of the main decoded picture to provide a continuous indication of the user's position in their own camera shot.

The physical layout of the VC9000-001 codec board is shown in Fig. 15.10. The board measures just 18×23 cm, and includes analogue processing to accept video in 625- or 525-line Y,Cr,Cb format, and to output video in R, G, B and PAL/NTSC formats. The encoder can generate pictures at up to 30 picture/s, but the decoder is limited to a maximum of 15 picture/s at 64 kbit/s.

Fig. 15.10   The VC9000 codec.

15.5.3   Codec software development

From the estimates of processing power required it was clear that the use of high-level language compilers or real time operating systems would not make it possible to deliver the efficiency required, so all the software was written in C30 assembly language. Appreciating that assembly language is intrinsically much less readable and self-documenting than high-level languages, the overall structure of the software was specified in terms of 'design modules' which defined what each associated 'code module' would do, its inputs and outputs, and roughly how the function was to be performed. The latter was usually specified in pseudo-code which allowed free use of C-type constructs and English language descriptions as appropriate.

The lower level subroutines were completely debugged and tested using the simulator tools available from Texas Instruments, which also allowed monitoring of the state of the C30's pipeline to check for unforeseen pipeline blockages. These could usually be removed by judicious re-ordering of the assembly language statements, or rewriting the code in a slightly different way. This refinement process was critical in being able to speed up the most time-consuming parts of the code such as the DCT, spatial filter and motion estimator.

As an example of the results of this process, consider the extract below, which is part of the spatial filter subroutine. As noted in section 15.4.2, the filter operation can be separated into two one-dimensional FIR filters of three taps each. The extract shown performs the second filter stage, with tap weights of 1, 2, and 1, and arithmetically shifts the result to normalise it back to an 8-bit result (the rounding value required having been added at a previous stage):

```
       .
       .
       .
    ASH      R5,R3,R1
    MPYI3    R6,*+ +AR0,R0      ; Store 2 times pixel n + 1 in R0
||  ADDI3    R2,*+ +AR1,R3
    ADDI3    R0,*-AR0,R2        ; Add pixel n to R0, store in R2
||  STI      R1,*AR2+ +(IR1)

    ASH      R5,R3,R1
    MPYI3    R6,*+ +AR0,R0
||  ADDI3    R2,*+ +AR1,R3      ; Add pixel n + 2 to R2, store in R3
    ADDI3    R0,*-AR0,R2
||  STI      R1,*AR2+ +(IR1)

    ASH      R5,R3,R1           ; Divide R3 by 16, store in R1
    MPYI3    R6,*+ +AR0,R0
||  ADDI3    R2,*+ +AR1,R3
    ADDI3    R0,*-AR0,R2
||  STI      R1,*AR2+ +(IR1)    ; Store R1 to array in RAM
       .
       .
       .
```

In this example, AR0 and AR1 are both being used to address the same line of input data in an array, whilst AR2 addresses the output array. IR1 contains the value 8, to allow AR2 to increment down the columns of an 8×8 array. R6 contains a filter weight value of 2, and R5 contains a value of -4 which is the required arithmetic shift for rescaling.

The parallel bars indicate a parallel instruction where the current line and the previous line actually form one instruction executed in a single cycle. The calculation of a single filter output value is distributed over three groups of identical assembly language terms to allow results from one instruction to be available for inclusion in a subsequent calculation. This is necessary since the result from one half of a parallel instruction is not available as an input to the other half of the instruction.

The extract shown implements a 3-point FIR filter including rescaling in just 3 cycles per output pixel (though the process is not quite as efficient at the start and end of each row of eight pixels). This high efficiency is only possible because of the simple filter weights being used, but demonstrates how the processing power estimate of section 15.4.2 has been met, and indeed exceeded, since rescaling is also included.

15.6   FUTURE TRENDS

The VC9000-001 development has shown that it is indeed possible to implement an algorithm as complex as H.261 on real time video signals using single DSP devices. To achieve this, use has been made of the various options that H.261 allows:

- only working with QCIF pictures at 64 kbit/s data rates;
- having a maximum picture decoding rate of 15 picture/s;
- using a reduced complexity motion estimation process.

The last option limits the quality achieved to slightly below that possible with a full hardware implementation, but at a fraction of the cost, size and power dissipation. It is likely that future generations of DSP devices, with faster cycle times, improved architectures and more internal parallelism, will enable higher picture quality to be achieved with single DSP devices, such that the quality is limited by the capabilities of H.261 rather than the implementation.

Similarly it may also be possible to implement a full CIF codec, or one operating at much higher bit rates. Such codecs have already been produced using multiple DSP devices but, as has been argued in section 15.1, this route often leads to a hardware complexity not much different from a full-blown hardware implementation. However, the trend (started by the INMOS transputer, and now followed by other devices such as the Texas Instruments TMS320C40) of enabling multiple processors to be easily linked together is likely to continue and will, to some extent, alleviate this problem.

Nevertheless, now that Recommendation H.261 has been ratified by the CCITT, the way is open for custom chip designers to produce full hardware implementations in very compact, even single chip, form, which will ultimately displace DSP device-based codecs from their 'low-cost and compact' niche. Similar considerations are likely to apply to other image compression standards such as JPEG (Joint Photographic Experts Group) for still picture applications [6] and the future MPEG (Moving Picture Experts Group) for moving pictures [7]. As these become ratified as standards, dedicated encoder and decoder chips, possibly able to deal with more than one of the standards due to their many common features, will become available at prices that undercut general purpose DSP device-based designs.

Image compression algorithms have until now been based largely on the removal of statistically redundant information, and approached from a mathematical viewpoint using techniques such as the DCT, variable length encoding, and optimized predictors. This form of encoding seems to be asymptotically nearing the limits of achievable compression, and new methods are being studied: these are based to a greater or lesser extent on using knowledge of how images are formed as projections of real world objects, and modelling these objects to give a more abstract and hence more highly compressed representation of the image.

One example of this approach is Mussman's work [8] in which objects are modelled as overlapping planar patches of arbitrary outline, which can move and distort to track changes in the objects in the image, but with the back-up of a conventional transform-based algorithm when the model fails.

Another example is Welsh's work [9] in which a human head and shoulders is modelled as a three-dimensional 'wire frame' on to which real image data is mapped. The model can be rotated to produce general head movements, and distorted locally to produce facial expressions. The image synthesis side of this work has already been demonstrated, using a C30 as the processing engine to perform the required calculations and image mapping operations in real time [10], but the image analysis problem is still under study.

These two examples indicate the current trend in image compression algorithms towards greater complexity, and in types of processing that cannot easily be translated into a pipelined hardware implementation.

In summary therefore, it would appear that the main application areas for the use of programmable DSP devices for image compression are likely to be:

- smaller-volume markets where the cost of custom IC development is not commercially justified;

- applications where one of the cheaper DSP devices is adequate to the task without using custom ICs;

- applications where the flexibility to alter or update algorithms is a key issue, such as algorithm development platforms, or leading-edge activities where standards are not yet defined;

- model-based codecs where the algorithm complexity is more suited to a software than a hardware-based codec.

15.7   CONCLUSIONS

It has been demonstrated that superficial estimates of the processing power required to implement image compression algorithms such as CCITT Recommendation H.261 can give rise to figures of the order of 9000 MOPS. But, by recognizing the limitations imposed by accepting the reduced picture resolution option of QCIF, at a maximum working bit rate of 64 kbit/s, and, by careful structuring of the algorithm, the processing requirement can be as low as 23 MOPS. Based on these observations, a video codec using just one Texas Instruments TMS320C30 for the encoder, and one for the decoder has been developed.

In the near future, standard algorithms such as H.261 will be implemented as custom ICs which will potentially undercut the price of DSP device-based codecs. However, there is still likely to be a place for programmable DSP devices to solve image compression problems where they are still cost-effective due to the small size of the market or the low cost of general-purpose DSP devices, or where the flexibility offered by a software approach is important, or where the algorithm complexity is beyond the reach of even advanced custom IC designs. In the latter case the use of multiple DSP devices with enhanced interconnection ability for parallel processing is likely to become a common feature.

REFERENCES

1.   CCITT Recommendation H.261: 'Video codec for audiovisual services at p × 64 kbit/s' (1990).

2.   Clarke R J: 'Transform Coding of Images', Academic Press, London (1985).

3.   Jayant N S and Noll P: 'Digital coding of waveforms', pp 465-485, Prentice-Hall (1984).

4.   Huffman D: 'A method for Construction of Minimum Redundancy Codes', Proc IRE, pp 1058-1101 (September 1952).

5.   Chen W H, Smith S H and Fralick S C: 'A fast computational algorithm for the discrete cosine transform', IEEE Trans, COM—25, pp 1004-1009 (September 1977).

6.   Wallace G K: 'The JPEG still-picture compression standard', Communications of the ACM, 34, No 4, pp 30-44 (April 1991).

7.   Le Gall D: 'MPEG: a video compression standard for multimedia applications', Communications of the ACM, 34, No 4, pp 46-58 (April 1991).

8   Mussman H G, Hotter M and Osterman J: 'Object oriented analysis-synthesis coding of moving images', Signal Processing: Image Communication 1, pp 117-138 (1989).

9.   Welsh W J, Searby S and Waite J B: 'Model-based image coding', BT Technol J, 8, No 3, pp 94-106 (July 1990).

10. Welsh W J, Simons A D, Hutchinson R A and Searby S: 'Synthetic face generation for enhancing a user interface', Proc Image Com 90, Bordeaux (November 1990).

16

DIGITAL IMPLEMENTATION OF NEURAL NETWORKS

D J Myers

16.1 INTRODUCTION TO NEURAL NETWORKS

Neural networks are assemblies of simple processor nodes with many inter-node connections. They are inspired by what is known about biological brain structures, and for this reason the nodes are often referred to as neurons and the interconnections as synapses. The neurons generally perform some simple operation such as forming the saturating (nonlinear) sum of their inputs. The inputs are outputs from other neurons, suitably weighted according to the synaptic strength of the connection, or weighted versions of elements of an externally applied vector.

Neural networks can be divided into categories depending on whether they are supervised or unsupervised. In unsupervised nets a set of training data is provided, which consists of representative samples of input data that the net will encounter in use. The network attempts to discover structure in the input, clustering inputs appropriately and associating classes of inputs with particular outputs. Such networks can be used as associative memories and in applications such as vector quantization. An example of an unsupervised net is the Kohonen self-organizing network [1].

In supervised neural networks a set of training data is provided, which consists of representative samples of input data that the net will encounter in use, and associated with them the desired outputs of the net in response to those inputs. In an initial training phase the training data set is applied to the net, the actual responses of the net are compared to the desired responses, and the internal synaptic weights of the network are then modified in accordance with some training algorithm in order to reduce the error or discrepancy between the actual and desired responses. A further data set, consisting of representative input data on which the net has not been trained, is used to test the network to give a measure of how it will perform on unseen data. The training data is typically presented to the net many times during training. When the discrepancy between the actual and desired responses has fallen below some predetermined value, the net is said to have converged. Supervised networks can be used to model nonlinear systems, and in applications such as pattern recognition. The multilayer perceptron (MLP), trained using the backpropagation algorithm [2], is perhaps the most popular neural network architecture in use today.

In the 1950s and early 1960s neural networks received a lot of attention. In particular networks such as the perceptron were studied by Rosenblatt [3] and others, and related work by Widrow resulted in the development of algorithms for training adaptive filters [4] which constituted one of the significant and enduring results of this period of neural network research. However, effort in this area virtually ceased after the publication by Minsky and Papert [5] of a critique which showed that the simple (single layer) perceptron had serious theoretical limitations. It could not, for example, solve pattern recognition problems that were not linearly separable. It was known that a multilayer perceptron (MLP) with nonlinearities in each layer overcame the limitations of the single layer net, but no way of training such networks was known.

That the subject has been revived in recent years is due in part to the development of algorithms, such as the backpropagation algorithm mentioned above, which allow the training of multilayer nets. A further stimulus for the revival of interest in neural networks is the hope that the neural network paradigm will prove useful in solving problems in areas such as speech recognition and image understanding, areas where conventional computing approaches have produced disappointing results.

Current research in the field includes the investigation of neural network algorithms, applications and implementation technology. Neural networks map poorly on to conventional Von Neumann computer architectures, because of their intrinsically parallel structures. The mismatch between the parallelism required for neural networks and the performance of sequential computer architectures is exacerbated as networks increase in size. As a consequence, a number of technologies are being investigated which allow the explicit mapping of neural network connectivity into hardware. At BT Laboratories these include optoelectronic [6] and amorphous silicon [7] technologies. Such approaches show promise and technology demonstrators have been produced; however a number of questions still remain. These include the problems of input-output (I/O), how to interface such sub-systems to conventional digital systems in such a way that their parallelism can be fully exploited, how to implement trainable nets, and the adequacy of the computational accuracy and resolution attainable.

Applications-oriented neural network research in the areas of speech recognition and image feature location at BT Laboratories has tended to focus on the use of the MLP architecture, trained using the backpropagation algorithm. Initial development has been undertaken on Apollo workstations, with simulations written in C or PASCAL using floating-point computation, but with the eventual aim of producing real time hardware-application demonstrators. The networks used are of moderate size, and in view of its maturity digital technology has been chosen as the means of implementing these hardware demonstrators.

This chapter looks specifically at the MLP architecture and the backpropagation training algorithm and discusses its computational requirements in general terms. Then two specific problems are considered (isolated word speech recognition and image feature location) in terms of network size, real time operating constraints, the effects of finite wordlength arithmetic and weight range limitation. The implementation of each of these example systems is described; because of their very different requirements, one system is implemented using off-the-shelf digital signal processing (DSP) device technology, whilst the other requires a custom very large scale integrated (VLSI) circuit.

16.2   COMPUTATION REQUIREMENTS OF THE MLP

The multilayer perceptron architecture is shown in Fig. 16.1. It consists of an input layer, one or more hidden layers and an output layer. All connections are between adjacent layers; there are no intra-layer connections. The input layer has no computation associated with its nodes. Their only function is to distribute each element of an input vector $\underline{Y} = \{y_1,..y_i,..y_N\}$ to all of the neurons in the layer above.

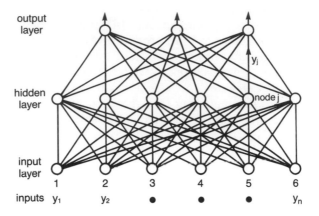

Fig. 16.1   Multilayer perceptron structure.

The computational requirements of the MLP fall into two categories:

- evaluation of the output of the net (forward pass);
- evaluation of an updated set of weight values for the net, as defined by the training algorithm.

A range of training algorithms exists. The training is often a more computationally intense activity than the evaluation of the output of the net, and requires a richer set of operations. The computational requirements (i.e. the nature of the computations) for the evaluation of the net output are relatively restricted and unchanging from net to net.

16.2.1   Evaluation of the net output

For the forward pass the computation required at each node varies with the neuron model being used. However, the model being considered in this chapter is the simple, widely used model shown in Fig. 16.2. Referring to the figure, for neuron j this requires the following computations:

- weighting of inputs y_i by weights w_{ij};
- summation of weighted inputs plus a threshold t_j of neuron j to form o_j:

$$o_j = \sum_i w_{ij}.y_i + t_j \qquad \qquad \dots (16.1)$$

- evaluation of the activation function output $y_j = f(o_j)$.

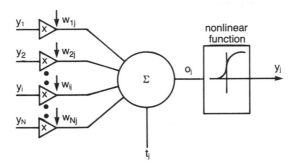

Fig. 16.2   Neuron model.

The model also requires the ability to store N weights values.

The activation function is a nonlinear function that may take a number of forms. The function most frequently made use of is the sigmoidal function:

$$y_j = 1/(1 + e^{-o_j}) \qquad \ldots (16.2)$$

This has the attraction that it is continuous and differentiable, an important consideration in the development of the backpropagation training algorithm. Its first derivative is given by:

$$y_j' = y_j.(1 - y_j) \qquad \ldots (16.3)$$

Therefore a simple digital processor dedicated to the evaluation of the neuron function could consist of N words of weights memory, a fast parallel multiplier to form the products $w_{ij}.y_i$, and an accumulator to form the sum of products o_j, followed by a circuit to evaluate y_j, the output of the activation function. The inputs y_i to the processor are provided sequentially.

It is also interesting to note that, neglecting the evaluation of the activation function, the single neuron is identical to the core computation involved in evaluating the output of a linear finite impulse response (FIR) transveral filter. Digital signal processing devices such as the Texas Instruments TMS320C25 [8] are optimized for the computation of the FIR function, via the highly pipelined MACD (multiply-accumulate and data move) instruction. The operation of this instruction requires that the weights are organized as a contiguous block in on-chip data memory, and a pointer to the memory location from which to fetch the next weight is incremented during the execution of the MACD instuction. When used in this way with the repeat instruction RPT or RPTK, the MACD instruction is executed repeatedly at a single clock cycle rate.

16.2.2   The activation function

The digital implementation of the activation function of equation (16.2) can be done in a number of ways. Perhaps the obvious method of evaluating it is by table look-up, in which the finite wordlength value o_j is used as an address to look up an appropriate value of y_j, stored in memory.

An alternative hardware implementation which maps compactly into an integrated circuit realization is based on a seven segment piece-wise linear approximation [9]. Each of the segments has a gradient which is a power of two. This circuit also gives the value of the first derivative of the activation function $y_j.(1-y_j)$ (see equation (16.3)) which is required when implementing training using the backpropagation algorithm.

16.2.3   The backpropagation training algorithm

As previously stated, a number of training algorithms exist. Each has its specific computational requirements, which will impact on the network hardware requirements for a trainable net. The backpropagation algorithm for training MLPs requires that an error signal be propagated back through the layers from the output layer. This affects the communications requirements of the net. By comparison, simple Hebbian learning [10] requires only 'local' information and therefore may be more suitable (in this regard) for hardware implementation. However the backpropagation algorithm has superior performance. The backpropagation algorithm is considered to be difficult to implement in hardware, for reasons which will be described.

The simplest form of the backpropagation algorithm is as follows.

- For the jth node in the output layer, an error term δ_j is formed, given by:

$$\delta_j = (e_j - y_j).y_j.(1 - y_j) \qquad \qquad \text{... (16.4)}$$

where y_j is the output of node j, and e_j is the desired output of node j. The weight connecting the ith node in the layer below to the jth node in the output layer is then modified as follows:

$$w_{ij}(n+1) = w_{ij}(n) + v.\delta_j.y_i \qquad \qquad \text{... (16.5)}$$

where v is a constant known as the learning rate.

- For the hidden layers the form of equation (16.5) remains the same, but it is not intuitively obvious what the desired output of a hidden node should be. For the jth node in the hidden layer, δ_j is given by:

$$\delta_j > y_j.(1-y_j) \cdot \sum_k \delta_k.w_{jk} \qquad \qquad \ldots (16.6)$$

where the summation is over all the nodes in the layer above. This calculation is performed first in the layer below the output layer, and then in succeeding hidden layers until the input layer is reached.

From these equations it is clear why the algorithm is known as backpropagation; the δ_k values for hidden nodes are obtained by propagating back and summing δ_k values from layers above, weighted by the strengths of the node connections. The reason that the algorithm might be considered difficult to implement in hardware can also be seen. The calculation of equation (16.4) requires information local to each output node, but calculation of equation (16.6) requires distributed information. What is worse, the summation in equation (16.5) is the exact opposite of the summation required in the calculation of the output of the net (equation (16.1)) in that it requires the weights values associated with the output of a particular node, whereas the forward pass summation at each node requires the weights associated with the inputs to the node. If hardware consists of simple digital node processors, as described in section 16.2.1, with a single processor dedicated to each neuron, then each of the weights values required to evaluate equation (16.6) will reside on a different processor. In the case of an implementation on a TMS320C25 processor the fact that the values required in equation (16.6) do not reside in a contiguous memory space means that the pipelined nature of the MACD instruction cannot be exploited.

A number of extensions to the backpropagation algorithm have been proposed, which have an impact on the hardware requirements. In a popular variant a momentum term is included in the weight updating equation, giving it the following form:

$$w_{ij}(n+1) = w_{ij}(n) + v.\delta_j.y_i + m.\Delta w_{ij}(n) \qquad \qquad \ldots (16.7)$$

where $\Delta w_{ij}(n) = w_{ij}(n) - w_{ij}(n-1)$, and m is a constant known as the momentum coefficient. The inclusion of the momentum term has important implications for the hardware; it becomes necessary to store either $\Delta w_{ij}(n)$ or $w_{ij}(n-1)$, significantly increasing memory requirements.

16.3   CASE STUDY — ISOLATED WORD SPEECH RECOGNIZER

In the first case study considered, an MLP has been trained to recognize spoken letters of the alphabet [11] independent of the speaker. The MLP was implemented using PASCAL and trained off-line using a variant of the backpropagation algorithm on an Apollo workstation. The final weights values obtained after training were used to code up a pre-trained MLP running on a Motorola 56001 DSP device, operating on data acquired in real time.

16.3.1   Description of the problem

The data for this problem consists of three utterances of each of the letters of the alphabet collected using a silence cabinet and a wideband telephone handset, from each of 104 speakers. The input was bandpass filtered between 170 Hz and 8 kHz and sampled at 20 kHz. This is an isolated word recognition problem which is quite challenging because of the potential confusability of words such as 'b','c','d','e','g','p','t' and 'v' which are only differentiated by a relatively short initial consonant.

16.3.2   Development of the neural network based solution

Each utterance in the data described above was automatically endpointed, and then checked manually and adjusted if necessary. The data was then split into frames which were preprocessed to provide mel frequency cepstral coefficients (MFCCs) [11]. Each utterance is of a variable length, which is inconvenient for MLPs which have fixed dimension inputs. Therefore the input data was time normalized using linear interpolation to represent each utterance by 15 frames of 8 coefficients each, giving rise to a 120 element input vector. The MLP has 26 outputs, one for each letter of the alphabet. The data from 52 speakers was used to train the net, and the data from the remaining 52 speakers was used to test it.

After a number of experiments, a net with a single hidden layer of 50 nodes was selected, as shown in Fig. 16.3. After training with an enhanced backpropagation algorithm, which included a momentum term, adaptive learning rate and momentum coefficient [11] and limitation of weight magnitudes to ≤ 1.5 [12], this achieved a recognition accuracy of 87.6% on the test set.

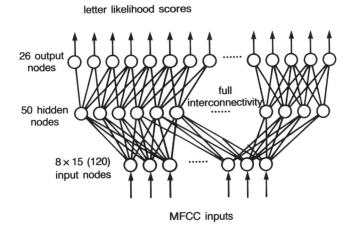

Fig. 16.3 Network for alphabet recognition.

16.3.3 Implementation of the system

The pre-trained MLP was implemented on a Motorola 56001 DSP device mounted on a development board inside a PC, as part of an overall system for performing real time spoken alphabet recognition, the output of which was fed into a program which looks up entries in the BT Laboratories internal telephone directory by taking the classification outputs of the MLP representing the first five letters of the surname required.

Input to the board was via a wideband telephone handset and filter of the same bandwidth as that used to obtain the training data. The board performs A/D conversion on the input signal. The 56001, which operates at 20.5 MHz with zero wait-state random access memory (RAM), is also used to automatically endpoint the speech and to perform MFCC and time normalization preprocessing on the data. Because the DSP device is optimized for the type of calculation (fast multiply accumulate) required to evaluate the forward pass output of the net, the MLP is executed in around 8000 processor cycles, taking about 780 μs. When preprocessing is taken into account, the execution time is still less than 1 ms. The MLP code is less than 20% of the overall program size, and requires about 7.5 k $\times$ 16-bit words to store the weights. The nonlinear activation function is implemented by table look-up with linear interpolation. A block diagram of the overall system is shown in Fig. 16.4.

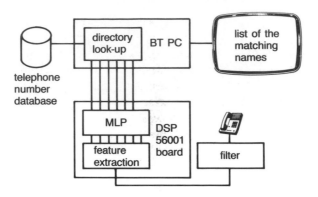

Fig. 16.4   Complete alphabet recognition demonstrator.

Although performance of the 56001 based MLP has not been quantitatively assessed, it appears to perform only marginally less well than the original Apollo simulation. This is due in part to errors introduced by the automatic endpointer and because the system is not used in ideal silence cabinet conditions.

In this example of a digital neural network implementation it was not necessary to devote effort to considering the problems of, for example, accuracy of weight representation in the 56001 based system. This is because experiments have shown that if the weights values after training are coarsely quantized, to as few as 3 bits in some cases, this does not cause the performance of the net to degrade appreciably [12]. Therefore the 16-bit weight representation used in this implementation is more than adequate.

16.4   CASE STUDY — IMAGE FEATURE LOCATOR

The second case study is in the area of feature location in images. MLPs have been trained to locate features in images at both low [13] and high resolution [14]. As in the case of the previous case study, this work was carried out on an Apollo workstation, the code for the MLPs being written in the C programming language. In order to produce a hardware demonstrator which operates in real time, a digital VLSI chip is being designed which performs the MLP function.

16.4.1   Description of the problem

The aim of the system under development is to locate and track facial features (eyes, mouths, etc) in 8-bit grey scale image sequences. The data used to train the MLPs was taken from a set consisting of 60 human head-and-shoulder still images — 30 male and 30 female. The images were available at the full resolution of 256×256 pixels, and also at the lower resolution of 16×16 pixels, obtained by pixel averaging and subsampling. The data used to test the system consists of moving head-and-shoulders sequences.

The ability to track features robustly has potential application in areas such as surveillence and enhanced performance videotelephony.

16.4.2   Development of the neural network based solution

In this application a hierarchical approach to feature location was taken [15], as illustrated in Fig. 16.5. MLPs trained to determine candidate locations of left eyes, right eyes and mouths in the low resolution image are used to reduce the search area of MLPs trained to accurately locate these same features in the high resolution image. The candidate areas detected by the low resolution MLPs are reduced by heuristic rules (e.g. left eye cannot be to right of right eye) prior to being used to restrict the search space in the high resolution image. This hierarchical approach improves robustness and decreases the computation required.

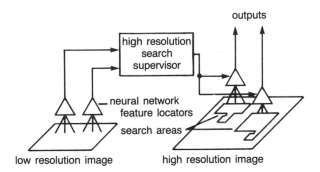

Fig. 16.5   Multi-resolution approach to feature location.

At each resolution the MLPs were trained to detect a particular feature by providing as input an $n \times m$ window of pixels from the image, such that the feature of interest can fit within the window. The MLPs therefore have $n \times m$ inputs and one output. They are trained to output a high value when the window is centred on the feature, and a low value otherwise. For the high resolution MLPs a 17×29 input window was chosen, because this is large enough to cover the eye completely in all the images. Each high resolution MLP therefore has 493 inputs, one output and a single hidden layer of 16 neurons. For the low resolution MLPs a 5×5 input window was chosen. The neurons in the hidden layer, however, have the 25 pixel values and the squares of these values presented to them allowing them to implement curved decision boundaries. Therefore each low resolution MLP has 50 inputs, one output and a single hidden layer of two neurons.

System simulations show that the low resolution stage of the feature locator reduces the search areas for the high resolution MLPs to about 5% of the total image area [13,15].

16.4.3   Implementation of the system

Assuming a modest frame rate of 15 frames/s, it can be shown that the low resolution stage of the feature locator system requires just under 40 000 multiply-accumulate operations per second, whilst the high resolution stage requires about 10^9 multiply-accumulate operations per second. Therefore, although it should be easily possible to implement the low resolution stage using a single general purpose DSP device, this is clearly not possible for the high resolution stage. A multiprocessor approach based on general purpose DSP devices is a possibility; however, it would be bulky and expensive.

In order to build a real time hardware demonstrator, the approach that has been adopted is to build a digital VLSI chip which implements the MLP function [16]. In order to make the chip more flexible and attractive for other applications, it has on-chip implementation of the backpropagation learning algorithm.

The chip architecture is based on a linear array of neural processors (NPs) which communicate locally with processors to their immediate left and right. Each NP consists of a weights memory (RAM), a fast parallel multiplier and accumulator, an activation function circuit based on a piece-wise linear approximation [9], I/O registers and a control unit. The control unit contains registers which can be loaded on initialization to provide information on the location (layer, position in layer) of the NP in the net to be implemented.

Two communication registers are provided with each NP, an input and an output register. Forward pass operation is achieved with this architecture by passing the input data serially through the network via the input registers as shown in Fig. 16.6. As each input arrives at a node processor in the first layer it is multiplied by the appropriate connection weight and the result is added to the accumulator. When all input values have been presented to each NP in the layer, all NPs in that layer evaluate their activation values and place them in their output registers. These are then passed on to the next layer. The process continues until the outputs from the top layer emerge from the pipeline, via any unused NPs.

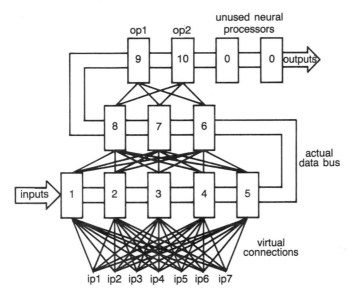

Fig. 16.6 MLP implemented using HANNIBAL architecture.

The use of separate input and output registers permits the layers to operate in a pipelined manner. The two NPs on either side of a layer boundary have to be connected such that the output register of the NP in the lower layer is connected to the input register of the NP in the upper layer. The structure can be maximally pipelined (i.e. inputs can be supplied continuously) if the number of inputs is greater than the number of units in the first layer, and the number of NPs in each succeeding layer monotonically decreases.

In order to allow training using the backpropagation algorithm, the architecture has been extended to common up the communication shift-registers as a double width feedback path during training. The summation required by equation (16.6) and discussed in section 16.2.3 is performed in a distributed way by threading the feedback path through the accumulators of the NPs where appropriate and forming partial sums of the $\delta_k w_{jk}$ terms which are fed back until the complete summation has been performed. The backpropagation algorithm implemented does not allow a momentum term as this requires a doubling of on-chip memory requirement.

The chip, known as HANNIBAL (hardware architecture for neural networks implementing backpropagation algorithm learning), consists of four node processors, each with the option of either 256×16-bit words or 512×8-bit words of weights memory. Because the weights memory is on-chip, the chip has a low pin count and large neural networks can be constructed by cascading chips, with little support circuitry required.

A hardware description model of the architecture was used to ensure that the choice of wordlengths internal to the chip did not result in poor performance [17]. Simulations of this model on a range of problems resulted in the selection of weights wordlengths, input, output and backpropagation communication wordlengths, and other wordlengths internal to the chip.

Some features of the chip are as follows:

- operation at 20 MHz on 8-bit input data, with 16-bit data paths for feeding back partial error terms — this gives a performance of up to 80 million synapses per second for a single chip;

- weights can be downloaded to run the net in pre-trained mode;

- reconfigurability at the chip level to provide the options of two NPs with 1024 weights per NP or 1 NP with 2048 weights;

- up to four levels of multiplexing at the chip level in pre-trained mode;

- automatic weight scaling option at each NP [17];

- provision of an option to bypass the activation function, allowing the external provision of alternative activation functions;

- the backpropagation training algorithm has an option in which pseudo-random noise (dither) is added to the updated weight in the accumulator, prior to its truncation when being rewritten to memory, thus allowing training to be performed successfully with 8-bit weights [17].

Using the HANNIBAL chip as a building block to implement a hardware version of the feature location system described above would require at most six of these chips for the low resolution stage MLPs and 18 chips for the high resolution stage MLPs, and should allow real time operation at a 15 Hz frame rate.

16.5   CONCLUSIONS

Although new technologies for the implementation of neural networks are the subject of much research effort, it has been seen that digital technology, and even general purpose DSP devices can be used to implement the medium size nets which are the basis of current applications prototypes. Such implementations may not have some of the postulated benefits of dedicated implementations with explicit connectivity (e.g. graceful degradation in the face of some types of hardware damage). However they have the advantage of a mature technology with the traditional benefits of the DSP approach such as arbitrary specifiable accuracy, and simple interfacing to host systems.

REFERENCES

1. Kohonen T: 'Clustering, taxonomy and topological maps of patterns', Proc Int Conf on Pattern Recognition, 1 , pp 114-128 (October 1982).

2. Rumelhart D E, Hinton G E and Williams R: 'Learning internal representations by error propagation', in Rumelhart D E and McClelland J I (Eds): 'Parallel distributed processing', 1 , MIT Press, Cambridge, Mass (1986).

3. Rosenblatt F: 'Principles of neurodynamics', Spartan Books, New York (1962).

4. Widrow B and Hoff M: 'Adaptive switching circuits', IRE WESCON Convention Record (1960).

5. Minsky M L and Papert S: 'Perceptrons', MIT Press, Cambridge, Mass (1969, revised and reissued 1988).

6. Barnes N M, Healey P et al: 'High speed opto-electronic neural network', Electron Letts, 26 , No 11, pp 1110-1112 (1990).

7. Reeder A A et al: 'Application of analogue amorphous silicon memory devices to resistive synapses for neural networks', Proc 2nd Int Conf on Microelectronics for Neural Networks, Munich, pp 253-259 (October 1991).

8. TMS320C25 User's Guide, Texas Instruments (1988).

9. Myers D J and Hutchinson R A: 'Efficient implementation of piece-wise linear activation function for digital VLSI neural networks', Electron Letts, 25 , No 4, pp 1662-1663 (1989).

10. Hebb D O: 'Organization of behavior', Science Editions, New York (1961).

11. Woodland P C: 'Isolated word speech recognition based on connectionist techniques', BT Technol J, 8, No 2, pp 61-66 (April 1990).

12. Woodland P C: 'Weight limiting, weight quantisation and generalisation in multilayer perceptrons', Proc 1st IEE Int Conf on Artificial Neural Networks, pp 297-300, London (October 1989).

13. Vincent J M: 'Facial feature location in coarse resolution images by multilayer perceptrons', in Kohonen T et al (Eds): 'Artificial neural networks', Proc of 1991 Int Conf ICANN-91, 1, pp 821-826, North Holland (1991).

14. Waite J B: 'Training multilayer perceptrons for facial feature location: a case study', in Linggard R, Myers D J and Nightingale C (Eds): 'Neural networks for vision, speech and natural language', Chapman & Hall (1992).

15. Vincent J M, Myers D J and Hutchinson R A: 'Image feature location in multi-resolution images using a hierarchy of multilayer perceptrons', in Linggard R, Myers D J and Nightingale C (Eds): 'Neural networks for vision, speech and natural language', Chapman & Hall (1992).

16. Myers D J et al: 'A high performance digital processor for implementing large artificial neural networks' BT Technol J, 10, No 3, pp 134-143 (July 1992).

17. Vincent J M: 'Finite wordlength, integer arithmetic multilayer perceptron modelling for hardware realization', in Linggard R, Myers D J and Nightingale C (Eds): 'Neural networks for vision, speech and natural language', Chapman & Hall (1992).

17

DSP IN NETWORK MODELLING AND MEASUREMENT

A Lewis, P Branch, P Barrett, M Ogden and P R Benyon

17.1   INTRODUCTION

In the same sense that Westall and Ip remarked (see Chapter 1) about digital signal processing (DSP) as a whole, the phrase 'network modelling' means different things to different people, depending on the level at which the subject is viewed. To some, it means impulse responses and eight thousand signal samples every second, while most telecommunications engineers will think of erlangs and unavailable hours per year. The ability to encompass and integrate these disparate viewpoints and move as seamlessly as possible across conceptual layers may become increasingly important in future, for reasons that will be described.

The concept of DSP in network modelling forms a nexus, or symbolic connection, of multidisciplinary themes, that are under study in universities and research organisations world-wide. These themes include computer aided engineering (CAE) tools, parallel processing, network emulation, network design, subjective assessment and DSP algorithms. Just as this book has linked seemingly unconnected areas of work, so DSP in network modelling and measurement potentially links the efforts and interests of many different development and research teams at BT Laboratories.

DSP has the reputation of being something of a black art, of use only in designing products for sale and therefore of peripheral interest to a network operator. The authors aim to show that such a view is narrow and misleading, by demonstrating the fertile potential of DSP to foster and support network evolution, helping to meet the requirements and expectations of future customers.

Sections 17.3 and 17.4 explain the role of DSP in network modelling and measurement. Because the subject and its ideas may be unfamiliar to some readers, the reasons and risks that create the potential need for this role will be described first.

17.2   TRENDS IN NETWORK EVOLUTION

17.2.1   Crystal gazing

Over recent years, the widespread digitization of telephone networks has created a medium that (with the exception of delay) is largely free from the accumulation of impairment with distance. Digital techniques mean that most questions about network operation have a binary answer: 'Yes, it's working' or 'No, it's not'. But is this new and comfortable sense of security permanent?

Relax for a few moments on a wild flight of fantasy, and imagine two radically different kinds of future for telecommunications. One is called 'bit-bountiful', the other 'bit-bare'.

In the first imaginary future, optical networks have re-cabled the world, right down to its local loop roots. Wavelength division multiplexing and photonic switching have provided more bandwidth at less cost than anyone can handle. A fresh optical fibre can be pneumatically-blown to any customer within the hour. Personal mobile communication has been declared a contravention of human rights and is illegal.

In this 'bit-bountiful' future, A-law speech encoding sounds unpleasantly distorted and customers insist on wideband telephony to hi-fi standards. Speech and video compression are regarded as laughably old-fashioned, dead-end technologies. There is a huge variety of different networks, but each exists only to transport bits transparently and all the value is added in the terminal equipment. Network quality is a binary variable and if it works it's perfect. In this utopia, no network operator needs DSP to measure or model anything.

In an alternative future, today's fixed circuits are all dead and gone. Communication nirvana[1] has been attained, with OneNetwork™

---

[1] State of beatitude, attained by the extinction of individuality.

(BISDN-7) that is all things to all people. Everyone is issued with their telephone number at birth. All telephones are personal and can be used anywhere on Earth. Every connection is either wireless or packet-switched and virtual. Calls are priced per bit, regardless of distance.

In this 'bit-bare' future, radio spectrum is the most precious commodity on the planet. All speech and video signals must be compressed close to their information limit, under severe penalty of international law. Bit transparency has been replaced by bit truancy, for a tenth of the price. Telecommunications networks exist to squeeze, massage, stretch and reconstruct far more bits than they ever actually transport and all the value is added in the network. The quality of speech and video transmission depends on more things than any customer feels comfortable thinking about. Every network operator relies on DSP to measure and model everything in this brave new world.

Reality, of course, won't be so extreme. The authors suspect it will turn out to be a mixture of elements from both these fantasy futures, as well as elements of neither. In the medium term, most market trends seem to have a 'bit-bare' direction.

17.2.2   Recent technical developments

DSP techniques are revolutionizing networks, products and services in telecommunications. Economic implementation of speech encoding at low bit rates [1-3] has enabled new mobile telephone services, such as the groupe speciale mobile (GSM) cellular radio system, the digital European cordless telephony standard (DECT) and the personal communications network (PCN) [4,5]. It is technically feasible to provide local loop access by similar techniques [6].

New transmission technologies, such as the synchronous digital hierarchy (SDH), asynchronous transfer mode (ATM) and the broadband ISDN are increasing network flexibility [7,8]. These evolving technologies promise a single, flexible and universal network, that can support a wide range of services without imposing a structure on the signal streams that are transported. Such services might include narrowband ISDN, metropolitan area networks and variable rate or packetized streams for speech, data, videoconference, multimedia, messaging, facsimile and document retrieval applications [9,10].

17.2.3   Market pressures

Network operators value the greater flexibility and freedom of network dimensioning and management promised by the developing SDH and ATM network standards. These technologies offer great potential for maximizing the utilization of network capacity and increasing the financial return on investment in new equipment. Customers are probably less concerned about technology, but no doubt want lower tariffs and good quality of communication. New transmission and channel-coding technologies are one way of satisfying these wants and squeezing more out of a finite resource.

The UK is one of the most deregulated telecommunications market-places in the world. This deregulation is accelerating the pace of change, so that new systems and services are increasingly likely to be specified, designed, developed and installed as stand-alone systems. Pressure for rapid change means that less time is available to test the many possible combinations of new equipment and services, before installation.

17.2.4 Digital distortions

Digitization has swept away many of the limitations of analogue, frequency division multiplex (FDM) networks. Indeed the very word 'digital' has come, in common speech, to mean high performance or near-perfect operation. So low bit rate or packetized digital encoding of speech must surely be perfect? In a sense it is, because DSP can implement distortion and nonlinearity just as perfectly as linear filtering. The tongue-in-cheek fantasies of section 17.2.1 had a serious purpose — to remind the reader that the perfect transmission of digital signals does not guarantee perfection in speech quality or the freedom to interconnect networks and equipment in arbitrary order.

Appropriate design, in low bit rate speech encoding, aims to exploit the psychoacoustic limits of human perception to conceal nonlinearities. The result is good quality speech with fewer bits. The compression of speech and video signals that can now be achieved [1-3], with little audible or visual loss, is indeed remarkable — but this compression is usually a very nonlinear process.

Nonlinearity means not just amplitude-dependent distortion, but also the effects of noise, timing changes, time-variable gain and signal-dependent quantization. It is also meant to include traffic-dependent modulations, which may occur in future networks. Not every nonlinearity is of equal subjective significance in speech communication.

However, it is well known that strongly nonlinear systems can display unexpected interactions when connected together. The transmission impairments of cascaded nonlinear network elements do not necessarily add linearly. Sometimes ten elements are no worse than one, while in other cases two can be ten times worse than one element. The infuriating answer is: 'Well, it depends.' Networks in a 'bit-bare' future might be less cascadable, or less freely interconnected, than are the networks of today.

Such interconnections are a root cause of the quality risk [11] that new speech encoding and transmission standards may raise — they are more nonlinear (and more 'bit-bare') than the A-/μ-law encoded 64 kbit/s systems of today. This risk to quality is the principal reason for using DSP in network modelling and measurement. Identifying the incompatible interconnections after installation will be much less comfortable than discovering them early in the specification or design phases.

17.2.5   Changing the channel

New world-wide transmission standards are not only increasing the flexibility and altering the topology of networks, they are even redefining what is meant by a speechband channel. In particular, these developments threaten to change the one-to-one mapping of logical channels to physical circuits that is now common.

At present, the plesiosynchronous digital hierarchy (PDH) provides a transmission system with channels that are mutually independent. Speech signals in the UK are converted to digital signals at 64 kbit/s, passing through the PDH in several stages of digital multiplexing, before conversion to analogue signals near their destination (Fig 17.1). Neglecting the very remote chance of error, bits from the source run up and down a complex cascade, multiplexing merrily, but emerging unscathed. The output is an exact replica of the input — that is the meaning of bit transparency. The channel, once allocated, uses a fixed network capacity continuously, until cleared at the

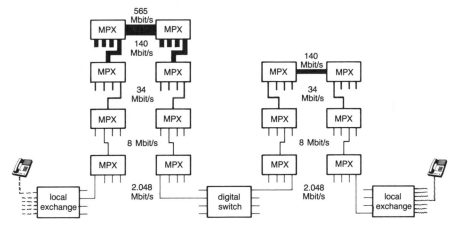

broken line — analogue, 2-wire connection   solid line — digital, 4-wire connection

Fig. 17.1   A connection through the plesiosynchronous digital hierarchy.

end of the call. Routeing of an individual channel is by physical circuit switching, which can occur only at the lowest level of the multiplex. In this world, channels remain pristine because they are inflexible.

In future networks, speech channels cannot be assumed to be bit transparent at 64 kbit/s. They may pass through several stages of speech encoding, echo cancellation, channel multiplication or packetization in cascade. In this case, bits inserted at the source will not emerge at their destination in the same pattern. Channels will not necessarily occupy a fixed amount of network capacity. Routeing may occur by virtual switching at higher levels of the multiplex. This world will have gained great flexibility, at the expense of reduced bit transparency of the channel.

International networks have already experienced some such developments, with the spreading use of digital channel multiplication equipment (DCME). Figure 17.2 shows a hypothetical example, where a GSM mobile radio customer in London is connected, via an ATM network in the USA, to a customer in Chicago. The colloquial descriptions of signal processing in Fig. 17.2 echo words from the 'bit-bare' fantasy future.

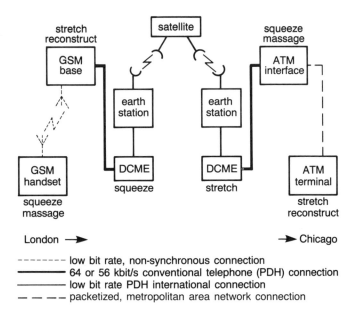

Fig. 17.2   Hypothetical future speech connection from London to Chicago.

17.2.6   Network design — bits in pieces

Designs of equipment for network, switching, transmission and terminal applications have traditionally been considered as independent engineering disciplines. This convenient subdivision into separate pieces was possible because simple interfaces between these disciplines could be defined and because each dealt with largely linear systems. The resulting independence of academic outlook has proven very successful in the design of tele-communications networks throughout the world. However, it may not be an adequate approach to the design of future networks.

A cross-disciplinary approach might be required to deal with the interactions and dependencies, between signals and protocols, that could arise in the more flexible but less transparent networks of the future. In particular, a subdivision of design tasks may be required that is favourable to a customer-facing viewpoint, rather than the organizational convenience of engineers or network operators. The use of DSP in network modelling is a good example of such a hybrid approach.

Speech is likely to remain the major purpose of telecommunications networks and the customer will be the judge of quality. If all those involved in the specification and design of speech networks succeed in their aims, then any loss of transparency that may occur in future networks will not cause a significant loss of quality [12].

17.2.7   A new role for DSP

The previous sections have explained why the authors believe that technical and market developments will combine to increase the risk of subtle, more varied and less predictable impairments to speech transmission quality. Such risks may especially occur when new systems and services are interconnected with existing networks. This can happen even though none of the equipment is faulty and each gives satisfactory speech quality when used in isolation. Consequently, there is a growing need for a more powerful kind of network modelling and measurement that reflects the customer's view of quality.

More powerful and panoptic network modelling can be created by intimately joining sample-driven DSP simulation with event-driven, statistical network modelling. This is a challenging but rewarding task, spanning and linking engineering disciplines that, by convention, are separate. A short feasibility study of the power of panoptic modelling is described in section 17.4, where two traditional approaches were mixed in one structure.

Such mixed modelling can blend the best of both worlds, simulating an entire communications channel from an overall viewpoint, instead of taking several unconnected views of individual elements. This broader view of network modelling could help predict subtle incompatibilities, especially at the interfaces between elements, and minimize the technical and business risks that incompatibilities create. It could aid the specification, verification and planning of new systems and services. Errors and omissions in specifications could be detected early in the design cycle [13], when correction is not only easier but cheaper, too.

In particular, mixed modelling could be used to reveal and investigate interactions between the protocols and the signal coding algorithms of future packet switched networks over a range of conditions that would be impractical or inconvenient to create in a real network.

In future, transmission quality may be affected by many factors, some of which may interact or depend on traffic volume, or be beyond the control of the network operator. Conventional measurements of transmission performance, made at night-time, would not be an adequate guide to day-time quality. Live measurements could become the only way of assuring customer satisfaction with conversational quality.

New applications of DSP technology can not only monitor live performance automatically, but can also characterize existing networks in much greater detail than the measurement methods in common use today. Because such detailed characterizations are the foundation of the power and accuracy of mixed modelling, the use of DSP in network measurement will be described next.

17.3   DSP AND SPEECH CHANNEL MEASUREMENT

17.3.1   Background

BT is publicly committed to improving quality. Major investment in new digital transmission and exchange equipment has significantly reduced the failure rate, the time to connect, the transmission loss and the noise level of the average call. But perceptions of quality can change over time, so that what was acceptable ten years ago may not be acceptable in future. Audio quality in consumer electronics has improved over recent years and customers may expect improvements in telephone speech quality.

Such customer expectations, with the technical developments and market pressures described earlier, could focus the attention of network operators on transmission quality. There are two strategies for transmission quality control — the first is by out-of-service measurement and the second is by in-service monitoring. Currently both methods are in use in BT, although by totally different physical means and with more frequent use of the first strategy. For out-of-service measurements, signals are injected into one end of an idle channel and their properties measured at the other end. For in-service monitoring, a carefully-trained human observer monitors brief samples of calls in the network and records an opinion of quality.

The results obtained from the two methods are complementary rather than alternative. Out-of-service measurements can be fast, rigorous and repeatable. In-service monitoring is inherently a less rigorous process, but it can be closely linked to customers' opinions. The two kinds of measurement can each be used in different ways — on individual channels they are an important aid to fault diagnosis and location, whilst on groups of channels they provide a statistical picture of the customers' perceptions of performance and quality of service. In telecommunications networks of all kinds, the trend will be towards a continuous assessment of quality.

17.3.2   Present measurement methods

Instrumentation and standards for audio measurement are well established [14], typically using individual analogue equipment such as oscillators, level meters, psophometers, frequency-analysers and bridges. Many commercial instruments are available, with varying degrees of sophistication, intended for making out-of-service measurements on telecommunications channels. Automatic types use a master/remote-slave configuration and are often specific to one application, such as a.c. and d.c. testing of the local line and telephone [15], or single-tone transmission loss and noise tests in networks [16]. These instruments are intended to confirm that a channel is functioning correctly but do not produce enough information to fully characterize its performance or nature.

BT Laboratories has developed an automated analyser that operates in a unique way. The T1000 series [17] is currently the only commercially-available instrument, known to the authors, that can objectively measure a connection in a way that closely correlates with customer opinions. It takes account of the electrical properties of the channel and of the nominal acoustic characteristics of the telephone. DSP techniques are used in a PC-based unit (Fig. 17.3) to calculate a range of parameters such as send, receive, sidetone and overall loudness ratings [18], line impedance, psophometrically-weighted noise [19], echo return loss and delay. Results can be presented in graphical form or stored for future comparison.

Fig. 17.3   T1000 network transmission performance analyser, from Fulcrum Communications Ltd.

17.3.3   New DSP methods for measurement

DSP technology is able not only to duplicate conventional measurements with precision, but can also integrate many functions into a single instrument [20,21]. Furthermore, the principles of digital adaptive filtering can create comprehensive new ways of measuring channels. Expertise in adaptive filtering, high-speed modem design, speechband network modelling and subjective assessment is being applied to develop new DSP methods for measuring and interpreting speech channel impairments.

BT Laboratories has used the latest generation of DSP microprocessors in a new range of high-speed datacommunication modems (see Chapter 7). These modems use several adaptive filters (Fig. 17.4) to maximize the rate of data transmission over public telephone networks, compensating for channel impairments by equalization and echo cancellation. This technology has led to an experimental instrument (Fig. 17.5) for out-of-service measurements of national and international channel impairments [22]. This equipment consists of a signal processing unit, which can be mains or battery powered, and a portable personal computer.

A laboratory prototype has been used at an international exchange and a network management centre. These initial trials have proven the feasibility of using adaptive filters, in a system identification role, to accurately measure parameters such as the delay, echo return loss and frequency response of both local and distant echo sources. Future plans include measurements of group delay, frequency offset and quantization distortion and the addition of a digital interface to A-/μ-law encoded channels.

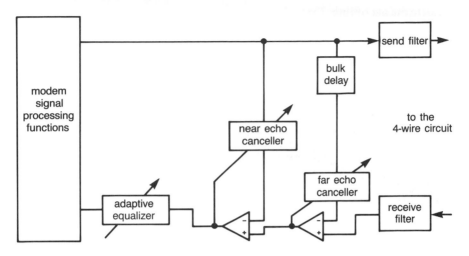

Fig. 17.4   The channel-modelling adaptive filters in a V.32 high-speed modem.

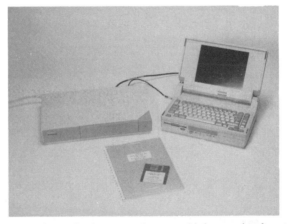

Fig. 17.5   An experimental echo-measuring set, with its associated portable PC.

17.3.4   Assessing quality in-service

As described earlier, speech quality in future networks may depend on traffic volume, so that night-time testing would not be a reliable guide to daytime performance [11]. Many circuits must be measured quickly for statistical accuracy in a large network. The cost of making such measurements out-of-service could be significant, since channels cannot carry revenue-earning traffic during testing. Extensive use of manual in-service observation can also be costly. Furthermore there are fundamental limits to the conclusions drawn from human observations, since the observer may not receive the same speech level as the customers. South [23] has suggested a possible solution to these problems, by means of an 'electronic observer' at a single-point in the digital network (Fig. 17.6).

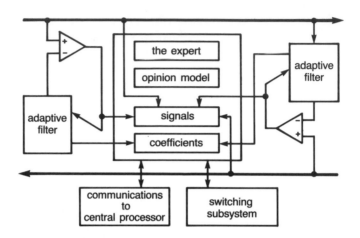

Fig. 17.6   Elements of an 'electronic observer' for transmission quality monitoring.

The BT digital adaptive filter chip-set (see Chapter 4) has been used to prove that conversational speech signals can converge a contra-directional pair of finite impulse response (FIR) least mean square (LMS) filters, as shown by the results in Figs. 17.7 and 17.8.

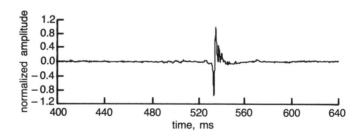

Fig. 17.7   Echo impulse-response of a UK to Auckland channel, measured with speech signals.

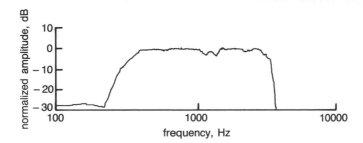

Fig. 17.8   Echo frequency response, derived from the measurement shown in Fig. 17.7.

This has demonstrated the feasibility of making accurate objective measurements for both directions of transmission in-service. If these measurements could be reliably mapped to customer perceptions and the need for a human expert was reduced or eliminated, then automated in-service measurements could estimate customers' opinions of call quality. A current initiative at BT Laboratories is studying the feasibility of such a system [24].

17.3.5   Measurements with adaptive filters

In the following discussion, the term 'adaption' refers to the process of adapting filter coefficients, while the word 'adaptation' means the final adapted state of the filter.

Using a conventional adaptive filter to make measurements with conversational speech signals requires accurate speech detection, to disable adaption of the filter modelling the path containing the speech source. Otherwise the filter coefficients diverge and accuracy is rapidly lost. Level-based speech detection is simple but can be ineffective in the presence of echo and background noise, especially when signal levels are widely asymmetric due to line loss, tolerance in telephone sensitivities and variation in the vocal level of the talkers [25]. However, a modified form of adaptive filter, with a so-called foreground/background architecture, eases the requirement for accurate speech detection (see Chapter 4).

The value of each adaptive coefficient displays an inherent dither, due to noise and nonlinearity in the channel, the statistical properties of input signal (see Chapter 4) and finite precision effects. It can be alleviated by increasing the DSP word size or slowing the speed of adaption. However, this dither can be useful because its variance can be related to quantization distortion in the channel. This technique can potentially distinguish between channels with A-/μ-law encoding and those using lower-rate schemes. It could detect time-dependent or traffic-related modulations in linearity, as in the embedded ADPCM packetized speech network examined in section 17.4.

Other algorithms and architectures offer faster adaption, improved accuracy or reduced sensitivity to signal statistics (see Chapter 3). Autoregressive moving average (ARMA) filters, with both recursive and feedforward terms, can model local line responses with fewer coefficients, although ensuring reliable and stable adaptation of such filters is more difficult than with FIR types.

A measuring probe, using the latest multiple device DSP hardware, is being developed at BT Laboratories for further investigation of both out-of-service and in-service measurement techniques, and to establish performance and interface specifications.

17.3.6 Parameters for assessment

Measuring the individual properties of, and the correlation between, the send and receive speech signals can provide a wealth of information about the quality of the connection. Such properties include the active speech level, the long-term level [26], the double-talk and talk-spurt durations and the psophometrically-weighted total noise [19]. Measuring these properties requires a suitable voice-activity detector [27-29].

However, it is the correlation between an incident signal and its reflection that is believed to provide the most information relevant to transmission quality. Reflections typically occur at the 2-wire to 4-wire converter, owing to a mismatch between the impedance of the local line and the fixed balance network (see Chapter 4). In a 4-wire connection there can still be coupling of acoustic energy between the transducers in a handset. Depending on the transmission delay, the customer hears this coupled signal as echo or as sidetone. The level of the echo loss can be calculated from the echo path impulse response, using Parseval's theorem. Similarly, the echo path delay can be estimated from the peak amplitude of the echo path impulse response. Given that measurements are made in both directions, the propagation delay of the entire connection can be determined.

Measurement of one parameter, such as frequency response or echo loss, is often adequate for fault diagnosis. However, many parameters must be measured to assess transmission quality, since it is the complex interaction between all the network parameters that determines customer perception. For example, a customer may not find a low value of echo return loss annoying if the overall loss is sufficiently high.

17.3.7   Mapping the subjective

The problem of mapping objective measurements to estimates of subjective speech quality in telephony has been extensively studied at BT Laboratories. In the 1970s, Richards [25,30] and his team did much original and formative work in this field. The work assessed the performance of end-to-end connections, establishing a relationship between subjective responses and physically definable quantities such as sound pressure, signal level, noise level, transmission loss and waveform distortion. The subjective responses had intuitively simple dimensions, such as loudness, quality, ease of understanding and objectionability of unwanted signals. Further work led to computer-based subjective models, such as the computer aided network assessment programme (CATNAP) for the prediction of transmission quality [31].

Current studies include the feasibility of using a knowledge-based system to extract statistical information from in-service objective measurements. For example, the vocal behaviour of the talker or the acoustic environment of the call might be estimated from an analysis of the probability density functions of the speech signals. New subjective models are also being investigated, based on a composite of several artificial intelligence techniques. Rule-based expert systems may be suitable at an abstract level, where interpretive decisions need to be made on limited data, as when mapping the opinions of trained observers on to combinations of objective measures. Neural networks might be used at lower levels for signal classification, voice-detection or trend analysis.

There are three directions that could be pursued to improve the present subjective models. Firstly, work is needed to map objective measurements of nonlinearity on to subjective opinions. Secondly, the information available in the network opinion models [31], used by several administrations to link objective data to subjective opinions, could be analysed. The third avenue for exploration is the mapping, under controlled conditions, of the opinions of expert observers on to those of the participants in the conversation for each set of non-intrusively measured impairments.

The design of such an interpretive subjective model is difficult and raises as yet unanswered questions. However, the overall structure of a network quality monitoring system that uses interpretive techniques can be identified, as described below.

17.3.8   Monitoring a network

Network performance monitoring equipment usually provides automatic scanning of many channels with remote reporting of results. An elaboration of these facilities will be used in future DSP measurement and monitoring systems, for out-of-service and in-service use.

To obtain statistically reliable results, a large number of channels must be monitored simultaneously, generating vast amounts of data at high speed. Applying this data to a single analytic and interpretive process would demand not only very large interconnection bandwidth but very high processing capacity. This heavy computational load can be distributed to DSP hardware and software in a per-channel measuring probe. Such a probe could perform speech level measurements, spectral transforms, cross-correlation and probability distribution analyses and other data reduction or feature extraction tasks, prior to subjective modelling or objective interpretation.

Figure 17.9 shows the important functional processes between the user and the per-channel probes. In this structure, interpretive software performs objective transformations and statistical analysis, monitoring options allow the unattended capture of fault signatures or the programming of alarm conditions, while hierarchically organised displays give clear visualization of key information without presenting confusing detail to the user. Virtual instrument techniques, with a graphical user interface, could allow the appropriate information to be shown on request, effectively providing a gamut of instrumentation in user-selectable layers of a single display screen.

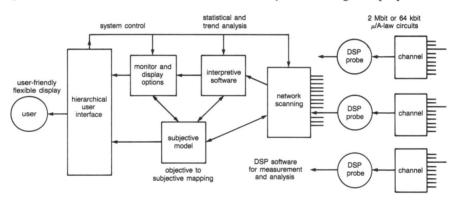

Fig. 17.9   The process links between the user and the DSP measuring probe.

For in-service monitoring and assessment of subjective quality a major process is added, that must perforce deal with large amounts of data to obtain a statistically reliable result. This subjective model takes its key inputs from an expanded interpretive process and from the user interface, feeding its outputs not only to the user and the options process but also to the probe for potential modification of its DSP software.

17.3.9   The potential benefits

The optimum design of the structure in Fig. 17.9 is challenging, because the complexity of this approach far exceeds that currently used to ensure functional confidence in network operation. Many complex factors are involved, in areas including DSP hardware and software design, technical interpretation, psychoacoustics, subjective assessment, expert systems, human factors, network topology and operational issues.

However, the authors believe that such a panoramic, yet comprehensive, precise, yet flexible, approach to service monitoring would have considerable benefits. It could assist the network operator and the customer alike, giving a live measure of overall quality and the means to rapidly detect and diagnose defects in a maintenance role. It could also provide appropriate data to police, both statistically and specifically, the network as it evolves.

Under controlled conditions of termination, a detailed characterization of individual network elements could be calculated, under operational conditions. This would allow simulations of new and unrealized systems to be interconnected with emulations of existing networks, for both objective and subjective assessments (see Chapter 13). The detailed characterization of existing networks elements underpins the future use of DSP in network modelling.

17.4   DSP AND NETWORK MODELLING

17.4.1   Conflicts of approach

Network modelling has traditionally used an event-based approach, expressed in text-based specialist languages. This type of model concentrates on a high-level view of the logical interaction of network architectures and protocols. It typically uses statistical paradigms to represent the macroscopic behaviour of traffic generation and flow. The key virtue of an event-based approach is that it reduces the computational complexity of the task and allows the simulation of large networks, by ignoring the channel signal waveforms. DSP simulations use a sample-based approach, concentrating on a low-level view of the deterministic manipulation of signal waveforms in the channel. In this approach, the computations of the model typically occur afresh each time a signal sample occurs.

Integrating these disparate approaches in a common value of time increment may seem a simple and satisfactory road to mixed modelling, but for many reasons this is impractical. The fundamental reason is that sample and event models treat time in a different way.

Most DSP simulations are synchronous processes, where time is a contiguous but discrete variable. This means that adjacent samples occur sequentially at fixed times and that intermediate values of time are disallowed, having no meaning. Event-based network simulations are usually asynchronous processes, where time is a discontiguous but continuous variable. This means that adjacent events may occur at arbitrary times (closely similar or widely different) and that intermediate activity is disallowed, having no meaning. Fortunately, mixing sample and event-based approaches does not need a common value of time increment.

17.4.2   Mixing samples with events

BT Laboratories has made a short feasibility study of mixed modelling [32], integrating sample flow structures of physical layer signal processing with event-driven, statistical models of higher layer protocols. This integration was successfully achieved by the simple but effective method of dividing the simulation model into a number of separately executed programme fragments, with files used to transfer parameters between fragments.

This non-real time, multipass approach was based upon an informed choice of the key parameters linking the fragments. It joined the two disparate environments of samples and events, simulating the effect of time-dependent network conditions upon the transmission of an encoded speech signal. It thereby provided an end-to-end viewpoint not only of speech encoding and network protocol effects but, more importantly, of their interactions.

As well as demonstrating feasibility, this study revealed useful information about an undetected minor deficiency in proposed CCITT specifications (section 17.4.4). Most usefully of all, it gave a clear insight into the potential use of DSP in network modelling and measurement. The following sections present a brief outline of what was simulated, before reviewing the implications for mixed modelling and suggesting some directions for future work.

17.4.3   Computer aided tools

DSP simulations have traditionally been written in high level languages for a mini-computer or in assembly code for a DSP microprocessor device. As described in Chapter 1, these methods lack speed, structure, reusability and ease of error checking. In recent years, CAE tools have emerged to make both DSP simulation and network modelling a less daunting task, better understood and more easily documented (see Chapters 1 and 5).

Graphical, block-diagram environments can be a powerful and user-friendly visualization interface. They allow complex simulation programmes to be constructed in a rapid, hierarchical and automatic way from predefined libraries of elementary blocks or from user-written blocks. These tools allow the design to be expressed at a high conceptual level and yet encompass low-level detail, without requiring lengthy programming work.

The penalty is primarily one of execution efficiency, through the slower execution speed of the resulting automatically constructed programme. In some cases, a DSP device can be used to accelerate this process and, in any case, the saving in design time and the pace of improvement in affordable computing power makes such drawbacks minor.

17.4.4   A study of mixed modelling

Two graphical, block diagram CAE tools were used in the mixed modelling study. SPW[TM] [33], intended for synchronous, sample-driven DSP, was used to simulate embedded ADPCM, variable bit rate speech encoding. BONeS[TM] [34] was used for asynchronous, event-driven modelling of a frame relay, packetized voice protocol (PVP) network, subject to the effects of congestion and bit errors.

CCITT Recommendation G.EMB (pre-cursor to G.727) describes low bit rate speech encoding by embedded ADPCM at 2-5 bits per sample [35]. The encoder (Fig. 17.10) consists of a 2-pole, 6-zero predictor and a 5-bit, non-uniform, one word memory adaptive quantizer. The quantization and prediction algorithms incorporate a strategy to limit the effect of channel errors.

Embedded encoding of speech has a significant advantage over non-embedded encoding in packetized transmission [7,8]. At times of network congestion, the least significant bits of the signal data can be intentionally discarded at the decoder (Fig. 17.11) without causing gross mistracking between the encoding and decoding algorithms. The result is that speech quality declines gracefully, rather than catastrophically, as the number of discarded bits increases. This embedded property is achieved by using only the two most significant bits of the encoded signal in the adaptive algorithms of both the encoder and decoder.

The CCITT draft Recommendation G.PVP [36] defines a protocol for voice-band transmission over a wideband packet network. Speech is transmitted in packets of 128 encoded signal samples. Each speech packet is divided into blocks, containing bits of the same significance. Network congestion can be ameliorated by discarding up to 3 blocks of low-significance bits from each speech packet [37,38]. Network capacity is therefore made usefully elastic at the expense of a temporary reduction in speech quality.

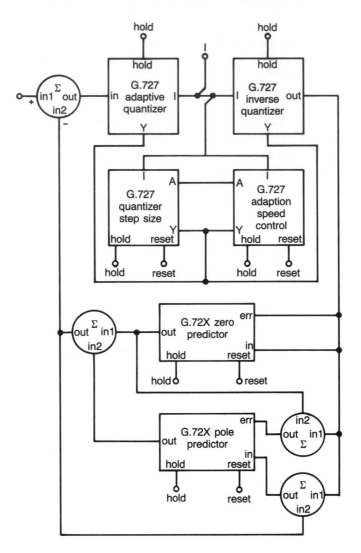

Fig. 17.10   Block diagram in SPW of embedded ADPCM encoder.

A measure of the background noise level is encoded in each packet header and speech packets are sent only when speech is present. Synthetic noise of a matching level is injected at the decoder during periods of silence, to improve the perceived signal continuity. Error protection is applied only to speech packet headers, not signal data, since speech perception is considered tolerant of occasional errors. A lost-packet strategy was not defined in the draft PVP Recommendation. Repetition of the previous packet and noise injection are possible strategies.

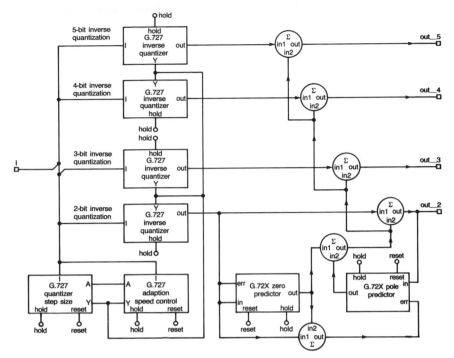

Fig. 17.11   Block diagram in SPW of embedded ADPCM decoder.

17.4.5   Mixed model processing

Two sentences, with a total duration of 23 secs, were taken from a CCITT compact disc [39] and processed at a variety of congestion levels and channel error rates. The complete mixed modelling simulation took place in four separate fragments.

Fragment 1:   For consistency, the speech signal power was estimated in SPW and normalized to 22 dB below the peak A-law value.

Fragment 2:   An SPW simulation extracted the speech-burst length and noise parameters from each speech signal and wrote them to a transfer file.

Fragment 3:   Several frame relay network simulations were made in BONeS (Fig. 17.12), using the transfer file to determine when packets were sent. The congestion level and error rate influenced which packets were delayed, truncated or discarded. This packet-handling data was written to a second transfer file.

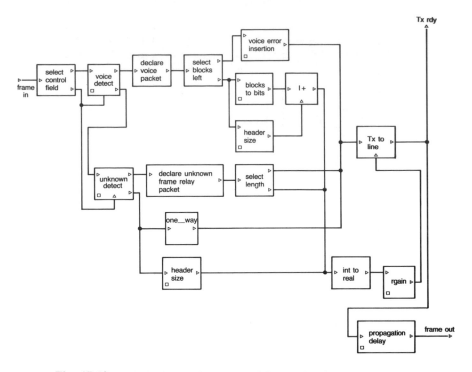

Fig. 17.12   Block diagram in BONeS of frame relay fast-packet network.

Fragment 4:   The speech signal was encoded, passed through a channel model and decoded in an SPW simulation (Fig. 17.13). The second transfer file controlled the injection of channel errors, discarded blocks and the selection of silence or decoded speech. Simultaneous decoding at 2-, 3-, 4- and 5-bit precision was used, to simulate discarded blocks by selecting the appropriate output.

Several simplifications proved expeditious in this simple study:

- a floating-point simulation of embedded ADPCM encoding was used;

- channel errors were simulated as a burst at a random point in the packet;

- traffic congestion and packet delay were simulated as random distributions;

- the cumulative error of time-stamp values was ignored;

- noise injection was omitted and the lost-packet strategy was the injection of silence.

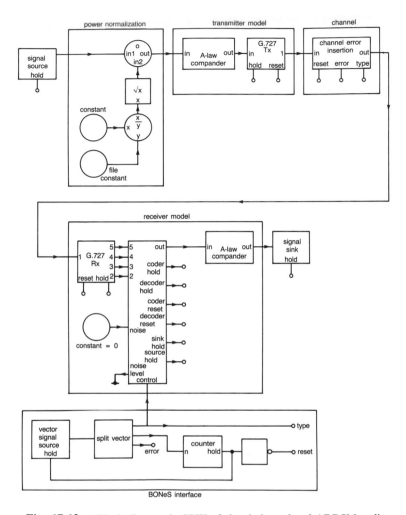

Fig. 17.13   Block diagram in SPW of signal channel and ADPCM coding.

17.4.6   Results from the model

Figure 17.14 shows the waveform of a portion of low level speech that was part of the input signal to the model. Figure 17.15 plots the waveform of the output signal, corresponding to this input, under conditions of slight network congestion. A comparison of Figs. 17.14 and 17.15 demonstrates how the encoding bit rate changes with time in response to temporary network congestion. The addition of a changing amount of quantization noise can be seen, especially near the start and the end of Fig. 17.15.

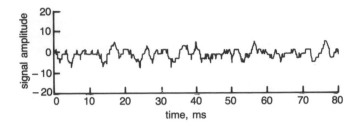

Fig. 17.14   Waveform of a low-level portion of the speech input signal, A-law encoded.

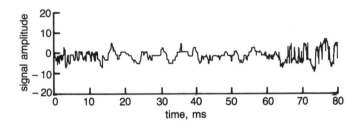

Fig. 17.15   Waveform of the output signal, corresponding to the input of Fig. 17.14.

Figure 17.16 shows the output waveform of the same input signal, during more severe congestion in the network, where the transmission bit rate is 2 bits per sample.

Figure 17.17 shows a larger portion of input speech. Figure 17.18 shows the corresponding output waveform during severe congestion, where a bit error in the channel has corrupted a packet header and resulted in a lost speech packet. At the output of the decoder, an interval of silence is followed by a burst of distorted speech, caused by a temporary loss of tracking between the adaptive prediction filters in the encoder and decoder.

A comparison of the original and processed speech signals showed an interesting feature. The audio delay could vary between utterances, because the draft Recommendation G.PVP [36] did not contain any mechanism to maintain absolute time synchronization during silence intervals. The variation in delay, of up to 1 ms, would be sufficient to disturb the operation of an echo canceller, having such a network in its echo path, at the start of each utterance. This was a consequence of the protocol as defined, although it was not clear whether this consequence was intended but considered irrelevant.

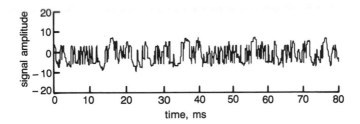

Fig. 17.16   Waveform of the output signal, during network congestion.

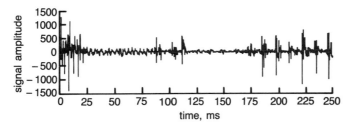

Fig. 17.17   Waveform of a second portion of speech input signal, A-law encoded.

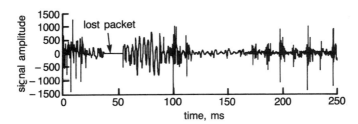

Fig. 17.18   Waveform of the output signal, showing the effect of a lost speech packet.

Prior to the mixed modelling study, this potentially deleterious interaction had not been obvious by inspection. The draft Recommendation G.PVP did not state that the system it described was unsuitable for connection to echo cancellers in this way.

17.4.7   Model timing

Timing was an important aspect of the fourth simulation fragment. The BONeS frame relay model simulated packet-delay jitter by altering the length of the silence intervals, causing the fourth fragment to discard signal samples.

A similar technique was used to handle lost packets. The SPW structure, chosen for simplicity, in the fourth fragment meant that the number of cycles needed to process a silence interval or a lost packet was about double that of a speech packet. The fourth simulation fragment was therefore the most computationally complex and had the slowest execution speed. The use of a DSP cross-compiler was examined in an attempt to increase this speed.

The SPW fragments were executed on two platforms — a Sun Sparc 1 + workstation and a Loughborough Sound Images TMS320C30 DSP board in a PC. Two SPW program generation methods were used on the Sun — the simulation program builder (SPB) and the code generation system (CGS). The latter transformed the block diagram simulation into a platform-independent, generic C program. This was also transformed into TMS320C30 code, using a Texas Instruments cross-compiler. Table 17.1 shows the approximate execution speed of these simulations. File access on the PC was via a shared network, which significantly slowed the speed of the TMS320C30 program. However, for the computationally-intensive fourth fragment a useful increase in speed resulted.

Table 17.1   Execution speed for different platforms and program generation techniques.

Simulation fragment	Approximate execution speed in seconds		
	Sparc 1 + /SPB	Sparc 1 + / generic C	TMS320C30
1	82	12	15
2	102	24	27
4	317	169	110

Neglecting file-access effects, the TMS320C30 execution speed was considerably less than that typical of a hand-written, assembly-level program. It was found that the C cross-compiler could produce inefficient code, partly due to limited utilization of the parallel architecture of the DSP device. However, by careful design of the custom code in the simulation of a simpler, non-embedded ADPCM encoder, a significant increase in efficiency proved possible [40]. The real time sampling rate of the TMS320C30 code was increased from 4.5 to 11 kHz by this means.

17.4.8   Future trends

Many other interesting and practically relevant systems can be usefully studied using mixtures of DSP and network modelling. For example, DSP could simulate the time-dependent behaviour of a radio channel in mobile radio applications, while network modelling could represent the call set-up, roaming or cell hand-over protocols.

The mixed modelling in this feasibility study was inherently a non-real time process, since the model was executed in separate fragments. The use of files as a parameter transfer interface between the two simulation domains proved to be a workable method, but not a particularly efficient or flexible one. File access slowed the execution of the simulation and would have represented a major overhead if a large number of parameters had been transferred.

A different strategy would be required to execute mixed model simulation close to real time. Instead of running multiple passes, the two simulations would have to execute concurrently. This would remove the need for signal storage in files and eliminate the overhead of file access. Alternative parameter transfer interfaces include the use of Unix^{TM} pipes for passing data or merging the DSP and network simulations into a single programme.

The key issue is that synchronization between the two simulation domains must be maintained, which implies an interlock mechanism to join synchronous and asynchronous processes. Creating such a mechanism as part of a CAE environment, in a way that is both rigorously correct and easy-to-use, is a topic of current academic debate [41].

Large simulations will require much greater computational power than that of the current generation of either DSP or general-purpose processor devices, if nearly real time simulation is desired. An alternative solution for increasing execution speed is distributed processing. DSP and network simulations typically have modular structures, that readily lend themselves to partitioning into subsets of interconnected processes. By allocating processes to separate DSP devices, significant improvements in execution speed should be possible. However, the overall speed would depend strongly not only on the software and hardware partitioning but also on the efficiency of parameter transfer and process scheduling. Creating such scheduling within a CAE environment, in a way that is both efficient and easy-to-use, is the subject of current academic research [42,43].

17.5 CONCLUSIONS

In decades past, telecommunications networks changed slowly. Sinewaves at 1 kHz were an adequate measuring tool, and channels were physical and mutually-independent. Network planners did not need to know about developments in switching and terminal equipment. A period of extremely rapid world-wide change has now begun, driven by an irresistible combination of technological and market forces. New services, like personal telephones, are introduced and create a prodigious demand seemingly overnight. Other services are introduced only to fail and be quietly withdrawn. The risks as well as the stakes are growing ever higher.

An important tool for success in this new environment will be the close integration of high-level network modelling with low-level DSP simulation. This will allow the specification, design and performance of new services to be verified, and any errors or omissions rectified, as early as possible. The mixed modelling feasibility study was necessarily brief and restricted in complexity. Yet it gave a surprisingly powerful insight into the future. Mixing samples with events means linking methods of simulation that are totally different in conceptual approach, to create a hybrid with fresh powers.

The key to accurately predicting transmission performance by mixed modelling is the comprehensive and precise measurement of existing networks. Fortunately, DSP promises to revolutionize the measurement of telephony transmission performance, through automatic in-service monitoring. The ability to quickly and economically collect accurate and comprehensive performance measures, that map closely to customer perceptions, will be an essential tool in managing future telecommunications networks effectively.

This new use of DSP promises both short and long-term benefits to network operators by:

- more precise, far-sighted and timely specification of systems and services;

- accurate prediction of speech and image performance in complex virtual networks;

- automatic and low-cost estimation of the customer's perception of quality.

DSP in network modelling and measurement is a multidisciplinary hybridization of statistical modelling, digital signal processing, artificial intelligence, parallel computing and telephonometry. It exemplifies a socio-technical trend described in Chapter 1, crossing and blurring many of the current boundaries between computing and engineering in telecommunications. It is an ambitious and embryonic subject, but one that may prove prodigious, posing major challenges to both DSP hardware designers and CAE tool developers. This chapter has provided but a hint of the difficulties and opportunities ahead.

REFERENCES

1. Jayant N S: 'High-quality coding of telephone speech and wideband audio', IEEE Comms magazine, 28 , No 1, pp 10-20 (January 1990).

2. Southcott C B et al: 'Low bit rate speech coding for practical applications', BT Technol J, 6 , No 2, pp 22—40 (April 1988).

3. Boyd I and Southcott C B: 'A speech codec for the Skyphone service', BT Technol J, 6 , No 2, pp 50—59 (April 1988).

4. Groves I S: 'Personal mobile communications — a vision of the future', BT Technol J, 8 , No 1, pp 7—11 (January 1990).

5. Jagoda R and de Villepin M: 'Personal communication services in Europe', Telecommunications, Int edition, p 62 of supp (August 1991).

6. van der Hoek H: 'Cordless access in the local loop', Telecommunications, Int edition, p 70 of supp (August 1991).

7. Giorcelli S: 'Network evolution towards an ATM-based B-ISDN', CSELT Technical reports, XVII , Pt 6, pp 403-415 (December 1989).

8. Gilbert H, Aboul-Magd O and Phung V: 'Developing a cohesive traffic management strategy for ATM networks', IEEE Comms magazine, 29 , Pt 10, p 36 (October 1991).

9. Gallagher I, Ballance J and Adams J: 'The application of ATM techniques to the local network', BT Technol J, 7 , No 2, pp 151-160 (April 1989).

10. Kitawaki N et al: 'Speech coding technology for ATM networks', IEEE Comms magazine, 28 , No 1, pp 21-27 (January 1990).

11. Wolf S et al: 'How will we rate telecommunications system performance?', IEEE Comms magazine, 29 , Pt 10, pp 23-29 (October 1991).

12. Appleton J M: 'Performance related issues concerning the contract between network and customer in ATM networks', BT Technol J, 9 , No 4, pp 57-60 (October 1991).

13. Harvey C: 'Performance engineering as an integral part of system design', BT Technol J, 4 , No 3, pp 142-147 (July 1986).

14. ANSI/IEEE Std 743-1984: 'IEEE standard methods and equipment for measuring the transmission characteristics of analogue voice frequency circuits' (November 1984).

15. CL680 line test system, Vanderhoff Communications Ltd.

16. Testnet 1000 network transmission test system, Teradyne Inc.

17. T1000 series network transmission performance analyser, Fulcrum Communications Ltd (1990).

18. CCITT Recommendation P.79: 'Calculation of loudness ratings', Melbourne (1988).

19. CCITT Recommendation O.41: 'Psophometer for use on telephone-type circuits', Melbourne (1988).

20. HP3560A portable dual-channel dynamic signal analyser, Hewlett Packard Co (1991).

21. PCM-4 channel measuring set, Wandel & Goltermann GmbH & Co (1984).

22. Alley D M, Lewis A V and Clarke A J: 'SCHEMAS echo measuring set user guide', Internal BT memorandum (May 1991).

23. South C R: 'In-service, non-intrusive assessment of telephony channels — an introduction', CCITT Int Symp on transmission quality of networks and telephone terminals, Brasilia (3—5 September 1991).

24. Branch P P: 'JANUS: An in-service non-intrusive measurement device for the assessment of network call quality', Internal BT memorandum (January 1992).

25. Richards D L: 'Telecommunications by speech', Butterworths (1973).

26. CCITT Recommendation P.56: 'Objective measurement of active speech level', Melbourne (1988).

27. Jankowski J A: 'A new digital voice-activated switch', COMSAT Technical Review, 6 , Pt 1, pp 159-176 (Spring 1975).

28. Freeman D K et al: 'The voice activity detector for the pan-European digital cellular mobile telephone service', Proc IEEE ICASSP, Glasgow, 1 , pp 369-372 (May 1989).

29. CCITT Recommendation G.763: 'Digital circuit multiplication equipment using 32 kbit/s ADPCM and digital speech interpolation', Melbourne (1988).

30. Richards D L: 'Transmission performance of telephone connexions having long propagation times', Third international symposium on human factors in telephony, Hague (1967).

31. CCITT Recommendations, P series supplement no 3: 'Models for predicting transmission quality from dynamic measurements', Melbourne (1988).

32. Barrett P A: 'An end-to-end network modelling environment for packetised speech transmission', Internal BT memorandum (1990).

33. 'Signal processing WorkSystem^{TM}', release 2.7, Comdisco Systems Inc (August 1990).

34. 'Block oriented network simulator^{TM}', release 1.5, Comdisco Systems Inc (November 1990).

35. CCITT draft Recommendation G.EMB, temp doc XV/2-E (July 1990).

36. CCITT draft Recommendation G.PVP, report COM XV-R 21 (January 1990).

37. Jayant N S and Christensen S W: 'Effects of packet losses in waveform coded speech and improvements due to an odd-even sample interpolation procedure', IEEE Trans Comms, 29 , Pt 2, pp 101-109 (February 1981).

38. Li S Q: 'Study of information loss in packetised voice systems', IEEE Trans Comms, 37 , Pt 11, pp 1192-1202 (November 1989).

39. Track 6 of CCITT CD: 'Speech data base for telephonometry', CRM-1059, NTT (1988).

40. Benyon P R: 'An exercise in SPW C code optimization', Internal BT memorandum (January 1991).

41. Buck J et al: 'Multirate signal processing in Ptolemy', Proc IEEE ICASSP, 2 , pp 1245-1248 (1991).

42. Lee E A and Messerschmitt D G: 'Static scheduling of synchronous data flow programs for digital signal processing', IEEE Trans Comms, 36 , Pt 1, sec V, pp 24-35 (January 1987).

43. Lee E A and Messerschmitt D G: 'Pipeline interleaved programmable DSPs: synchronous data flow programming', IEEE Trans ASSP, 35 , Pt 9, sec III-B, pp 1334-1345 (September 1987).

Index